NATIONAL GEOGRAPHIC

ILLUSTRATED GUIDE
TO
Nature

NATIONAL GEOGRAPHIC

ILLUSTRATED GUIDE

TO

Nature

FROM YOUR BACK DOOR
TO THE GREAT OUTDOORS

Wildflowers | Trees & Shrubs | Rocks & Minerals
Weather | Night Sky

NATIONAL GEOGRAPHIC
WASHINGTON, D.C.

Lightning illuminates the sky at Three Lakes Wildlife Management Area in Florida.

PRECEDING PAGES: *The moon rises over Canyonlands National Park from Green River Overlook, Utah.*

CONTENTS

Introduction 6

Fall colors surround oak trees in the
Shawangunk Mountains of New York.

Knowledge, Pleasure & Wonder
Exploring the World of Nature

"The world is too much with us," wrote the poet William Wordsworth, and by that, he meant the world of bustle and busyness, deadlines and expectations, noise and hubbub. That world has only grown louder and more predominant in the intervening two centuries, built up with trucks and highways, skyscrapers and traffic, television and smartphones. Our ears and minds and days are crammed full of information. We yearn for forest silence, for the lapping of ocean waves on beach sand, for early morning birdsong.

Studies conducted at the University of Kansas show the benefit of spending more time in nature. Testing the theory that "nature has specific restorative effects on the prefrontal cortex"—the part of the brain responsible for creativity and planning—the researchers gave a standardized problem-solving test to backpackers ages 18 to 60 before and after a four-day sojourn into the wilderness. Campers returning scored 50 percent higher.

Why? "Nature is a place where our mind can rest, relax, and let down those threat responses," says Ruth Ann Atchley, lead investigator. When we spend time in nature, Atchley proposes, "we have resources left over—to be creative, to be imaginative, to problem-solve—that allow us to be better, happier people who engage in a more productive way with others."

This book is designed to invite people, young and old, to step with greater pleasure into the out-of-doors. We intend the book to both educate and entertain. Selecting one realm of nature—the world of wildflowers, trees and shrubs, rocks and minerals, the weather, or the night sky—and getting to know it more intimately can bring lifelong satisfaction. Or just open up this book occasionally to learn the names of clouds above or flowers at your feet.

Working with experts in the various fields of natural history, we have selected 781 species and phenomena from five of the great realms of nature and offer images, identifying features, and brief descriptions of each for you to learn and enjoy. This is an arbitrary number, and relatively low, given that the book's geographic range covers the entire continental United States and Canada. This illustrated guide is thus a starting point from which every nature enthusiast will proceed to fuller, longer, and more detailed topic-specific field guides as well as a perennial treasury to which you and others in your household will return again and again, weaving the natural world more intimately into your daily life.

"The more high-tech we become, the more nature we need," writes Richard Louv, author of *The Nature Principle* and *Last Child in the Woods*. May this volume bring enjoyment, relaxation, productivity, and wonder as you step out into the world of nature that shares this amazing planet of ours.

Native wildflowers blanket the crest of Rowena Plateau along the Columbia River in Oregon.

1 | Wildflowers

Nature's Exterior Decoration

Well before the trees have leafed out, wildflowers signal the new season. Woodland spring ephemerals poke through the moist leaf litter, reaching for the sun before the tree canopy blots it out. The wildflower parade continues throughout the growing year—long or short depending on the species and its location—filling fields, woods, roadsides, desert washes, and myriad other habitats throughout the continent with a kaleidoscopic flowering array.

■ What Is a Wildflower?

Basically, a wildflower is a noncultivated flowering plant that is not a shrub or tree. Most wildflowers are classified as herbaceous plants—flowering plants with nonwoody stems. (We tend to think of herbs as plants with flavorful leaves that enhance our culinary efforts, but that is too narrow a definition.) Under certain conditions, some wildflower species can grow into shrubs or even small trees, so the distinction is not absolute.

In botanical terms, a wildflower is an angiosperm, a seed-producing flowering plant. All wildflowers bear flowers, although they may be small, inconspicuous, or so short-lived that we may know the plant mainly by its leaves, or foliage. Wildflowers can be annuals, going through a flowering and seeding period in a single year; biennials, setting a rosette of leaves one year and flowers and seeds the next; or perennials, living and flowering for three or more years.

Wildflowers come in all shapes and sizes, meeting the basic needs of survival and reproduction in a mind-boggling variety of flower configurations,

leaf shapes, and fruits (the part of the flower that houses the seeds). Equally intriguing are their numerous strategies for pollination (fertilization) and seed dispersal. Flowers can have the tidy ring of petals in a buttercup, the banners and folds of the Pea family, or the elegant, tapered tepals (look-alike petals and sepals) of many lilies. The leaves also take a multitude of shapes—simple and compound, with variations upon variations. The leaves at the base of a wildflower plant may differ greatly from those on the stem. Fruits are likewise diverse, in forms such as seed heads, pods, capsules, and berries. Learning all these distinctions heightens appreciation of nature's wildflower bounty.

■ Wildflower Versus Weed

Technically, every weed deserves to be called a wildflower, but whether some wildflowers deserve to be called weeds is a judgment call. One person's wildflower is often another person's weed. Wildflowers that thrive where they're often not welcome—in a turf lawn, a flower bed, or a vegetable plot—tend to be called weeds. The dandelion is a classic example. Viewed against the standard of neatly groomed, cultivated flower species, many wildflowers look unkempt and unruly, so they often end up in the weed category. Of course, there are wildflower species of indisputable beauty by any standard. But all wildflowers deserve a look, and on their own terms. Every wildflower, no matter how homely or unwelcome in the eye of a viewer, has a story to tell about its life, survival strategies, and place in the world of living organisms.

◼ Exotics & Invasives

Humans have a long history with wildflowers. Aesthetics figure into this relationship, but so does practicality: potential for use as food, beverage, medicine, cosmetics, or decoration. Wildflowers have often been exploited for these purposes to the point of extermination, such as some wild orchids for their beauty and rarity.

Over time, humans have become very attached to certain wildflowers, and as people take up roots and settle in different regions or on different continents, they want their familiar plants to accompany them. Nostalgia for vine-covered cottages and buildings,

for instance, helped bring English Ivy to North America. Highly successful in its new environment, the species can outcompete native species and establish itself as a monoculture, thwarting biodiversity. Introduction of nonnative species can occur unintentionally as well. For example, unwanted seeds can enter crevices in footwear and vehicle tires, and can contaminate grain seed shipments.

An introduced, nonnative plant is called an exotic. When it runs rampant and impinges on other species (which not all introduced species do), it is designated as invasive. Invasive exotics have significantly changed North American plant communities. Certain native plants can become invasive, but they tend to have a lesser impact. Many local conservation groups are leading efforts to remove exotic invasives and restore native plant communities. Most offer opportunities for volunteers to aid these efforts.

◼ Identifying Wildflowers

The wildflower species included in this chapter are just a small sampling of the thousands of flowering plants in North America north of Mexico.

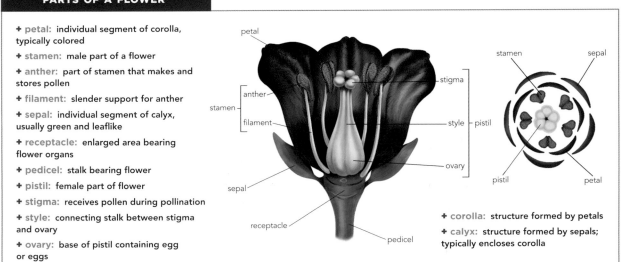

PARTS OF A FLOWER

+ **petal:** individual segment of corolla, typically colored

+ **stamen:** male part of a flower

+ **anther:** part of stamen that makes and stores pollen

+ **filament:** slender support for anther

+ **sepal:** individual segment of calyx, usually green and leaflike

+ **receptacle:** enlarged area bearing flower organs

+ **pedicel:** stalk bearing flower

+ **pistil:** female part of flower

+ **stigma:** receives pollen during pollination

+ **style:** connecting stalk between stigma and ovary

+ **ovary:** base of pistil containing egg or eggs

+ **corolla:** structure formed by petals

+ **calyx:** structure formed by sepals; typically encloses corolla

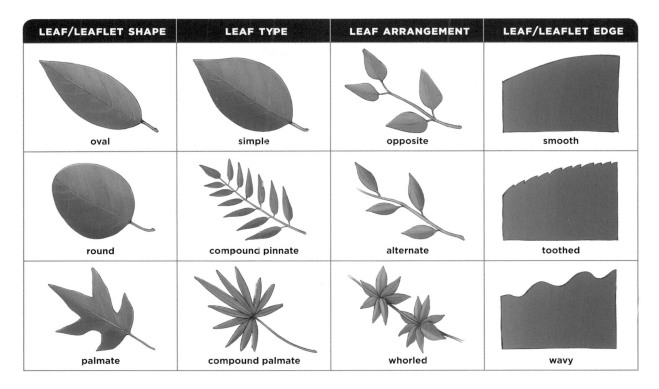

LEAF/LEAFLET SHAPE	LEAF TYPE	LEAF ARRANGEMENT	LEAF/LEAFLET EDGE
oval	simple	opposite	smooth
round	compound pinnate	alternate	toothed
palmate	compound palmate	whorled	wavy

The 160 species represent different types of common wildflowers from different regions of the continent, showing a range of flower shapes and sizes, plant structures, and distribution. Each entry offers characteristics that will help you identify a species and compare it to similar flowers; additional information about habitat, range, and bloom period; and other details useful for understanding and appreciating the species. The species are presented by color groups—initially the most useful distinguishing characteristic for identification—and within those groups, by similarities of flower shape, foliage, and general appearance. Though some entries mention culinary or medical uses of various wildflowers, readers should not eat any part of any plant unless they are 100 percent certain that it is 100 percent safe. Other species are described as being toxic; if a species is not so described, it should not be deduced that any part is safe to eat. Many books and online references offer detailed information on which plants are unsafe.

A weed is no more than
a flower in disguise,
Which is seen through at once,
if love give a man eyes.
—JAMES RUSSELL LOWELL

■ Wildflowers at Home

When you make native perennial wildflowers part of your home landscape, everyone wins: you, the wildlife, and the local ecosystem. Native wildflowers add beauty and feed birds, butterflies, and other animals in ways that most imported annuals cannot begin to match, and they allow other native plant species to gain a foothold. Where can you find native species for home planting? Though you should never collect plants on public or private lands without permission, you might find specimens for sale through local nature centers, botanical gardens, and native plant societies. These organizations often have demonstration gardens that will give you ideas about plant combinations and light and moisture requirements. Friends with extensive plantings are another source, as are growing numbers of commercial suppliers. Always try to find local ecotypes of the plants you want; they're best suited to your growing conditions and the surrounding ecosystem.

You can take this endeavor one step further by making native wildflowers the foundation of a wildlife habitat that provides the basic necessities for all manner of creatures, from insects to birds to mammals. Wildflowers provide nectar and pollen for insects; nectar, berries, and seeds for birds; and those things plus flowers and foliage for a wide variety of mammals. Vining and shrubby plants provide shelter and places for nesting; and a birdbath, fountain, or small pond provide crucial water. Your yard can satisfy both you and the wildlife.

OFFICIAL FLOWERS OF THE U.S. STATES AND CANADIAN PROVINCES

+ **Alabama: Camellia** (*Camellia japonica*)
+ **Alaska: (Asian) Forget-me-not** (*Myosotis alpestris asiatica*)
+ **Arizona: Saguaro Cactus blossom** (*Carnegiea gigantea*)
+ **Arkansas: Apple blossom** (*Malus pumila*)
+ **California: California Poppy** (*Eschscholzia californica*)
+ **Colorado: Rocky Mountain (Colorado Blue) Columbine** (*Aquilegia coerulea*)
+ **Connecticut: Mountain Laurel** (*Kalmia latifolia*)
+ **Delaware: Peach blossom** (*Prunus persica*)
+ **District of Columbia: American Beauty Rose** (*Rosa* "American Beauty")
+ **Florida: Orange blossom** (*Citrus × sinensis*)
+ **Georgia: Cherokee Rose** (*Rosa laevigata*)
+ **Hawaii: Pua Aloalo** (*Hibiscus brackenridgei*)
+ **Idaho: Syringa Mock Orange** (*Philadelphus lewisii*)
+ **Illinois: Violet** (*Viola* spp.)
+ **Indiana: Peony** (*Paeonia lactiflora*)
+ **Iowa: Wild Prairie Rose** (*Rosa arkansana*)
+ **Kansas: Sunflower** (*Helianthus annuus*)
+ **Kentucky: Goldenrod** (*Solidago* spp.)
+ **Louisiana: (Southern) Magnolia** (*Magnolia grandiflora*)
+ **Maine: Eastern White Pine tassel and cone** (*Pinus strobus*)
+ **Maryland: Black-eyed Susan** (*Rudbeckia hirta*)
+ **Massachusetts: Mayflower (Trailing Arbutus)** (*Epigaea repens*)
+ **Michigan: Apple blossom** (*Malus pumila*)
+ **Minnesota: Pink-and-white Lady's Slipper** (*Cypripedium reginae*)
+ **Mississippi: (Southern) Magnolia** (*Magnolia grandiflora*)
+ **Missouri: Hawthorn** (*Crataegus* spp.)
+ **Montana: Bitterroot** (*Lewisia rediviva*)
+ **Nebraska: Goldenrod** (*Solidago gigantea*)
+ **Nevada: Sagebrush** (*Artemisia tridentata*)
+ **New Hampshire: Purple Lilac** (*Syringa vulgaris*)
+ **New Jersey: (Common Blue) Violet** (*Viola sororia*)
+ **New Mexico: (Soapweed) Yucca** (*Yucca glauca*)
+ **New York: Rose** (*Rosa* spp.)

+ **North Carolina: Flowering Dogwood** (*Cornus florida*)
+ **North Dakota: Wild Prairie Rose** (*Rosa arkansana*)
+ **Ohio: Scarlet Carnation** (*Dianthus caryophyllus*)
+ **Oklahoma: Mistletoe** (*Phoradendron leucarpum*)
+ **Oregon: Oregon Grape** (*Berberis aquifolium*)
+ **Pennsylvania: Mountain Laurel** (*Kalmia latifolia*)
+ **Rhode Island: Violet** (*Viola* spp.)
+ **South Carolina: Yellow Jessamine** (*Gelsemium sempervirens*)
+ **South Dakota: Pasqueflower** (*Anemone patens* var. *multifida*)
+ **Tennessee: (German) Iris** (*Iris germanica*)
+ **Texas: Bluebonnet** (*Lupinus* spp.)
+ **Utah: Sego Lily** (*Calochortus* spp.)
+ **Vermont: Red Clover** (*Trifolium pratense*)
+ **Virginia: Flowering Dogwood** (*Cornus florida*)
+ **Washington: Coast Rhododendron** (*Rhododendron macrophyllum*)
+ **West Virginia: (Big) Rhododendron (Great Laurel)** (*Rhododendron maximum*)
+ **Wisconsin: Wood Violet** (*Viola papilionacea*)
+ **Wyoming: Indian Paintbrush** (*Castilleja linariifolia*)

Canadian Provinces and Territories
+ **Alberta: Wild Rose** (*Rosa acicularis*)
+ **British Columbia: Pacific Dogwood** (*Cornus nuttallii*)
+ **Manitoba: Prairie Crocus** (*Anemone patens*)
+ **New Brunswick: Purple (Marsh Blue) Violet** (*Viola cucullata*)
+ **Newfoundland and Labrador: (Purple) Pitcher Plant** (*Sarracenia purpurea*)
+ **Northwest Territories: Mountain Avens** (*Dryas integrifolia*)
+ **Nova Scotia: Mayflower (Trailing Arbutus)** (*Epigaea repens*)
+ **Nunavut Territory: Purple Saxifrage** (*Saxifraga oppositifolia*)
+ **Ontario: (Large-flowered) White Trillium** (*Trillium grandiflorum*)
+ **Prince Edward Island: (Pink) Lady's Slipper** (*Cypripedium acaule*)
+ **Quebec: Blue Flag** (*Iris versicolor*)
+ **Saskatchewan: Western Red Lily** (*Lilium philadelphicum*)
+ **Yukon Territory: Fireweed** (*Epilobium/Chamerion angustifolium*)

WILDFLOWERS

||||||||||||||||||||||||||||||||||||||||||||||||||||||||||||||||||||||||||||||||||||||||||||||||||||||||

Yellow Trout Lily

Erythronium americanum H 5–10 in (13–25 cm)

A scaly bulb gives rise to a single stem bearing the blossom of the woodland Yellow Trout Lily. Bloom and bulb shape once gave it a misleading common name of Dogtooth Violet.

KEY FACTS

The 6-part pendent yellow flower has tepals (look-alike petals and sepals) that are bent back; 2 mottled leaves clasp the flower stem.

+ **habitat:** Moist woods, swamps, and meadows

+ **range:** Eastern U.S. and Canada, except for Florida

+ **bloom period:** March–June

From a basal pair of long, elliptical, brown-mottled leaves, the Yellow Trout Lily blossom tops a single stalk, although there may be many such blooms in the species' large colonies. The leaf markings resemble those of a Brown or Brook Trout and give the native perennial its common name. The bent-back petals and sepals provide a signature look for this lily, a spring ephemeral. The flowers are yellow on the outside and brownish within, and display six brownish stamens with prominent anthers. Yellow Trout Lily often occurs in large colonies that include many nonflowering plants with solitary leaves.

Greater Yellow Lady's Slipper

Cypripedium parviflorum var. *pubescens* H 6–28 in (15–70 cm)

A fat, bright-yellow lip petal gives the Greater Yellow Lady's Slipper its name. Usually, only one of the striking flowers crowns the top of a leafy stem, but at times there are two or three.

KEY FACTS

The blossom is yellow and brownish. The stem has up to 6 pointed leaves.

+ **habitat:** Bogs, marshes, and moist woods

+ **range:** Throughout much of the U.S. and Canada, except for far northern areas, and parts of southern and western U.S.

+ **bloom period:** May–July

The Greater Yellow Lady's Slipper, a member of the Orchid family, shows a typical orchid structure with some unique features. The bulbous lower petal of the native species has a rounded opening at the base. The two lateral petals are long, thin, and very twisted. Broad, bright-green leaves have pointed ends and prominent lengthwise ribs. The leaves sheathe the stem at their bases and grow alternately along it. There are both large and small varieties of Yellow Lady's Slippers, and some are quite fragrant. Like most orchid species, their current populations struggle as a result of indiscriminant and often illegal collecting.

Swamp Buttercup

Ranunculus hispidus var. *nitidus* H 12–36 in (30–90 cm)

Despite its name, this native buttercup emerges in both wet and dry habitats. Once classified as a separate species *(R. septentrionalis)*, it now stands as a variety of the Hispid Buttercup.

KEY FACTS

The flower, up to 1 in (2.5 cm), has 5 yellow petals; 3-lobed, hairless, deeply divided leaves are usually as wide as they are long.

+ **habitat:** Swamps, marshes, other wet habitats, woods, and grasslands

+ **range:** Throughout much of central and eastern U.S. and Canada, except in northern areas

+ **bloom period:** April–July

The Swamp Buttercup produces leaves that serve as a base for its flowers, and also alternate leaves on stems that spread along the ground. The yellow flowers rise on stalks, with petals that are bright and shiny most of their length, but become a pale greenish yellow toward the base. Fine lines on the petals serve as guides to the nectar. Bees, flies, butterflies, and beetles visit the flowers to feed on both nectar and pollen. The Swamp Buttercup resembles the Early Buttercup *(R. fascicularis)*, which true to its name gets an earlier start and has more pointed leaves longer than they are wide. It is also significantly hairier than the Swamp Buttercup.

Black Mustard

Brassica nigra H 12–96 in (30–240 cm)

A poultice of Black Mustard leaves placed on the chest was once the go-to remedy for colds, bronchitis, and similar ailments. The seeds are used to make the iconic yellow condiment.

KEY FACTS

Narrow flower clusters have 4-petaled yellow flowers. Upper leaves are toothed and narrow; lower leaves are deeply lobed.

+ **habitat:** Grassy hills, fields, roadsides, and wasteland

+ **range:** Naturalized throughout the U.S. and Canada, except in far northern areas

+ **bloom period:** June–October

The Black Mustard plant is capable of growing very tall and lanky to the point that it flops sideways. At its tallest, the lower leaves can measure up to 10 inches (25 cm) long and may wilt in dry heat. Each flower turns into a slender seedpod that grows close to the flower stalk. A native of Eurasia, the prolific, weedy Black Mustard has adapted well to most of North America. It shares a genus with vegetables such as Brussels sprouts, broccoli, cauliflower, and cabbage and has its own extensive list of culinary and medicinal uses. Many songbird species favor the plant's tiny black seeds.

Silverweed

Argentina anserina L runners to 36 in (90 cm); H to 8 in (20 cm)

The low-growing Silverweed gets its name from the silvery, hairy undersides of its compound basal leaves, which grow on hairy leaf stalks.

KEY FACTS

Solitary, 5-petaled flowers grow on long, hairy stalks; pinnate compound leaves are composed of mostly alternating toothed leaflets.

+ **habitat:** Pond and stream banks, shores, marshes, wet meadows, and roadsides

+ **range:** Much of northern and southwestern U.S. and far northern areas into Canada

+ **bloom period:** June–August

The Silverweed's bright yellow blossoms with blunt petals and a cluster of short yellow stamens strongly suggest the flower's place in the Rose family. The flowers may appear to be at a distance from the plant's foliage, but they are connected; both the flower stalks and the compound leaves spring from nodes on often inconspicuous runners. Long used for food and for medicinal purposes, the Silverweed's cooked root has the taste of parsnips or sweet potatoes. A coastal variety is less hairy or even hairless and is found hugging the Pacific coast and the Atlantic coast from New England northward.

Puncture Vine

Tribulus terrestris L creeper to 36 in (90 cm)

Similar to the Silverweed but far less benign, the Puncture Vine can injure grazing animals with the sharp spines on its fruits as well as a toxic substance that causes sun sensitivity.

KEY FACTS

The 5-petaled yellow flowers grow on short stalks that emerge from the axils of the opposite, compound leaves.

+ **habitat:** Fields, open ground, and also disturbed areas; often in sandy or gravelly areas

+ **range:** Naturalized throughout most of the U.S. and parts of Canada

+ **bloom period:** June–September

The Puncture Vine, a European import, sprawls along the ground on creeping stems that send up compound pinnate leaves with leaflets that are opposite but often of unequal lengths. The seed case breaks into five sections, each with a pair of strong, sharp spines. These can cause great damage to livestock when gathered into a bale of hay used for fodder, and they make the species a noxious weed in the eyes of farmers and ranchers. They also give the plant its common names, including Goathead and Caltrop—the latter a reference to the spiked weapon deployed on the ground to impede horses, troops, and vehicle tires.

Common St. John's Wort
Hypericum perforatum H 12–30 in (30–75 cm)

A prolific European import, Common St. John's Wort wasted no time getting established in North America. It now outcompetes many native wildflower species.

KEY FACTS

The small dark glands edge the bright yellow, 5-petaled flowers with many long and yellow stamens.

+ **habitat:** Open woods, meadows, and along roadsides

+ **range:** Introduced and naturalized throughout much of the U.S. and Canada

+ **bloom period:** June–September

The Common St. John's Wort, a hardy perennial, is named for St. John the Baptist. When its flower petals are pinched, they turn red from the oil in their glands; this unusual property associated them with the saint's beheading. Tiny, stiff, linear-to-oval-shaped leaves grow opposite each other on numerous stems. Oil glands also cover the leaves, visible as tiny, translucent dots. The upper stems culminate in multiple flower clusters. The plant as a whole gives off a faint odor of turpentine or balsam. Part of nature's pharmacopoeia for millennia, Common St. John's Wort has pronounced antiviral and antidepressive effects among its many benefits.

Downy Yellow Violet
Viola pubescens H 6–18 in (15–45 cm)

The Downy Yellow Violet is named for its soft, hairy leaves and stems. Leaves near the bottom of the stem are heart-shaped and have scalloped edges.

KEY FACTS

The petals are bright yellow in color; the lateral ones are bearded, and the lower petal is heavily veined with purple.

+ **habitat:** Woods, woodland edges, and thickets

+ **range:** Eastern U.S. and Canada

+ **bloom period:** May–June

The large flowers of the Downy Yellow Violet appear on slender stalks from the axils of the plant's upper leaves. In addition to the notable veins on the lower petal, other petals may have one or two thin lines. A beard of hairs on the lateral petals encourages visiting insects to brush up against the stigma and anthers, where they pick up pollen on their way to sipping nectar from the spur. In summer, most Downy Yellow Violets grow specialized flowers without petals on short stems growing from the plant's root. These never open, but nonetheless become self-fertilized inside the bud.

Common Evening Primrose
Oenothera biennis H 12–72 in (30–180 cm)

The Common Evening Primrose bides its time during the day, waiting until dusk to unfurl its yellow flowers—each destined to last only one night—and allow moths to enter for pollination.

KEY FACTS

The 2-in (5-cm) flower has 4 rounded and lemon yellow petals.

+ **habitat:** Open areas with sandy soil, such as embankments, meadows, and along roadsides

+ **range:** Central and eastern U.S. and Canada

+ **bloom period:** June–September

After a night in bloom, the Common Evening Primrose flowers wither by morning—at least before noon. The plant reveals its blossoms gradually during the blooming period: Lower ones open first and by the time upper flowers bloom, the lower ones have become seedpods. The plant's tall stem is stout and covered with soft hairs. The long, lance-shaped leaves grow alternately on the stem. They and the stem share a coarse texture. Among its many medicinal applications, the Common Evening Primrose is harvested for an oil that is successfully used to treat hormonal imbalances and eczema.

Common Bladderwort
Utricularia macrorhiza H 4–16 in (10–40 cm)

The Common Bladderwort is a rootless, native, carnivorous aquatic plant. It captures prey by means of trigger hairs at the mouth of underwater "bladders," or inflated segments.

KEY FACTS

The 5-lobed flowers are yellow, or rarely pink or purple; the lobes are fused to form 2 lips, the lower lip with a sickle-like spur.

+ **habitat:** Lakes, ponds, slow-moving water, and ditches

+ **range:** Throughout U.S. and Canada

+ **bloom period:** June–August

Common Bladderwort flowers are distinctive, forming loose clusters on a stalk that clears the water; however, much of the interest lies underwater, where fine branches that fork between three and seven times display numerous bladders. The bladders are a transparent green when young, turning darker and brownish as they age. Once a bladder has captured its prey, including water fleas and mosquito larvae—an action that may take as little as $1/460$ of a second—enzymes within it begin to digest the prey. Some experts consider the scientific name of the Common Bladderwort to be *U. vulgaris macrorhiza*.

Common Mullein

Verbascum thapsus H 24–72 in (60–180 cm)

Many people don't realize that the Common Mullein, a denizen of empty lots and roadsides, grows flowers in its second year. They know the Eurasian import only by its grayish leaves.

KEY FACTS

The spikes of dense 5-petaled yellow flowers grow on the woolly stems.

+ **habitat:** Fields, roadsides, disturbed areas, and waste ground

+ **range:** Naturalized throughout U.S. and Canada, except northwestern Canada

+ **bloom period:** June–September

Common Mullein, a member of the Figwort family, is a biennial wildflower that takes off in its second season. A long stem covered with downy hairs emerges from a rosette of large basal leaves, and as the bright yellow flowers start to bloom, the erect spikes take on the appearance of candles or torches—an apt association on several counts. The leaves, when dried, are highly flammable and traditionally were used as candlewicks; the flower spikes were dipped into melted fat and set alight as torches. Even today, the flowers are brewed into a tea that treats respiratory congestion.

Common Monkeyflower

Mimulus guttatus H 6–36 in (15–90 cm)

Clearly defined hairy ridges on the lower lip of the distinctive blossom are a characteristic of the Common Monkeyflower, also known as Yellow Monkeyflower, a member of the Lopseed family.

KEY FACTS

The yellow flowers are sometimes solitary or in clusters, on long stalks, strongly 2-lipped, with red to maroon dots.

+ **habitat:** Wet places such as stream banks and other wet areas

+ **range:** Native to the western U.S. and in Canada; highly naturalized in the East

+ **bloom period:** March–August

The Common Monkeyflower can vary significantly in height and structure, appearing dwarflike with small leaves at times, and tall and imposing at others. The flowers are similar in size and appearance to their relative, the Common Snapdragon. The leaves of Common Monkeyflower also vary significantly in size and shape: Some are oval, some are roundish, some are kidney shaped, and some are heart shaped. Leaves at the bottom of the plant generally have stalks; those at the top often are stalkless and clasp the stem. The species can reproduce by means of a creeping stem, or stolon, that sends down new roots at intervals to form new plants.

Coastal Sand Verbena

Abronia latifolia H 1–3 in (2.5–8 cm)

The Coastal Sand Verbena sprawls along Pacific beaches and dunes, well adapted to the drying effects of wind and sun with its moisture-retaining stems and leaves.

KEY FACTS

The small, yellow, trumpet-shaped flowers form a globular head that grows on a succulent stem.

+ **habitat:** Sandy soils of beaches and dunes

+ **range:** Western coast of the U.S. and Canada to British Columbia

+ **bloom period:** May–August

The perennial native Coastal Sand Verbena forms low, extensive mats of succulent stems that often are buried in the sand and anchored by deep roots. The stems bear many smooth-edged fleshy leaves that are oval to kidney shaped and have slightly undulating surfaces. The flower cluster rises gracefully—*Abronia* comes from the Greek for "graceful"—on a stem that grows from a leaf axil. Two other Sand Verbenas grow in the same habitat along the Pacific coast. The Red Sand Verbena or Beach Pancake (*A. maritima*) is similar but has reddish purple flowers, while the flowers of the Beach Sand Verbena (*A. umbellata*) range from white to deep pink.

Greater Celandine

Chelidonium majus H 12–30 in (30–76 cm)

An introduced member of the Poppy family, the Greater Celandine is noted for a yellowish orange sap traditionally used to treat warts, source of another common name, Wartwort.

KEY FACTS

The bright yellow, 4-petaled flower less than an inch (2.5 cm) wide forms loose clusters.

+ **habitat:** Woods, fields, roadsides, waste ground, and disturbed areas

+ **range:** Native to Eurasia; naturalized throughout northern and central U.S. and Canada in disparate regions

+ **bloom period:** April–August

The individual flowers of the Greater Celandine grow on short stems and form a cluster at the end of a slender stalk up to 4 inches (10 cm) tall. The plant's ribbed leaves alternate on their branching stems and are deeply lobed into rounded segments with irregularly toothed margins. Leaves can be up to 14 inches (35 cm) long and typically are composed of five to nine lobes. In addition to serving as a wart remover, the plant's caustic yellow sap was also used to make eyedrops. A very aggressive import, Greater Celandine pushes out native wildflowers, especially woodland species.

Curlycup Gumweed

Grindelia squarrosa H 12–36 in (30–90 cm)

The flower head of this species is clothed by a distinctive cluster of overlapping bracts with recurved tips—the curly cup. The cup also is very resinous—the gumweed reference.

KEY FACTS

The flower head typically has both disk and ray yellow flowers, although the rays sometimes are absent.

+ **habitat:** Fields, prairies, and waste areas

+ **range:** Throughout the U.S. and Canada, except in far northern areas and southeastern U.S.

+ **bloom period:** July–September

The native daisy-like Curlycup Gumweed, a member of the Aster family, rises on a stalk with 2 to about 20 companions. The flower head, up to 1.5 inches (3.75 cm) long, bears many rays and disks, which are darker than the rays. Waxy, oblong, stemless leaves, often with toothed, curled edges and a pointy tip, alternately clasp the smooth, erect stalk that often is branched near the top. With roots up to 6 feet (180 cm) long, the plant does well under drought conditions. American Indians used the species extensively to treat respiratory ailments and stomach and liver problems, and even as a topical treatment for saddle sores.

Eastern Prickly Pear

Opuntia humifusa H to 18 in (45 cm) in clumping form

Sparse, slender spines and short, barbed bristles on fleshy stems make the Eastern Prickly Pear—eastern North America's main claim to flowering cactus fame—require careful handling.

KEY FACTS

The yellow- to orangish-tinged flowers are 2–3 in (5–7.5 cm) wide; multiple flowers grow on the fleshy segments.

+ **habitat:** Sandy, rocky, and hilly areas; sometimes planted in gardens

+ **range:** Central and eastern U.S. and Canada, excluding the upper Northeast

+ **bloom period:** May–August

The Eastern Prickly Pear takes several forms. In most parts of its range, it is a low clumping plant consisting of numerous flattened, ovoid, shiny green, and fleshy segments. These are covered with clusters of reddish brown barbed bristles called glochids and less numerous long, thin spires. The glochids can be as injurious as the cactus's spines, as they have barbed tips and lodge stubbornly in the skin. Low-growing plants can spread some 3 feet (90 cm) in diameter. In some parts of its range, notably in Florida, the Eastern Prickly Pear grows into an erect shrub, more than 6 feet (180 cm) tall.

Creeping Wood Sorrel
Oxalis corniculata H 4–8 in (10–20 cm)

The Creeping Wood Sorrel is a member of the genus *Oxalis,* so named for the Greek word for "sour," a reference to the sour-tasting oxalic acid that occurs throughout the plant.

KEY FACTS

The small yellow flowers with 5 petals and 5 stamens form an umbrella-like cluster; the petals are separate down to the base.

+ **habitat:** Lawns, flower beds, fields, nursery grounds, and disturbed areas

+ **range:** Naturalized throughout most of the U.S. and Canada

+ **bloom period:** June–September

A low, creeping wildflower with heart-shaped, green to purple leaves, the Creeping Wood Sorrel sends down roots and forms stems from nodes as it grows along the ground. It also spreads by very small brown seeds that explode from a ripe capsule. The angular seeds adhere to many objects and easily transport themselves to new areas. The name sorrel itself comes from the German for "sour," and sorrels were an old-time salad ingredient, rich in vitamin C, though we now know that ingestion of too much oxalic acid can block the absorption of calcium. *Oxalis* plants are often sold commercially as "shamrocks."

Black-eyed Susan
Rudbeckia hirta H 12–36 in (30–90 cm)

The Black-eyed Susan flower head rises singly on a long, erect stem and gets its name from a dark purple center composed of tubular florets that appear almost black from a distance.

KEY FACTS

The flower head is large and daisy-like with yellow rays and a dark, cone-shaped disk.

+ **habitat:** Fields, prairies, and the open woods

+ **range:** Throughout much of the U.S. and Canada, except in parts of the Southwest and the far northern areas

+ **bloom period:** June–October

In the Black-eyed Susan, a member of the Aster family, radiating green bracts surround the emerging blossom. As the flower head opens, its long, orange-yellow rays extend to meet the collar, some 2 to 3 inches (5–7.5 cm) across. The rays sometimes bear reddish brown splotches. In its first year, the plant forms a rosette of lance-shaped or oval hairy leaves; in the second year, tall hairy stems produce the conspicuous flower heads. The species flourishes in challenging conditions and provides nectar for bees, butterflies, and other insects, as well as seeds favored by a variety of birds.

Common Sunflower
Helianthus annuus H 3–10 ft (1–3 m)

We mostly know the Common Sunflower from the mammoth-headed, heavily seeded cultivars that nod from their weight. The wild flower heads are smaller, with smaller centers.

KEY FACTS

The flower heads have dense yellow rays and a brownish red central disk. Oval to heart-shaped leaves are sometimes toothed.

+ **habitat:** Fields, plains, roadsides, and waste areas

+ **range:** Throughout much of the U.S. and Canada, except in far northern areas

+ **bloom period:** June–September

A member of the Aster family, the Common Sunflower grows tall on an erect, hairy stem with many rough, hairy leaves. The stem branches at the top and produces multiple flower heads. The flower heads follow the sun during the day as it moves across the sky. American Indians utilized many parts of the species: They made dyes from the flower rays and seeds of its flat central disk; a tea from the flowers to treat lung ailments; and poultices from the leaves for snake bites. They also ground the seeds into flour and used seed oil for cooking and to dress hair.

Common Dandelion
Taraxacum officinale H 2–16 in (5–40 cm)

An introduced species from Eurasia that has spread from coast to coast, the Common Dandelion is easily recognized with its yellow flower heads, jagged leaves, and wispy seed heads.

KEY FACTS

The single yellow flower head composed of many ray-like toothed ligules is cupped by green bracts on a solitary stem with leaves all in a basal rosette.

+ **habitat:** Lawns, fields, roadsides, and disturbed areas

+ **range:** Naturalized throughout the U.S. and Canada

+ **bloom period:** Mainly March–September

The bane of lawn warriors, the Common Dandelion would figure on many top ten weeds lists. The name derives from the French for "lion's tooth," a reference to the distinctive edges of the leaves. This hardy wildflower, a member of the Aster family, blooms much of the year—a few stragglers even show up in the dead of winter. All parts of the plant exude a milky juice when broken. The species is highly edible and valued in traditional medicine for its support of liver and kidney function and diuretic properties. The roots are fermented into beer, and the flowers are used to make a potent wine.

Tickseed

Coreopsis species H 6 in–8 ft (15–240 cm)

Members of the large Aster family, tickseeds in North America encompass more than two dozen species that share mostly yellow, daisy-like flower heads of varying size and delicate leaves.

KEY FACTS

Flower heads are largely daisy-like, yellow, mostly with both disk and ray flowers; some are sunflower-like; a few have pink flowers.

+ **habitat:** Old fields, prairies, roadsides, open woods, swamps, and sandy areas

+ **range:** Throughout U.S. and Canada, except far northern areas; widely cultivated

+ **bloom period:** May–September

Just as there is wide variation among the flowers, tickseed leaves and leaflets can appear at the base of the plant or on the stem, they can be opposite or alternating or appear in whorls, and can be lance-shaped, threadlike, or may be lacking entirely on bare leaf stems. Tickseeds often form large colonies whose member plants bloom over a long period of time. The genus and common names for tickseeds refer to the bedbug, an image related to the appearance of the plants' flattened, seedlike fruits. Many tickseed species have escaped cultivation and are widespread outside their customary native ranges.

Goldenrod

Solidago species H 12–96 in (30–240 cm)

Goldenrods unfairly take the fall for another common wildflower—Ragweed—as a major allergen, but the genus traditionally is a potent healer, as its name—Greek for "make well"—testifies.

KEY FACTS

The goldenrod species have both disk and ray flowers, which are relatively small and are usually yellow.

+ **habitat:** Fields, prairies, open woods, thickets, along roadsides, and in salt marshes

+ **range:** Throughout the U.S. and Canada

+ **bloom period:** July–November

The many members of the genus *Solidago* bear flower heads that vary in shape, size, and arrangement on the stem, but typically are a distinctive golden yellow, although *S. bicolor* has white ray flowers, and *S. ptarmicoides* has white disks and rays. Flower head arrangement can vary from flat-topped rounds to loosely branching, arched clusters that sometimes are one-sided. Leaves are basal or alternate on the stem; they are succulent in salt marsh species. Goldenrods are late summer flowers that rely on insects for pollination. The real culprits in the seasonal sneezefest are Ragweeds and other wind-pollinated species. A number of goldenrod species will hybridize.

Common Jewelweed/Spotted Touch-me-not
Impatiens capensis H 24–60 in (60–150 cm)

Pendent yellow-and-orange flowers ornament this leafy annual with a preference for shade and moisture. Touching the swollen, seed-filled, ripe fruits causes them to burst explosively.

KEY FACTS

Drooping yellow and orange tubular flowers have brown, red, and orange spotting; 2 grow on each flower stalk, flowering one at a time.

+ **habitat:** Shady wetlands, woods, stream and riverbanks, and pond edges

+ **range:** Throughout much of northern, central, and eastern U.S. and Canada

+ **bloom period:** June–October

If not for the graceful, nodding flower—the species' obvious jewel—this plant would seem more of a weed: long, slender, succulent stems laden with oval leaves growing mostly alternately along the stem. The nectar-rich flowers are a favorite of hummingbirds, bees, and butterflies. Sap from the plant's leaves and stems thwarts rash development and soothes itching from Poison Ivy exposure, and often is made into soap for this purpose. The sap also serves as an effective fungicide and treatment for athlete's foot. The similar yellow Pale Jewelweed (*I. pallida*), found in eastern North America, shares these properties.

Western Wallflower/Prairie Rocket
Erysimum capitatum H 6–36 in (15–90 cm)

Western Wallflowers get their common name from a similar Old World relative in the Mustard family that often is seen embedded in drystone walls, a common landscape feature in Europe.

KEY FACTS

Orange to yellow flowers with 4 petals and 4 sepals grow in loose terminal clusters on short stalks.

+ **habitat:** Plains, foothills, deserts, open woods, cliffs, and coniferous forests

+ **range:** Throughout much of western and central U.S. and Canada

+ **bloom period:** March–July

Western Wallflowers often stand out in their habitats in dense clusters of usually erect stems bearing showy flowers with knobby stigmas. The plants are quite leafy—and hairy—with lance-shaped hairy lower leaves and smaller upper leaves, both growing alternately on a hairy stem that is sometimes branched. Sandy and rocky environments suit this hardy plant. It is also known as Prairie Rocket, not for its shape but for its relationship to Salad Rocket, another species in the Mustard family that also is known as Arugula. The Western Wallflower is highly variable, appearing in a number of other color palettes, including yellow, orange-brown, and maroon.

California Poppy
Eschscholzia californica H 6–24 in (15–60 cm)

Fields of California Poppies grace the landscape in the Golden State, where they have been the state flower since 1903. Bright sun opens the flowers; they close at night and on cloudy days.

KEY FACTS

Yellow to brilliant orange flowers have 4 satiny petals that often have an orange spot at the base.

+ **habitat:** Coastal dunes, open forests, meadows, plains, and desert edges

+ **range:** West Coast of the U.S. from southern California into southern Washington

+ **bloom period:** February–September

You can distinguish a California Poppy from other poppy species in its genus by its disk-like collar that sits at the base of the blossom. The flower has many stamens; often a dozen or more crowd the center. A closed flower takes the shape of a nightcap. The California Poppy's leaves have three deeply divided, feathery lobes and typically are blue-green with a grayish cast. Depending on the location, the species can be low and spreading or tall and erect. Also depending on environmental circumstances, it can be either an annual or perennial. The poppy contains several compounds with potential for treating cancer.

Common Orange Daylily
Hemerocallis fulva H 24–72 in (60-180 cm)

This Asian native has long been cultivated in North America as an ornamental plant. Escapees have naturalized widely, making the ephemeral flowers a common roadside attraction.

KEY FACTS

Orange flower is composed of 3 petals with wavy edges and 3 sepals with smooth ones.

+ **habitat:** Roadsides, meadows, woodland edges, stream banks, and disturbed areas

+ **range:** Naturalized in much of U.S. and Canada, except southwestern U.S. and western Canada

+ **bloom period:** May–July

True to its name, the flower of the Common Orange Daylily lasts only one day. But a number of the funnel-shaped blossoms form on each leafless stalk to prolong blooming. The flowers face out, not downward like other lilies, and the petals curve backward. The multibranched stalks, or scapes, rise directly from the root and are taller than the profuse strap-shaped leaves. The blossoms feature yellow centers and radiating thin stripes as well as pronounced veins. The species very rarely forms seeds; most Common Orange Daylily plants produce by cloning from fleshy roots or rhizomes. Another escaped cultivar produces doubled flowers.

Blackberry Lily/Leopard Lily

Iris domestica (Belamcanda chinensis) H 18–48 in (45–120 cm)

The Blackberry Lily spreads wide its distinctly spotted tepals (look-alike petals and sepals) as if to draw attention to its short-lived beauty, as each blossom lasts only one day.

KEY FACTS

The flower has 6 orange tepals that are heavily spotted with red; the flowers grow in multiples on naked, branched stems.

+ **habitat:** Roadsides, grasslands, meadows, open woods, rocky outcrops, and disturbed areas

+ **range:** Naturalized throughout central and eastern U.S.

+ **bloom period:** June–August

A native of China, the Blackberry Lily has escaped cultivation to become widely established in North America. Showy flower sprays appear in the midst of fan-shaped clusters of long, narrow, flat, medium-green leaves. Pear-shaped seedpods form in late summer. When ripe, they split to reveal a cluster of shiny blackberry-like seeds, the source of the plant's common name; the spots, of course, lend another name—Leopard Lily. A species of a different genus also goes by the name Leopard Lily; *Lilium pardalinum*, native to California, has somewhat similarly spotted tepals that curl. Its range does not overlap with that of *Iris domestica*.

Wood Lily

Lilium philadelphicum H 12–36 in (30–90 cm)

Despite its name, the Wood Lily flourishes equally in sandy and brush-covered habitats, with its striking blooms standing erect at the top of their stalks and not drooping like other lilies.

KEY FACTS

Orange-red blooms have tepals that lighten toward the tapered base and are spotted with brown.

+ **habitat:** Prairies, open woods, roadsides, power line cuts, dunes, barrens, and mountain meadows

+ **range:** In U.S. and Canada, except far northern areas and parts of western and southeastern U.S.

+ **bloom period:** June–August

From one to four funnel-shaped blossoms crown the summit of the Wood Lily plant. Narrow, lance-shaped leaves, pointed on the ends, are in whorls of three to six, and usually alternate on each tall, erect stem. American Indians ate the flavorful bulbs and sprinkled the nutritious pollen on numerous dishes. A medicinal tea treated stomach ailments and fever, and aided childbirth. The bulbs and leaves were used in poultices on wounds, bruises, and spider bites. The Wood Lily is losing ground within portions of its range due to development, overpicking, grazing, and browsing by increasing numbers of White-tailed Deer.

Butterfly Weed/Orange Milkweed
Asclepias tuberosa H 12–36 in (30–90 cm)

Like its cousin the Common Milkweed, the Butterfly Weed attracts insects with an abundance of tempting nectar, but this species serves it up in brilliantly colored orange flowers.

KEY FACTS

Clusters of small, orange flowers top erect, hairy stems; long, oblong leaves alternate on the stem.

+ **habitat:** Prairies, open woods, road-sides, and disturbed areas

+ **range:** Through-out most of the U.S., except the Northwest, and eastern Canada; highly cultivated

+ **bloom period:** May–September

The individual flowers of the showy, flat-topped clusters have five bent-down petals and a crown of five erect hoods. The clusters them-selves are about 2 inches (5 cm) across. Slits in the flower's central column contain pollen sacs, and the plant relies on visiting bees and butterflies getting a leg tangled in a slit and picking up and carrying off a load. Unlike the Common Milkweed, this species exudes a clear, not milky, sap when the leaves are bruised. If trying to attract butterflies to your garden, instead of planting imported purple *Buddleia* species, choose native Butterfly Weed, which supports species such as the Monarch and the Queen.

Fiddleneck
Amsinckia species H 4–30 in (10–75 cm)

Fiddleneck flowers develop along curled stems in a characteristic fiddlehead shape, unfolding as they grow into hairy, one-sided clusters of yellow trumpet-like flowers.

KEY FACTS

Small, 5-lobed trum-pets ranging from pale yellow to yellow-orange populate one side of the flower stem.

+ **habitat:** Dry areas: meadows, open woods, shrubby and disturbed areas

+ **range:** Native to western U.S. and Canada but were introduced east of the Rockies in scattered locations

+ **bloom period:** February–June

Fiddlenecks are a hairy lot. The flowers develop at the tips of bristly, green stems and branches, with bristly, lance-shaped leaves. As the stem uncoils, the flowers appear first from the bottom of the cluster. Later, each flower produces a fruit comprising four triangular nutlets that start out green before turning gray to black. Ten different species of the native annual fiddleneck popu-late western North America and locations to the east where they have been introduced. Differences among them are often very subtle, making iden-tification difficult. Some species have red splotches in the throats of their flower tubes.

Hoary Puccoon
Lithospermum canescens H 4–20 in (10–50 cm)

The word "Puccoon" in this wildflower's common name is of American Indian origin and refers to plants that yield pigment for dye or paint, in this case usually a deep yellow.

KEY FACTS

Flat-topped, 5-lobed, yellow-orange flowers grow in clusters. The leaves are long and narrow, and largely stalkless.

+ habitat: Fields, open woods, grassland edges, dry and sandy areas

+ range: Throughout much of central and eastern U.S. and Canada

+ bloom period: April–June

The leaves and stems of the Hoary Puccoon are densely covered with soft, grayish white hairs, both color and fuzziness suggesting the adjective "hoary." The half-inch-long (1.25 cm) flowers grow in flat or curled clusters at the ends of stems with alternating, lance-shaped leaves. A similar species, the Hairy Puccoon (*L. caroliniense*), sports hairier leaves with longer hairs. In addition to making dye and paint from the roots of the plant, American Indians made an herbal tea from its leaves and used it to wash feverish and convulsive patients. Related species of puccoon produce red or purple pigments.

Orange Milkwort
Polygala lutea H 4–16 in (10–40 cm)

Like other species in the Milkwort family, the Orange Milkwort seems misnamed. A milky juice does not ooze from leaves, stems, or elsewhere when crushed.

KEY FACTS

Bright orange flower has 5 sepals, 2 of which form "wings," and 3 fused petals that form a tube with a fringe. Flowers grow in compact clusters.

+ habitat: Moist habitats, including pine barrens, damp sandy or peaty soil, meadows, and ditches

+ range: Coastal states from mid-Atlantic to Gulf Coast

+ bloom period: June–October

The dense, brilliant orange flower heads of the Orange Milkwort grow on single or branched stems heavily populated with alternating, spoon-shaped leaves. Leaves at the base of the plant are the same shape but significantly broader and form a rosette. Milkworts get their name from the traditional belief that consuming the plants increased milk production in cows and nursing human mothers. The species' scientific name also seems to contain a misnomer. *Lutea* means "yellow"— clearly not the case with plants in bloom, but appropriate when the blossoms dry out. The Milkwort flower's complex structure makes pollination a tricky business for insects.

Nodding Onion
Allium cernuum H 6–24 in (15–60 cm)

A pronounced bend in the stem just under the flower head gives the Nodding Onion its distinctive shape. This perennial species carries the typical onion odor and taste.

KEY FACTS

Tiny, bell-shaped flowers form an umbel, or cluster, at the top of a leafless stem.

+ **habitat:** Prairies, glades, rocky embankments, meadows, stream banks, and moist, cool soils at higher elevations

+ **range:** Throughout most of U.S. and Canada, except much of Southeast and far northern areas

+ **bloom period:** June–October

Up to several dozen tiny bells composed of six pointed white or pale pink tepals (look-alike petals and sepals) attach by slender stalks to the tip of the scape, supported by green bracts. The umbel nods on a bent scape, a stem that is leafless and arises directly from the plant's underground bulb. Three to five thin, flat, bladelike leaves up to 16 inches (40 cm) long emerge at the base, and sheathe the scape. American Indians, trappers, and settlers found food and medicinal value in this species, the most widely distributed wild onion, and the similar, erect Wild Autumn Onion (*A. stellatum*).

Lesser Purple Fringed Orchid
Platanthera psycodes H 6–40 in (15–100 cm)

The pronounced beauty of the Lesser Purple Fringed Orchid has been its undoing. This native orchid species is listed as threatened or endangered throughout much of its range.

KEY FACTS

White, pink, or purple blooms have a highly fringed, 3-lobed lip, and a long, tubular spur at base.

+ **habitat:** Moist woods, stream banks, meadows, slopes, marshes, old fields, and roadsides

+ **range:** Throughout much of eastern U.S. and Canada, except Southeast and far northern areas

+ **bloom period:** June–September

This spectacular native orchid bears many fringed flowers loosely or densely arranged in a long cluster toward the tip of the stem. Alternate oval or lance-shaped leaves sheathe the smooth, green stem. Those at the bottom of the plant are up to 8 inches (20 cm) long; leaves at the top are much smaller and bract-like. Unlike their tropical counterparts, which commonly are epiphytes (air plants), most North American orchids take root on land. This species is superficially similar to its larger counterpart, the Large Purple Fringed Orchid (*P. grandiflora*), except for the size of the flowers.

Lady's Thumb
Persicaria maculosa H 8–36 in (20–90 cm)

A prolific European import, Lady's Thumb is generally considered a weed. The name comes from a dark spot in the middle of the leaf that was seen to resemble a woman's thumbprint.

KEY FACTS

Tiny pink to purple flowers, composed of 5 tepals joined at the base, form slender, elongated spikes at the end of pinkish stems.

+ **habitat:** Moist areas including roadsides and disturbed areas

+ **range:** Throughout the U.S. and Canada, except in some far northern areas

+ **bloom period:** March–November

The delicate, pink flower spikes of Lady's Thumb make subtle arches on the ends of jointed stems. The stems support lance-shaped leaves, arranged alternately, that can be smooth or covered with short, stiff hairs, and have a papery sheath around the stem at their bases. Lady's Thumb can grow straight up or it can sprawl over the ground. It is opportunistic, often taking up space at the expense of native wildflower species. The plant requires very little substrate to gain a foothold; with enough moisture, it will sprout easily in a thin layer of organic debris in a gutter. The species is often referred to as Smartweed.

Virginia Spring Beauty
Claytonia virginica H 6–12 in (15–30 cm)

An early herald of spring, the Virginia Spring Beauty's delicate blossoms pop up on forest floors, bearing the name of 18th-century American botanist John Clayton.

KEY FACTS

The blossoms have 5 whitish to pink petals veined with deeper pink and 5 stamens; they grow in loose clusters.

+ **habitat:** Moist woods, clearings, seeps, bluffs, lawns, and roadsides

+ **range:** Throughout much of central and eastern U.S. and Canada

+ **bloom period:** January–May

The low-growing, perennial Virginia Spring Beauty often makes an appearance while it is still winter, pushing up from underground potato-like tubers. The plant has two levels of leaves, a basal pair and a pair of opposite stem leaves that are pointed and sometimes fused where they meet. The loose clusters of small, candy-striped flowers make a striking display among the emerging green of the forest, especially in a large stand. But by the time its glossy seeds have ripened, the rest of the plant has all but disappeared above ground. Both American Indians and colonists relished the tuber, which tastes like a chestnut.

One-seeded Pussy Paws

Cistanthe monosperma H 4–20 in (10–50 cm)

The top-heavy, upturned "cat feet" of the One-seeded Pussy Paws may lie on the ground or rise above the plant's basal rosette. This perennial wildflower thrives in a dry, rocky environment.

KEY FACTS

Small flowers ranging from white to pink have 4 petals and uneven sepals, giving the flower a ruffled appearance.

+ **habitat:** Sandy, rocky, and gravelly areas in mid to high elevations

+ **range:** Western U.S. to British Columbia

+ **bloom period:** April–September

The blossoms of the One-seeded Pussy Paws often form one-sided, umbrella-shaped clusters that look like the pads of a cat's paw. Fleshy, spatula-shaped leaves create a rosette at the base of the plant. Two or more stems rise from the basal leaves; these can be erect or prostrate, sometimes varying with the time of day. Leaves higher up are smaller. The perennial One-seeded Pussy Paws are members of the Miner's Lettuce family (previously in the Purslane family), related to Spring Beauty and Bitterroot. The Latin species name, *monosperma,* refers to the seed capsule, which often contains only one shiny black seed.

Bitterroot

Lewisia rediviva H 1–3 in (2.5–8 cm)

In 1806, Lewis and Clark brought back a Bitterroot plant from Montana that received Lewis's name, and the name *rediviva,* meaning "reborn," after the dried specimen was coaxed into bloom.

KEY FACTS

The flower is composed of a variable number of oblong rose-pink petals, usually 10–19; 20–50 stamens crowd the flower's center.

+ **habitat:** Rocky, gravelly areas and wooded and brushy slopes

+ **range:** Western U.S. and Canada

+ **bloom period:** March–June

Bitterroot, Montana's state flower, has large flowers about 2 inches (5 cm) wide that can range from white to light pink to deep pink. Their height is exaggerated by the fact that they grow low to the ground on very short solitary stems. By the time they bloom, the fleshy, cylindrical green leaves that pushed up earlier have all but disappeared, leaving the flowers to grace patches of rock or gravel with moisture and good drainage. Northwestern American Indians routinely ate the nutritious fleshy taproots after boiling them long enough to remove the bitterness, which also causes them to become jellylike.

New England Aster

Symphyotrichum novae-angliae H 24–72 in (60–180 cm)

One of the largest and most imposing asters, with flower heads measuring more than an inch (2.5 cm) across, the New England Aster has a much wider range than its name suggests.

KEY FACTS

Dense petal-like rays that range from blue to rose-purple and rarely white surround an orange-yellow disk.

+ **habitat:** Wet meadows, moist fields, stream banks, roadsides, and thickets

+ **range:** Much of U.S. and Canada, except southernmost areas, desert lands, and far northern areas

+ **bloom period:** July–October

The striking flower heads of the New England Aster have a base of hairy bracts with curled tips and grow on hairy and sticky stalks that form a cluster at the top of the plant, making a dramatic statement. The plant's mainly smooth-edged leaves measure up to 4 inches (10 cm) long and alternate along the stout plant stems, clasping them at their bases. The species is also known as the Michaelmas Daisy because its blooms linger into the fall and often are still present at the time of the feast of St. Michael on September 29. This aster species is pollinated by long-tongued bees and butterflies.

Deptford Pink

Dianthus armeria H 6–24 in (15–60 cm)

Named for the London suburb where it once flourished, the Deptford Pink is an introduced member of the Carnation family admired for its handsome pink flowers.

KEY FACTS

Deep pink flower has 5 spreading, jagged-edged petals dotted with white; sepals form a tube.

+ **habitat:** Pastures, old fields, woodland edges, roadsides, disturbed areas, and waste areas

+ **range:** Introduced; naturalized throughout most of U.S. and Canada, except far northern areas

+ **bloom period:** May–September

The Deptford Pink resembles a simplified carnation, its highly saturated pink blooms highlighted by the profusion of light green, leaf-like bracts that stand erect around the loose, flat-topped flower clusters. Narrow, erect leaves appear in opposite pairs at wide intervals on slender, branched stems. This Eurasian native was introduced and has escaped cultivation. It self-seeds prolifically, giving it a well-deserved reputation as a weedy invasive. The Deptford Pink is somewhat similar to another introduced species, the Sweet William (*D. barbatus*), which is bulkier with fatter petals and is widely cultivated, but less prone to escape.

Soapwort/Bouncing Bet
Saponaria officinalis H 12–30 in (30–75 cm)

When crushed, the leaves of the Soapwort exude a gluey juice that lathers in water, a trait that gives this member of the Carnation family its common name.

KEY FACTS

Pinkish or white 5-petaled, tubular flowers grow in a flat-topped cluster; the petals are notched or deeply cleft.

+ habitat: Waste ground, such as along roadsides and railroad beds

+ range: Naturalized throughout most of the U.S. and Canada, except in far northern areas

+ bloom period: July–September

The Soapwort stem is wide, smooth, and straight and rarely branches. Long, oval, veined leaves grow opposite each other on the stem. The Soapwort has two methods to aid its rapid spread: The hardy plant spreads by means of seeds and also by underground runners that send up new plants. Soapwort was one of the first flowers introduced by colonial settlers, for its useful cleaning properties as well as a host of medicinal uses. Its alternative common name, Bouncing Bet, may refer to the activity of barmaids, known as Bets, who cleaned ale bottles by filling them with water and a plant sprig and shaking them up.

Pacific Bleeding Heart
Dicentra formosa H 8–18 in (20–45 cm)

Clusters of pendent pink blossoms dangle from arched stalks to create a delightful natural valentine, an alluring benefit of a walk in damp western habitats.

KEY FACTS

The flowers are heart-shaped, puffy, and range from creamy white to rose-purple; tips of outer petals curve outward.

+ habitat: Damp woods, clearings, meadows, and stream banks

+ range: Native from British Columbia through California; widely cultivated in the U.S. and Canada

+ bloom period: March–July

Pacific Bleeding Heart blossoms, about three-quarters of an inch (2 cm) long, appear in small, branched clusters on a stalk that rises directly from the crown of the plant's root. The deeply cut, fern-like green basal leaves, often with a bluish cast, develop in the same manner. After flowering, this native perennial produces kidney-shaped seeds in a long conical pod. Some Northwest Indian tribes chewed the raw, succulent taproots of the species to alleviate toothache and also made a decoction of the roots that was taken to purge worms. The shade-loving plant is widely cultivated as a garden ornamental.

Spotted Geranium/Wild Cranesbill

Geranium maculatum H 12–24 in (30–60 cm)

The Spotted Geranium builds on the principle of five: sepals, petals, leaf lobes, and often blooms in a cluster. The alternate common name refers to the shape of the beaked fruit capsule.

KEY FACTS

The pink, violet, or white flower has 5 fine-lined petals and 5 pointed green sepals underneath.

+ **habitat:** Moist woods, meadows, thickets, and woodland edges

+ **range:** Throughout most of central and eastern U.S. and Canada

+ **bloom period:** March–July

Spotted Geranium flowers appear in small clusters of three to five above green palmate leaves with five deep lobes. The lobes themselves are lobed or toothed on the margins. The species reproduces by means of seeds and also by rhizomes that generate new growth, but it does so in moderation and does not overtake other plants in its environment. The genus name, *Geranium*, derives from the Greek for "crane." American Indians commonly used tea made from boiled plants to treat diarrhea, and they also bathed mouth sores with an infusion that capitalized on the plant's astringent properties.

Swamp Rose

Rosa palustris H 36–96 in (91–244 cm)

A member of the sizeable Rose family, the Swamp Rose usually grows into a medium-size upright shrub, seeking out moist habitats with rich, loamy soil.

KEY FACTS

The flower—up to 3 in (7.5 cm)—has 5 pink petals, a ring of yellow stamens, and a central flattened mass of stigmas.

+ **habitat:** Swamps, marshes, ditches, stream banks, and pond and lake edges

+ **range:** Native throughout most of the eastern U.S. and Canada; widely cultivated

+ **bloom period:** May–June

Swamp Rose blossoms punctuate the shrub's bushy, leaf-filled branches and prickly stems for only a month or two each year. The leaves are odd-pinnate, usually composed of seven leaflets, and the prickles are curved. Red hips, or fruits, replace the flowers; as in other rose species, they are rich in vitamin C. A Swamp Rose bush is like a wildlife grocery, supplying food for scores of different species of insects, birds, and mammals including White-tailed Deer and American Beaver. It is also like a condo, providing sheltered nest sites for many bird species, including warblers and the Northern Cardinal.

Purple Loosestrife
Lythrum salicaria H 12–60 in (30–150 cm)

Introduced from Europe, Purple Loosestrife makes a dramatic statement in large stands, at a cost to native aquatic plants and wildlife. As such, it is classified as a noxious weed in many states.

KEY FACTS

Lavender or rarely magenta to red flowers with 5–7 wavy petals occur in small clumps on long flower spikes.

+ **habitat:** Wet areas, including meadows, ditches, and floodplains

+ **range:** Throughout much of the U.S. and Canada, except in far northern areas and parts of southeastern U.S.

+ **bloom period:** June–September

The terminal flower spike of the Purple Loosestrife takes up about the last foot (30 cm) or more of the plant's long, stout, angled stem. Lance-shaped leaves without stalks appear opposite each other or in whorls along the stem, and rounded or heart-shaped ones grow at its base. Dozens of stems can rise from a single rootstock, and the flowers on these can produce millions of tiny seeds; abundant nectar encourages pollination. The genus name derives from the Greek for "blood," a reference to the plant's color or possibly to the styptic (blood-staunching) quality of some loosestrife species.

Large Beardtongue
Penstemon grandiflorus H 18–48 in (45–120 cm)

The Large Beardtongue is a *Penstemon*, a genus of more than 200 species that is endemic to North America. These attractive wildflowers have been developed into popular cultivars.

KEY FACTS

A large broadly tubular pink to lavender corolla flares into lobes, forming an upper lip with 2 lobes and lower lip with 3 lobes.

+ **habitat:** Dry prairies, plains, woods, and thickets

+ **range:** Throughout most of the central U.S.; escaped from cultivation elsewhere

+ **bloom period:** May–June

The Large Beardtongue bears large, inflated flowers, up to 2 inches (5 cm) long, that grow horizontally from the axils of leafy bracts. The bracts are smaller but similar in appearance to the upper stem leaves. Fine purple lines in the flower's throat serve as nectar guides for pollinators such as long-tongued bees. Leaves at the base form a rosette in the plant's first year; in the second and subsequent years, flower stalks emerge. *Penstemon* means "five stamens," which all the species in this genus have. One of them typically is sterile and hairy—the source of the common name Beardtongue.

Virginia Meadow Beauty
Rhexia virginica H 6–36 in (15–90 cm)

The Virginia Meadow Beauty mostly confines itself to wet places where its classic looks stand out. The attractive native perennial also goes by the name Handsome Harry.

KEY FACTS

The flower has 4 widely spreading, pink to purple petals and 8 showy stamens with long, slender anthers.

+ **habitat:** Wetlands, wet meadows, pond edges, and ditches

+ **range:** Much of central and eastern U.S. and Canada

+ **bloom period:** May–October

The flowers of the Virginia Meadow Beauty form loose clusters at the ends of hairy, squared, slightly winged, erect stems. The blossoms have petals nearly as long as they are wide and long stamens with large, bent, bright yellow anthers. Oval leaves clasp the stem opposite to each other and have rounded bases and three prominent veins. The fruit of this species is a reddish, distinctively urn-shaped, four-pointed capsule that holds many tiny seeds. Members of the Melastome family typically are tropical, but a number of different *Rhexia* species are native to temperate North America.

Fireweed
Chamerion angustifolium H 24–96 in (60–240 cm)

The aptly named Fireweed often makes an appearance after a wildfire, when it moves into a burned-out forest area, taking advantage of sunny conditions to form large, impressive stands.

KEY FACTS

The flower is composed of 4 pink, rose-purple, or rarely white petals; it has 8 even-length stamens, and a 4-lobed stigma.

+ **habitat:** Open areas, slopes, and roadsides

+ **range:** Mainly western and northern U.S. and throughout most of Canada

+ **bloom period:** June–September

A member of the Evening Primrose family, the Fireweed sends up reddish stalks with loose clusters of blossoms in terminal spikes. The blossom, though four-petaled, is less symmetrical than that of the Virginia Meadow Beauty. Its leaves are long, narrow, alternate, and willowlike, and its seeds form in slender pods that peel back from the top when ripe. The hair-tufted seeds disperse efficiently—and far—in the wind, helping this opportunistic species establish itself rapidly. Following the eruption of Mount St. Helens, it was one of the first plant species to populate the wasteland, adding an early burst of color.

Tongue Clarkia/Diamond Clarkia

Clarkia rhomboidea H 12–36 in (30–90 cm)

The Tongue Clarkia's unusual diamond-shaped petals with toothed bases help identify this annual member of the Evening Primrose family. It also is noted for its blue-gray pollen.

KEY FACTS

The flower is rose or lavender-pink, sometimes spotted, with 4 diamond-shaped petals each with 2 small teeth at base.

+ habitat: Open woods and pine forests; slopes to about 10,000 ft (3,000 m)

+ range: Western U.S. and northward into British Columbia

+ bloom period: June–July

Tongue Clarkia keeps a fairly narrow profile. It sends up a thin, erect, usually unbranched stem with sparse numbers of short, smooth-edged, lance-shaped leaves. The leaves attach alternately to the stem by means of short stalks. Few flowers adorn the upper portions of the stems. The distinctive flower buds are nodding, and often have pointed tips. This genus is named for William Clark, the explorer who accompanied Meriwether Lewis, and with him served as co-leader of the Corps of Discovery, on a transcontinental expedition to the Pacific Northwest from 1804 to 1806. The expedition collected many floral and faunal specimens.

Trailing Arbutus/Mayflower

Epigaea repens H 2–3 in (5–8 cm) L to 16 in (40 cm)

For those patient and lucky enough, searching among woodland leaf litter in early spring may reward with a glimpse of the Trailing Arbutus with its fragrant flowers among evergreen foliage.

KEY FACTS

The half-inch (1.25 cm) corollas are pink to white, more or less trumpet-shaped, and have 5 lobes and hairy throats.

+ habitat: Moist and dry woods and in rocky places

+ range: Eastern U.S. and Canada

+ bloom period: February–May

Creeping along the ground, the tangled, hairy stems of Trailing Arbutus, a low-growing member of the Heath family, are replete with leathery leaves and flowers that grow on stem ends or emerge from the leaf axils (the crook where the leaf joins the stem). Fleshy capsules replace the flowers. The species serves as a bellwether for environmental disturbance, being sensitive to changes in climate and habitat; it has become more rare over time. Eastern Indian groups, such as the Cherokee and the Iroquois, made decoctions of the plant to treat mainly stomach and kidney ailments.

Dark-throated Shooting Star
Dodecatheon pulchellum H 2–16 in (5–40 cm)

Less is more for the Dark-throated Shooting Star, also known as the Western Shooting Star. Usually only a few flowers emerge in a cluster from the tip of a single scape (a long leafless stalk).

KEY FACTS

The lobes of the 5-lobed magenta flower sweep back, allowing the stamens and style to protrude.

+ **habitat:** Moist, wet meadows, prairies, bogs, and stream banks from sea level to the timberline

+ **range:** Throughout the western U.S. and Canada

+ **bloom period:** April–August

The flowers of the Dark-throated Shooting Star point downward until they are pollinated, and then turn heads up, all the while impersonating their celestial namesake. Leaves that are mostly oblong grow in profusion at the base of the plant. The leaves can be toothed or have smooth edges, and can have smooth or hairy surfaces. There is also significant variation in the number of flowers on a scape of this native perennial. Some western variations seem to inter-breed and produce intergrades. An eastern species, *D. meadia*, has paler flowers that range from white to pale pink to lavender.

Pink Turtlehead/Lyon's Turtlehead
Chelone lyonii H 12–36 in (30–90 cm)

This member of the Plantain family sports a distinct flower that is unmistakably turtle-like. Its genus name is derived from the Greek for "tortoise."

KEY FACTS

The 2-lipped flow-ers are deep pink to purple with yellow bearded interior of lower lip.

+ **habitat:** Moist woods, stream banks, and coves

+ **range:** Native to southeastern U.S.; escaped cultivars naturalized in the Northeast

+ **bloom period:** July–October

Compact clusters of striking pink blossoms against deep green foliage make the Pink Turtlehead an unusually attractive wildflower. The shape of its flower adds a comical element. The flowers are borne mainly at the tops of erect stems with many opposite, oval or lance-shaped, toothed leaves. The species prefers a moist footing. Pol-linating bees enter the concave, upper, shal-lowly two-lobed lip of the turtlehead's "mouth," attracted by the ridged, hairy, slightly three-lobed lower lip with its yellow "signals." The species is named for John Lyon, a transplanted Scot who made important contributions to early 19th-century American botany.

Rose Pink
Sabatia angularis H 12–36 in (30–90 cm)

At the center of the Rose Pink flower, a member of the Gentian family, is a star formed by yellow marks at the base of each petal. Dark pink outlines the star.

KEY FACTS

Rose pink to white flower has 5 petals that are slightly joined at the base, and a greenish yellow center with yellow stamens and anthers.

+ **habitat:** Prairies, marshes, open woods, thickets, old fields, and roadsides

+ **range:** Throughout much of eastern and southern U.S. and Canada

+ **bloom period:** July–September

The biennial Rose Pink starts in the first year as a basal rosette of leaves. Its four-angled main stem shoots up the following year, branching near the top. Smooth oval leaves more than an inch (2.5 cm) long closely attach opposite to each other on the stems. A single blossom about an inch wide forms at the tip of the main stem, and others join it on side stems to form a cluster called a cyme. The fragrant flowers open during the day and close at night. After flowering, a single-chambered capsule develops that contains many light, tiny seeds. These can easily disperse on the wind.

Spreading Dogbane
Apocynum androsaemifolium H 6–48 in (15–120 cm)

The poisonous nature of this milkweed relative gives the species its common and genus names: *Apocynum* derives from Greek words meaning "away dog," referring to its toxicity to dogs.

KEY FACTS

Small, pink or white, 5-lobed bell-shaped flowers grow in a terminal cluster.

+ **habitat:** Woods, woodland edges, prairies, meadows, fields, and roadsides

+ **range:** Throughout much of the U.S. and Canada, except in far northern areas and some southeastern states

+ **bloom period:** June–August

Spreading Dogbane produces many forked, branching stems that are green to reddish and bear loose clusters of slightly nodding flowers at the ends. The interior of the fragrant flower has darker pink stripes that lead to the nectar sources. The opposite, lance-shaped leaves either spread out from the stem or droop noticeably; they tend to have smooth upper surfaces and slightly hairy lower ones. Like milkweeds, Spreading Dogbane exudes a milky sap when its stems or leaves are crushed. A related species, *A. cannabinum* or Indian Hemp, has smaller greenish white flowers. American Indians used its long stem fibers to make rope.

Rose Vervain
Glandularia canadensis H 5–10 in (13–25 cm)

A member of the Vervain family, the Rose Vervain either creeps horizontally, forming a mat, or grows upright. In either case, it sends up flat clusters of small, trumpet-shaped flowers.

KEY FACTS

The pink to lavender, trumpet-shaped flower has 5-lobed corollas, the lobes are sometimes notched at the tips.

+ **habitat:** Prairies, open woods, pastures, fields, lake edges, and rocky or gravelly areas

+ **range:** Central and eastern U.S. and Canada

+ **bloom period:** February–September

Small flowers, about a half inch (1.25 cm) wide, crowd the Rose Vervain's terminal flower spikes on erect or sprawling hairy stems. The leaves of this plant are lance-shaped overall, though deeply lobed with roughly serrated edges; they are positioned opposite each other on the stem. Stems often branch. As the plant ages, its flower clusters elongate. This native perennial is often confused with escaped cultivars of the genus *Verbena* that are introduced in many areas but are generally less hardy, especially in the winter. Flowers of purple or white might indicate a cultivar.

Common Milkweed
Asclepias syriaca H 24–72 in (60–180 cm)

The most abundant milkweed, the Common Milkweed yields a thick, bitter, milky liquid when its leaves or stem are bruised. The plant contains compounds used to treat heart disease.

KEY FACTS

Pink to purplish flowers growing in rounded clusters are slightly pendent and formed of 5 bent-back corolla lobes and a central crown of 5 curved lobes.

+ **habitat:** Roadsides, old fields, and disturbed areas

+ **range:** Native throughout most of central and eastern U.S. and Canada

+ **bloom period:** June–August

Common Milkweed blossoms grow from the axils of the upper leaves. The green leaves in general are broadly oval and hairy—sparsely above and fuzzier below, creating a grayish cast. In autumn, the flower heads give way to large, coarse seedpods filled with seeds tipped by silky hairs that eventually burst, dispersing black seeds on the breeze. The caterpillar of the monarch butterfly feeds exclusively on the foliage of milkweeds, whose toxic properties transfer to the bodies of both larva and butterfly and offer protection from predators. Milkweeds attract other butterfly species and other insects, including bugs, beetles, and bees.

Wild Bergamot
Monarda fistulosa H 12–48 in (30–120 cm)

A showy perennial in the Mint family, Wild Bergamot and its relatives attract butterflies and bees with dense globes of pink, lavender, white, or scarlet flowers.

KEY FACTS

Pink to lavender flowers have 2 lips; upper lip has 2 lobes, lower lip is broad with 3 lobes.

+ **habitat:** Fields, prairies, meadows, open woods, thickets, and woodland edges

+ **range:** Native throughout most of the U.S. and Canada east of the Rockies, naturalized throughout wider area

+ **bloom period:** May–September

A single Wild Bergamot flower cluster with conspicuous projecting stamens appears atop a whorl of leaflike bracts at the summit of a square, hairy stem. The blossoms attract butterflies, bees, and hummingbirds. The stem bears opposite, toothed, lance-shaped, and fragrant leaves often used to make a tea. Along with a close relative, *M. didyma* or Bee Balm, Wild Bergamot has been used by American Indians and others to treat a wide range of ailments, including stomach and intestinal troubles, respiratory problems, fevers, and skin eruptions. Wild Bergamot is not related to the orange subspecies *Citrus aurantium bergamia*, which is commonly used to flavor tea.

Rocky Mountain Beeplant
Cleome serrulata H 12–60 in (30–150 cm)

True to its name, the Rocky Mountain Beeplant attracts numerous bees. Many American Indian groups used the foliage and seeds for food and medicinal purposes.

KEY FACTS

Pink to purplish flowers have 4 petals and 6 long, projecting stamens attached along a terminal spike.

+ **habitat:** Prairies, open woods, washes, disturbed areas, and scrub; occurs to 5,000 ft (1,500 m)

+ **range:** Native to western U.S. and Canada; introduced eastward, mostly absent in the Southeast

+ **bloom period:** July–September

The showy pink clusters of the Rocky Mountain Beeplant, a member of the Caper family, grow on the ends of erect, branched stems with alternate palmate leaves, usually composed of three to seven leaflets. The flower spikes continue to elongate as additional flowers open, and it is not unusual for a single spike to display buds, flowers, and long, slender seedpods all at the same time. The species is also called Stinkweed in recognition of the unpleasant odor given off by the leaves when crushed. The similar Yellow Spiderflower (*C. lutea*) has yellow flowers and a more limited distribution.

Purple Passionflower/Maypop

Passiflora incarnata L vine to 30 ft (9 m)

This vining plant with showy blooms owes its name to features resembling symbols of Christ's passion. It grows from slender stems along vines that trail or climb by means of curly tendrils.

KEY FACTS

The purple-and-white blossom is 2–3 in (5–7.5 cm) across; 3-lobed leaves are toothed.

+ **habitat:** Fields, fencerows, open woods, sandy thickets, stream and riverbanks

+ **range:** Native to southeastern U.S. and parts of the Midwest; introduced elsewhere

+ **bloom period:** April–September

Missionaries named the intricate flower for the Christian symbolism they read into its parts: Three central stigmas signify crucifixion nails; five stamens correlate to Christ's wounds; the fringe of purple-and-white filaments resembles the crown of thorns; and the ten combined petals and sepals under the fringe represent the faithful apostles (minus Judas and Peter). A yellowish orange to yellowish green egg-shaped fruit, known as a Maypop, is filled with abundant small seeds embedded in yellow pulp. American Indians used Purple Passionflower to heal wounds and bruises. Today, it is valued for its mildly sedative properties and is used in teas.

Eastern Purple Coneflower

Echinacea purpurea H 12–36 in (30–90 cm)

Its large flower head pointed skyward, the instantly recognizable Eastern Purple Coneflower, also known as Echinacea, is one of the most widely used and studied medicinal plants.

KEY FACTS

The stiff, brownish central disk is surrounded by densely packed, notched, and drooping magenta rays.

+ **habitat:** Prairies, open woodlands, and dry forest margins; widely cultivated as an ornamental and for herbal medicine

+ **range:** Central and eastern U.S. and Canada

+ **bloom period:** June–October

The Eastern Purple Coneflower blossom rises on a stiff, hairy stem. Leaves that are also stiff and hairy alternate along the stem; the upper ones are smooth edged, but the lower ones are toothed, and each bears five distinct ribs. The scientific name for this perennial plant comes from the Greek *echinos*, which means "hedgehog," a reference to the numerous sharply pointed scales in the central disk of the flower head. For centuries, Echinacea was taken internally and applied externally by American Indians for a wide range of ailments. Today, its perceived properties as an immune system booster make it a popular remedy for fighting colds and flu.

Sweet Joe-Pye Weed

Eutrochium purpureum H 24–84 in (60–210 cm)

Projecting styles on the purplish florets of its rounded terminal flower clusters give the Sweet Joe-Pye Weed a fuzzy, out-of-focus look. Despite its name, the species is a popular garden addition.

KEY FACTS

The small pink to purplish 5-lobed disk flowers with long styles make up each of the many flower heads that form a cluster.

+ habitat: Open woods, woodland edges, wet meadows, and ravines

+ range: Throughout much of the eastern U.S. and Canada

+ bloom period: July–September

A member of the Aster family, the Sweet Joe-Pye Weed lacks the rays on its flower heads that are characteristic of many other species in this family. The plant has distinctive stems that are mostly green but become dark purple where leaves attach; the stems bear whorls of three to six coarse, toothed leaves with short stalks, broad bases, and sharply pointed ends. The leaves are up to 10 inches (25 cm) and may give off the scent of vanilla when crushed. After flowering, the clusters convert to balls of slender, five-sided brown seeds that are tufted with brownish hairs to help them disperse in the wind.

Common Ironweed/Prairie Ironweed

Vernonia fasciculata H 24–48 in (60–120 cm)

Ironweeds received their name from the toughness of their stems. The genus name, *Vernonia*, honors William Vernon, an English botanist who collected in Maryland in the late 17th century.

KEY FACTS

The tightly compressed, 5-lobed magenta disk florets with prominent divided styles form a compound flower.

+ habitat: Wet prairies, marshes, field edges, and sloughs along railroads

+ range: Native throughout much of central U.S. and Canada; introduced elsewhere

+ bloom period: July–September

Rugged round central stems support the dense clusters of Common Ironweed flower heads. Narrow, lance-shaped leaves with serrated edges grow alternately on the stem, which can be white, light green, or reddish purple. In addition to its strong constitution, the plant also tends to resist pests and disease and tolerates short bouts of flooding. Common Ironweed blooms attract many bees and butterflies, and its foliage hosts the caterpillar of the American Painted Lady. But its bitter foliage is not a favorite of mammalian herbivores, who tend to shun Ironweed until it is one of the last plants left in a heavily grazed pasture.

Dense Blazing Star/Gayfeather

Liatris spicata H 12–72 in (30–180 cm)

A stand of Dense Blazing Star rising above a damp meadow or prairie is a wondrous sight. The tall plants bear spikes of plumy purple flowers, the source of another common name, Gayfeather.

KEY FACTS

The flower is a rayless, rose-purple disk with long projecting styles that emerges above a base of overlapping bracts.

+ **habitat:** Prairies, damp meadows, open woods, and marsh edges

+ **range:** Eastern U.S. and Canada; cultivated widely elsewhere

+ **bloom period:** July–September

The feathery flower heads of the Dense Blazing Star attach without stalks and crowd a long spike, a foot (30 cm) or longer; they bloom from the top down. The plant stem grows from a base of grasslike leaves, also a foot long. Dense stem leaves appear alternately, becoming progressively smaller toward the top. The flower heads of the conspicuous terminal spike have no scent, though they are highly attractive to bees, butterflies, and moths; two moth caterpillar species feed specifically on the foliage, stems, and developing fruits. The perennial Dense Blazing Star is a member of the Aster family.

Bull Thistle

Cirsium vulgare H 24–72 in (60–180 cm)

The large and distinctive Bull Thistle, a Eurasian import, mounts many defenses to protect itself: Blossoms, stems, and leaves all have spines or prickles in various combinations.

KEY FACTS

The flower head up to 3 in (8 cm) wide is composed of slender purple disk flowers that are supported by spiny green bracts.

+ **habitat:** Roadsides, fields, pastures, disturbed areas, waste places

+ **range:** Introduced and naturalized in most of the U.S. and Canada

+ **bloom period:** June–September; year-round in mild climates

Bull Thistle flower heads emerge at the tips of thick, angular stems and branches armed with sharp, hairy, and prickly leaves and sticky bracts that discourage marauding insects. The spiny leaves are deeply lobed and toothed and the stout stems bear spiny "wings." The biennial plant's defenses are combined with a very efficient seed-dispersal system. The downy bristles of its fruit act like parachutes to transport the light, dry seed. The Bull Thistle is similar but much larger than its cousin the Canada Thistle (*C. arvense*), which also has a spineless stem but is classified as a noxious weed in many states.

Fuller's Teasel
Dipsacus fullonum H 18–72 in (45–180 cm)

The prickly Fuller's Teasel name derives from the use of the dried flower heads by fullers, or cloth cleaners, to tease up the nap of washed cloth, particularly wool.

KEY FACTS

Dense bands of 4-lobed purplish flowers encircle a prickly, ovoid flower head with spiny, upwardly angled bracts.

+ habitat: Old fields, roadsides, ditches, and disturbed areas

+ range: Naturalized throughout much of the U.S. and Canada, except far northern areas and the extreme Southeast

+ bloom period: July–October

The Fuller's Teasel comes armed with multiple defenses. In addition to the prickles and sharply pointed scales of the flower head, short prickles project from the erect stem and from the midrib of the lance-shaped leaves that clasp the stem and form a joined pair from opposite sides. The basal leaves are wider with wavy margins; these leaves die off in the biennial plant's second season. A native of Eurasia, Fuller's Teasel forms large expanses that crowd out many native species. The genus name, *Dipsacus*, means "thirsty," and is thought to refer to the water-catching cup formed by the joined leaf pairs.

Skunk Cabbage
Symplocarpus foetidus H 12–24 in (30–60 cm)

The stinky Skunk Cabbage is often the first wildflower to bloom each year. As early as February, the purplish spathes push up, their internal heat melting surrounding snow.

KEY FACTS

Tiny yellow to purplish flowers form on a spadix (dense spike) that is inside a hoodlike spathe.

+ habitat: Wet woodlands, stream banks, open swamps and marshes, and other wet areas

+ range: Mid-Atlantic and northeastern U.S. and Canada

+ bloom period: February–May

The spathe precedes the leaves of the Skunk Cabbage in seeking the light of day. It changes color as it ages, starting out pale green and barely streaked with brown, and ending up darker and heavily stained with purplish brown. The spathe shelters a thick fleshy spadix, on which large numbers of small flowers form. The almost hidden flowers give off a rank odor, intensified by the plant's internal heat, which can rise more than 30°F (17°C) higher than the outside temperature. When the flowers wither, the leaves emerge from the ground. They are large and green, resembling cabbage leaves.

Jack-in-the-pulpit
Arisaema triphyllum H 12–36 in (30–90 cm)

Tiny "Jack," the spadix, peeps out of his "pulpit," the spathe, in this native woodland perennial. The Jack-in-the-pulpit prefers rich soil and grows well in cultivation.

KEY FACTS

A curved, green spathe, striped with white or purplish brown, folds over a tube with a club-shaped spadix (spike) inside.

+ **habitat:** Damp woods, swamps, bogs, and marshes

+ **range:** Throughout central and eastern U.S. and Canada

+ **bloom period:** March–July

The hooded spathe, or sheath, of the Jack-in-the-pulpit rises from a tube on its own stalk and hovers over the small spadix, or spike, covered with tiny male or female flowers. The spathe emerges near a canopy provided by two (sometimes only one) large, palmate leaves rising on separate stalks. The leaves usually divide into three lobes. At maturity, the spadix produces shiny red berries that provide food for birds and mammals. American Indians collected the short, broad tubers of the plant to dry and cook; this practice supplied the alternate common name Indian Turnip. The roots and other plant parts can blister and irritate the skin.

Checker Lily/Mission Bells
Fritillaria affinis H 6–40 in (15–100 cm)

The Checker Lily may or may not be "checkered" (that is, mottled with brown, red, or purple). Its range of appearance made botanists wonder whether some variants were separate species.

KEY FACTS

A pendent yellow-green flower, composed of 6 lance-shaped tepals (look-alike petals and sepals), is often heavily mottled to appear dark purple.

+ **habitat:** Prairies, open woods, grassy and brushy slopes to higher elevations

+ **range:** Western U.S. and Canada

+ **bloom period:** February–June

Two to five bell-shaped flowers nod from the top of a solitary, erect, unbranched stem of this member of the Lily family. Narrow, lance-shaped leaves grow alternately on the upper portion of the stem and in whorls of three to five lower down. The Checker Lily sprouts from unusual, scaly bulbs that are surrounded by a mass of rice-size bulblets. The genus name comes from the Latin for "fritillary," which means "dice box," a reference to the checker pattern of these lilies and the butterflies known as fritillaries. Another common name for this species is Mission Bells.

Scarlet Rose Mallow
Hibiscus coccineus H 36–96 in (90–240 cm)

The Scarlet Rose Mallow, a type of hibiscus, beguiles with its large, showy red blossoms. Although the perennial species has a small native range, it is cultivated widely in climate-friendly zones.

KEY FACTS

The flower has 5 separated crimson petals and anthers; the stamens form a long tube around the styles and the stigmas.

+ habitat: Swamps, marshes, and ditches

+ range: Native to southeastern U.S.; also widely cultivated

+ bloom period: June–September

Scarlet Rose Mallow sets its blossoms from the upper axils of its palmated, highly divided, pointed, and toothed deep green leaves, which bear a strong resemblance to the foliage of the Hemp plant, *Cannabis sativa*. The large flowers, up to 8 inches (20 cm) wide, are star shaped and appear above undersized green bracts. The stamen column formed around the styles and stigmas is a distinctive Mallow feature.

The species forms clumps several feet wide and blooms later in the season than many other wildflowers. Some *Hibiscus* cultivars have more rounded, overlapping petals.

Common Blanketflower
Gaillardia aristata H 6–30 in (15–75 cm)

This member of the Aster family gets its common name from similarities between the cheerful flower and the colorful patterns woven into American Indian blankets.

KEY FACTS

Flower head is formed of yellow rays, usually 3-forked at the tip, with dark red bases around a center of dark red disk flowers.

+ habitat: Prairies, woodlands, meadows, and disturbed areas

+ range: Native in much of northwestern U.S. and Canada; highly cultivated and escaping farther south and east

+ bloom period: July–September

Blanketflowers often grow as single flower heads on stalks rising from a base of rough, hairy, gray-green leaves that alternate on the stem. The leaves, up to 6 inches (15 cm) long, may be lance shaped or lobed, with or without toothed edges. Occasionally, rays will be absent, leaving only the dark central disk flowers. Though not much favored by grazing animals, Blanketflower's presence in a pasture indicates that the preferred plants there have reached their grazing readiness. The species has had many uses in traditional medicine and is being investigated for potential anticancer and antibacterial properties.

Wild Ginger
Asarum canadense H 6–12 in (15–30 cm)

Wild Ginger's pair of large, heart-shaped leaves tower over the small, low-growing and solitary brownish maroon flower that is often overlooked, obscured by the woodland leaf litter.

KEY FACTS

The bell-shaped brownish maroon flower has 3 pointy lobes, about 1 in (2.5 cm) long, and it grows on a short stalk.

+ **habitat:** Moist, rich woods and wooded slopes

+ **range:** Central and eastern U.S. and Canada

+ **bloom period:** April–May

Unrelated to the zesty tropical spice, Wild Ginger gets its name from its similar underground stems and sharp taste. The brownish maroon flower shares its color and fetid smell with rotting meat, making it very attractive to its chief pollinators—various fly species. The plant's distinctive leaves rise on long stalks from the creeping stems, and the pair will often sway together in a gentle breeze. They have dark, woolly, and attractively veined upper surfaces and lighter undersides. The flowers emerge near the ground on their short stalks in the fork between the two leaf stalks.

Red Columbine/Wild Columbine
Aquilegia canadensis H 12–24 in (30–60 cm)

The genus name for columbines, members of the Buttercup family, derives from *aquila,* the Latin for "eagle," and is thought to refer to the talon-like appearance of the flower's distinctive spurs.

KEY FACTS

The flower's 5 yellow-faced petals form red tubular spurs and its red sepals project forward; the yellow stamens protrude.

+ **habitat:** Woods, hillsides, cliffs, bogs, and roadsides

+ **range:** Native to central and eastern U.S. and Canada; also highly cultivated

+ **bloom period:** March–July

The Red Columbine is a woodland wildflower with the whole package: colorful, distinctive, bell-like pendent blossoms that grow gracefully on branched stems adorned with delicate, three-lobed compound leaves. The spurred red flowers present a challenge for pollinators and rely on hummingbirds and long-tongued insects to get the job done. The perennial plant spreads by prolific self-seeding. A similar species, the Western Columbine (*A. formosa*), has petal-like sepals that spread out. Various native columbines hybridize among themselves. As cultivars, columbines are much admired for their hardiness, persistence, and shade tolerance.

Fringed Redmaids
Calandrinia ciliata H 2–16 in (5–40 cm)

Low-growing Fringed Redmaids got their common name from botanists who encountered the little red flowers on an 18th-century expedition to Peru and Chile.

KEY FACTS

The flower has 5 pale pink to rose-red petals, yellow-tipped stamens, and sepals with a hairy fringe.

+ **habitat:** Grasslands, fields, gravelly washes, desert, rocky slopes, and disturbed areas

+ **range:** Western U.S. to British Columbia; Mississippi; Massachusetts

+ **bloom period:** February–May

Fringed Redmaid flowers grow in loose clusters from the axils of leaflike bracts on erect or prostrate stems that bear lance-shaped, alternately arranged, fleshy leaves. The foliage varies somewhat, but the flower appearance remains consistent. This member of the Miner's Lettuce family (previously in the Purslane family) often keeps low to the ground and spreads prolifically. Fringed Redmaids yield huge quantities of black oil-rich seeds that are a favorite food of many songbirds and rodents. California Indians also collected the seeds, which they parched and ground into meal. They sometimes set grasslands on fire to encourage this fire-following species.

Red Clover
Trifolium pratense H 6–24 in (15–60 cm)

Red Clover, a Eurasian import, is often planted in pastures and as a cover crop. Like its cousin White Clover *(T. repens),* it can quickly become a widespread landscape feature.

KEY FACTS

Numerous pink to red tubular flowers composed of 5 narrow petals form dense egg-shaped masses to globes.

+ **habitat:** Meadows, pastures, old fields, lawns, and disturbed areas

+ **range:** Throughout the U.S. and Canada, except in far northern areas

+ **bloom period:** May–September

Red Clover, a member of the Pea family, sometimes has red blossoms, but more often they are pinkish, or rarely, white. This species has a compound leaf structure of three oval leaflets with smooth edges, with a V-shaped pattern in the middle of each. Flowering branches are leafier than those of White Clover, carrying one or several leaves just below the flower head. Red Clover flowers have a honey-like fragrance, and the foliage, especially in a dense patch, smells, well . . . like clover. Farmers and home gardeners often improve soil quality by planting this nitrogen-fixing species as a winter cover crop or in rotation.

Scarlet Gilia

Ipomopsis aggregata H 12–36 in (30–90 cm)

You can dress a Scarlet Gilia up, but you cannot hide its less attractive feature: Glandular foliage gives this member of the Phlox family a skunky odor.

KEY FACTS

The trumpet-shaped flower is composed of 5 fused petals with red, pink, or white spreading and pointed lobes, which are sometimes speckled.

+ **habitat:** Meadows, open woods, and rocky slopes

+ **range:** Western U.S. into British Columbia

+ **bloom period:** August–October

Many wildflowers deserve to be called striking, but for the Scarlet Gilia, it is part of the scientific name: *Ipomopsis* derives from the Greek for "striking appearance." The species' leaves don't compete with its flowers; deeply and narrowly lobed, they are sparsely scattered on the stem. Stems often spring from a basal rosette of leaves that may persist for several years after flowering has ended. This common western wildflower was previously placed in the genus *Gilia* and still retains that name in some horticultural literature. Its odor gives it another moniker: Skunkflower. Lewis and Clark brought back a specimen collected in Idaho in 1806.

Indian Warrior

Pedicularis densiflora H 6–24 in (15–60 cm)

This western species, a member of the Broomrape family (previously in the Figwort family), gets its common name from the shape of its flower cluster, which resembles a feathered plume.

KEY FACTS

The red or fuchsia, 2-lipped tubular flower has a straight, hooded upper lip and a smaller lower one.

+ **habitat:** Open woods, chaparral, and slopes to 6,000 ft (1,828 m)

+ **range:** Native to California and Oregon; introduced elsewhere

+ **bloom period:** March–May

The Indian Warrior's clublike flower stalks rise above toothed bracts at the end of a hairy stem. The fernlike leaves, up to 6 inches (15 cm) long, occur mainly at the base; those farther up the stem are much smaller. The species is part of a group of plants once thought to transmit lice to cattle, the reason for the genus name *Pedicularis*, derived from the Latin for "louse." Indian Warrior will opportunistically parasitize the roots of other plants, especially Manzanita shrubs. The species has many traditional medicinal uses, including as a muscle relaxant and nerve-pain reliever.

Scarlet Indian Paintbrush

Castilleja coccinea H 6–24 in (15–60 cm)

Brilliant red-tipped bracts surrounding tubular greenish yellow flowers in a long spike give the Scarlet Indian Paintbrush the appearance of a quick dip in a pot of scarlet paint.

KEY FACTS

The fan-shaped, 3-lobed, red-tipped to all-yellow leafy bracts appear with the long, tubular greenish yellow flowers.

+ **habitat:** Moist places such as prairies, meadows, and roadsides

+ **range:** Central and eastern U.S. and Canada

+ **bloom period:** May–July

Like the Poinsettia, the blaze of color in the Scarlet Indian Paintbrush belongs to the leafy bracts, not the flower petals. Flowers appear in the axils of the bracts, which form a dense spike at the end of a hairy stem. Often, the stigma-tipped style projects past the corollas. Oval leaves at the base of this native plant create a rosette, while those on the stem are narrowly divided. The Scarlet Indian Paintbrush leads a partially parasitic life, appropriating nourishment from other plants by fastening its roots onto theirs. Other *Castilleja* species are very common in western North America.

Seaside Petunia

Calibrachoa parviflora H to 12 in (30 cm)

Seaside Petunias come in a variety of colors, red being one of them. Despite the name, the species is also found inland in habitats with sandy soil.

KEY FACTS

The small 5-lobed trumpet-shaped flower appears in red, pink, purple, and other colors.

+ **habitat:** Dry streambeds, stream banks, washes, and wetlands

+ **range:** Native to southwestern North America; introduced elsewhere; also highly cultivated

+ **bloom period:** April–September

A member of the Nightshade family, the Seaside Petunia is a sprawling, low-growing species that forms a dense mat, sending down new roots at the leaf nodes as it creeps along the ground. The plentiful evergreen leaves are lance shaped, fleshy, and very sticky; the flowers grow from their axils. The species was first classified as a true petunia, but based on DNA evidence, it was put into the genus *Calibrachoa*, named for the 19th-century Mexican botanist Antonio de la Cal y Bracho, a switch that is not always recognized. Both *Petunia* and *Calibrachoa* are considered basically South American genera.

Cardinal Flower

Lobelia cardinalis H 12–60 in (30–150 cm)

Growing tall on erect stems, the bright red blossoms of the Cardinal Flower signal a bounty of nectar to the Ruby-throated Hummingbird, which is one of the plant's chief pollinators.

KEY FACTS

The bright red, tubular flowers have a 2-lobed upper lip and a 3-lobed bottom one.

+ habitat: Moist areas, especially along streams and ponds

+ range: Throughout much of the U.S. and Canada, except in the Northwest and in far northern areas

+ bloom period: July–September

The flowers of the Cardinal Flower, a member of the Bellflower family, grow in loose to dense, elongated spikes. A flower's stamens are fused together and protrude above the tube. Alternate, lance-shaped leaves line the tall stalk at close intervals. The lower leaves have stalks and are toothed; the upper leaves are smoother and clasp the stem. A perennial that prefers a moist footing, the Cardinal Flower often sends up new plants from its creeping, underground rootstalk. The closely related Great Blue Lobelia (*L. syphilitica*) is similar in almost all respects except the color of its flowers, and the two hybridize occasionally.

Blood Sage/Scarlet Sage

Salvia coccinea H 12–24 in (30–60 cm)

Square stems with showy whorls of flaming scarlet blossoms make Blood Sage impossible to overlook. A long blooming season adds to its prominence.

KEY FACTS

The red tubular flower has 2 lips: a small upper one with 2 lobes and a larger bottom one with 3.

+ habitat: Sandy sites in open woods, woodland edges, thickets, and disturbed areas

+ range: Mainly on Coastal Plain of southeastern U.S.; highly cultivated and sometimes escaping farther northward

+ bloom period: February–October

The striking flowers of Blood Sage, also commonly known as Salvia, appear in well-separated whorls, forming an elongate, loose cluster atop a straight stem. This member of the Mint family has bright green, roughly triangular leaves that grow opposite one another. Of the many native *Salvia* species, Blood Sage is uncommon in having red flowers, a feature that of course makes it very popular with hummingbirds. Its foliage is pungent, however, which discourages some herbivores such as the White-tailed Deer. *Salvia coccinea* has been developed into many flashy cultivars.

Virginia Spiderwort
Tradescantia virginiana H 8–24 in (20–60 cm)

The deep purplish blue blossoms of the Virginia Spiderwort live very short lives, blooming only for a day before the petals begin to contract and form gummy blobs by evening.

KEY FACTS

Flower has 3 broad oval petals and 3 hairy sepals above leaflike bracts; 6 orange-tipped stamens with long, hairy filaments provide contrast.

+ habitat: Woods, woodland edges, prairies, meadows, stream banks, roadsides, and disturbed areas

+ range: Eastern North America and California

+ bloom period: April–August

Clusters of Virginia Spiderwort flowers grow at the end of rounded stems that can be solitary or branched at the base of the plant. Long, narrow leaves with a pronounced midrib form an arched clump. The plants multiply mainly by underground runners that root and then send up new plants. There are a number of proposed explanations for this wildflower's common name. One is that the leaf growth pattern resembles a crouched spider; another suggests that the fine filaments of the stamens look like a spider's web. Yet a third attributes the name to a threadlike secretion that flows from the cut stem.

Eastern Bluestar/Blue Dogbane
Amsonia tabernaemontana H 12–36 in (30–90 cm)

The rather unwieldy scientific name of the Eastern Bluestar honors the 16th-century botanist Jacobus Theodorus Tabernaemontanus, considered by many the father of German botany.

KEY FACTS

Star-shaped flower has 5 blue, spreading lobes above a slender tube, with a white or yellow patch at the base of each lobe.

+ habitat: Open woods, thickets, meadows, low prairies, marshes, stream banks, and roadsides

+ range: South-central and eastern U.S., except in some northeastern areas

+ bloom period: March–July

Eastern Bluestar's easily recognized sky blue flowers appear in loose clusters on branching stems. The stems are dense with alternate, willowlike, lance-shaped, green leaves that turn yellow in the fall. After the flowers bloom, two cylindrical seedpods up to 5 inches (12.5 cm) long replace each of the pollinated flowers. At maturity, the pods split along the side to reveal a row of cylindrical seeds. The plant reproduces by self-seeding. The species, which contains a toxic, milky sap in its leaves and stems, also goes by the name Blue Dogbane and is a member of the Dogbane family.

Round-lobed Hepatica

Anemone americana H 2–6 in (5–15 cm)

What look like colorful petals on the Round-lobed Hepatica, a member of the Buttercup family, are actually its sepals. They can vary in number, but six is a frequent count.

KEY FACTS

The lavender, blue, pink, or white flower is formed of 5–12 sepals with many white stamens and is supported by 3 green bracts.

+ habitat: Woods, rocky slopes, and bluffs

+ range: Eastern North America, except in far northern areas

+ bloom period: February–April

The solitary, hairy flower stalk of the Round-lobed Hepatica rises from a base of leaves with three rounded lobes that inspired the name Hepatica, which refers to the liver. These leaves can have mottling above and a purplish cast underneath. They grow along the ground on their own stalks, appear mostly after the flowers have bloomed, and persist through the winter before withering. A similar species, the Sharp-lobed Hepatica (*A. acutiloba*), has both pointier bracts and leaves and occurs within roughly the same range. These two species historically were assigned to the genus *Hepatica*.

Pasqueflower/Prairie Smoke

Anemone patens var. *multifida* H 4–16 in (10–40 cm)

The Pasqueflower receives its common name from the fact that its blossoms often emerge during the celebration of Easter, which was known in an old French form as Pasque.

KEY FACTS

The blue to purple flower has 5–8 petal-like, pointed sepals and numerous yellow stamens.

+ habitat: Prairies, meadows, and rocky areas

+ range: Northern and central U.S. and Canada into western Canada

+ bloom period: March–June

The hairy, leafless flower stalk of the Pasque-flower rises from under the ground in early spring. Its large, solitary crocus-like flower—reaching 3 inches (7.5 cm) across when open—begins to bloom as the foliage starts to form. The basal leaves of the plant are silvery, deeply divided, and fernlike, and the bracts beneath the flower are hairy and whorled with linear lobes. The foliage and flower stalks continue to grow after the flower has bloomed. Attractive silky and feathery seed heads replace the blossoms, giving it the alternate name Prairie Smoke. A cultivated variety from Eurasia (*A. patens patens*) has a similar appearance but a wider range of flower colors.

Colorado Blue Columbine

Aquilegia coerulea H 12–24 in (30–60 cm)

This perennial member of the Buttercup family serves as Colorado's state flower and has been taken to the max in cultivation, including forms with double blossoms.

KEY FACTS

The flower has 5 deep blue petal-like sepals and 5 white petals, each with a backward-pointing blue spur.

+ **habitat:** Moist woods and meadows

+ **range:** Mid- to higher elevations of Rocky Mountain states in the U.S.; highly cultivated

+ **bloom period:** June–August

The showy blossoms of the Colorado Blue Columbine often face proudly upward. They appear in bushy clumps with fernlike foliage in the form of compound leaves that are deeply lobed and cleft. There is considerable color variation in the species. Flowers to the north and west of the native range tend to be paler, going even into white forms. Also, native flowers at higher altitudes tend to be a deeper blue than those at lower altitudes. The genus name derives from the Latin for "eagle," and refers to the flower's long spurs that are thought to resemble an eagle's talons. The species name refers to the flower's blue color.

Tall Blue Larkspur

Delphinium exaltatum H 48–72 in (120–180 cm)

The "spur" in the name Larkspur refers to the fact that two of the blossom's petals and one of its petal-like sepals form a single, cone-shaped prong, or spur, that projects backward.

KEY FACTS

Complex purplish blue flower has 4 petals that form a nectar tube at their base, and 5 hairy, petal-like sepals.

+ **habitat:** Open woods, rocky slopes, and limestone bluffs

+ **range:** Scattered native distribution in parts of eastern U.S.; highly cultivated

+ **bloom period:** July–September

Tall Blue Larkspur flowers, which form long clusters, have a lot going on. In addition to their distinctive spurs, they have notched upper petals and bearded, two-lobed lower ones. Lateral sepals have a greenish spot in the center, and lower and upper sepals have a greenish spot at the tips. The three-lobed leaves sometimes have divided basal lobes and are widely dispersed on the stem. This perennial member of the Buttercup family ranks as the tallest of the native larkspurs and typically presents more luxurious foliage than other species. The toxic alkaloids it contains present a danger to grazing cattle.

Great Blue Lobelia

Lobelia siphilitica H 12–48 in (30–120 cm)

Blooming late in the season, the striking flower clusters of the Great Blue Lobelia stand out in their rich, moist habitats.

KEY FACTS

The bright blue or lavender blue tubular flower has a 2-lobed upper lip and a 3-lobed lower lip, with 2 white bumps on the lower lip.

+ **habitat:** Woods, meadows, stream banks, swamps, and ditches

+ **range:** Throughout much of central and eastern U.S. and Canada

+ **bloom period:** July–October

The flowers of the perennial Great Blue Lobelia form long, slender clusters at the ends of tall, erect, leafy stems. The oval to lance-shaped leaves grow up to 5 inches (12.5 cm) long; they lack stalks and are alternate on the stems, which seldom are branched. This member of the Bellflower family is similar to the red Cardinal Flower (*L. cardinalis*) in most aspects except color and some details of flower structure, and sometimes hybridizes with it. Many bee species seek out its nectar and pollen. The plant's species name comes from the erroneous belief that it could cure syphilis; in fact, it contains several toxic alkaloids.

Wild Lupine/Sundial Lupine

Lupinus perennis H 8–24 in (20–60 cm)

Lupines get their name from the Latin for "wolf," a misnomer bestowed when it was believed these members of the Pea family devoured minerals from the soil.

KEY FACTS

The blue flower has 5 petals: a large upper banner petal, 2 forward-projecting side wing petals, and 2 lower ones called a keel that are fused at the base.

+ **habitat:** Dry open woods, woodland edges, and fields

+ **range:** Much of eastern U.S. and Canada; also highly cultivated

+ **bloom period:** April–July

Wild Lupine forms elongated flower clusters up to 10 inches (25 cm) long at the top of erect, branching stems. The clusters tend to be looser than those of the Texas Lupine (*L. texensis*) with its crowded flower stalks. The flowers are typically blue and less commonly white or pink. The Wild Lupine's leaves are palmately divided into seven to eleven leaflets that are hairy on the undersides, as are the stems. This species also is known as the Sundial Lupine. Far from depleting the fertility of the soil, lupines enhance it by fixing nitrogen in their roots.

Texas Lupine/Texas Bluebonnet

Lupinus texensis H 6–18 in (15–45 cm)

The Texas Lupine, better known as the Texas Bluebonnet, received its traditional common name because its flowers resemble the sunbonnets that pioneer women wore.

KEY FACTS

The 5-petaled pea-like blue flower has a white spot on its upper petal that turns dark purplish as the flower ages.

+ **habitat:** Prairies, open fields, and roadsides

+ **range:** Texas, Oklahoma, Louisiana, and Florida; also highly cultivated

+ **bloom period:** March–May

Large, elongated flower clusters dense with pealike flowers top the branching stems of the Texas Lupine, which rise from a basal rosette of leaves. The topmost flowers (not just the central spots) on the dense clusters are conspicuously white. The leaves of the plant are light green and velvety, divided palmately into five pointed leaflets. Long, green pods replace the flowers. When the seeds inside these mature, the plant will die back. The Bluebonnet is the state flower of Texas, but the designation represents all the native *Lupinus* species there, resulting in multiple state flowers, *L. texensis* being just one of the honored plants.

Hookedspur Violet/Western Dog Violet

Viola adunca H 3–6 in (7.5–15 cm)

Hookedspur Violet flowers do not grow from their stems or roots, like other violet species. Instead, they appear from the plant's leaf stalks.

KEY FACTS

Blue-violet flower has 5 petals; the side petals have bearded tufts and the lower petals have dark nectar guides and a long, curved spur.

+ **habitat:** Meadows, open woods, slopes, and stream banks

+ **range:** Throughout northern and western U.S. and Canada

+ **bloom period:** April–August

The flowers of the Hookedspur Violet grow on stalks from the axils of the plant's leaves. The leaves themselves are heart shaped to oval and finely but bluntly toothed and are alternate on the stems. The lower flower petal's long nectar spur may curve and appear above the top flower petals. A number of fritillary and other butterfly species and their larvae favor this violet for its nectar and foliage. Like some other Violet species, the Hookedspur Violet produces small, inconspicuous cleistogamous flowers that never open, but produce viable seeds by self-pollinating. The name Dog Violet apparently arose to separate it from the sweet-scented violets.

Birdfoot Violet

Viola pedata H 3–10 in (8–25 cm)

Birdfoot Violets are named for their three-lobed, deeply divided leaves with secondary lobes that widen toward the tips, a structure that resembles a bird's foot.

KEY FACTS

The blue, purple, or lavender (sometimes two-toned) beardless flower is composed of 5 petals and 5 stamens.

+ **habitat:** Open woods, prairies, rocky slopes, and along roadsides

+ **range:** Throughout much of central and eastern North America

+ **bloom period:** March–June

The flowers of the Birdfoot Violet rank as one of the largest among the blue violets—up to 1.5 inches (3.75 cm) across—their blueness offset by a bright orange center composed of a cluster of conspicuous stamens. The plant has no aerial stems; instead, its leaves and flower stalk emerge directly from the rootstock. A striking variation of the species has dark purple upper petals and pale violet lower ones. The Birdfoot Violet's coppery seeds produce a sugary, ant-attracting gel that encourages the insects to carry them off and disperse them. The species sometimes continues to flower into the fall.

Common Morning Glory

Ipomoea purpurea L vining to 15 ft (4.5 m)

The genus name of the Common Morning Glory derives from Greek words meaning "like a worm," referring to the twining growth of the plant.

KEY FACTS

The funnel-shaped blue, purple, pink, white, or variegated flower has 5 united petals and a style with a 3-parted stigma.

+ **habitat:** Fields, fence lines, along roadsides, and in disturbed areas

+ **range:** Introduced and naturalized throughout much of the U.S. and eastern Canada

+ **bloom period:** June–September

Common Morning Glory plants with their large, trumpet-shaped flowers climb over anything natural or built that provides support, or they otherwise sprawl over open ground. The plants have slender stems that are somewhat hairy, with large, usually heart-shaped, bright green alternate leaves on long slender stalks. Small flower clusters appear from the axils of some of the leaves. The blossoms open in the morning and last only one day, and are replaced by round seed capsules. Introduced from South America, the species has proven very adaptable and prolific, though not as aggressive in its spread as some other morning glories.

Distant Phacelia

Phacelia distans H 6–36 in (15–90 cm)

These pretty blue flowers of arid regions carry a toxic secret: Handling the plants without gloves can cause a severe case of contact dermatitis.

KEY FACTS

The pale blue, funnel- or bell-shaped corollas have 5 rounded lobes and 5 delicate protruding stamens.

+ **habitat:** Desert, washes, slopes, and sandy roadsides

+ **range:** Southwestern U.S.

+ **bloom period:** February–June

Distant Phacelia flowers emerge in a hairy, coiled terminal cluster, on only one side of the stem. The curl tends to straighten out as the plant transitions from flower to fruit. The leaves of the species are hairy, pinnate, and fernlike, and they grow on hairy, reddish stems that can be erect or lie along the ground. Distant Phacelia is one of about a hundred *Phacelia* species in the western United States, many of them quite similar. Due to the shape of its curled flower coil, a distinguishing feature, it also goes by the name Scorpionweed.

Wild Blue Phlox

Phlox divaricata H 10–20 in (25–50 cm)

Native perennial Wild Blue Phlox spreads on creeping stems that send up delicate, fragrant flowers and dense, green foliage on rich woodland floors.

KEY FACTS

The trumpet-shaped flower of blue, lavender, or white has 5 flat lobes that can be notched, and stamens and pistils contained within the slender tube.

+ **habitat:** Woods, fields, and stream banks

+ **range:** Throughout central and eastern U.S. and Canada; highly cultivated

+ **bloom period:** March–May

The hairy and sticky stems of Wild Blue Phlox have narrow, lance-shaped leaves that send down roots at their nodes and send up stems with loose, flat-topped terminal clusters of sweetly fragrant flowers. They can reproduce by cloning, forming extensive mats over time. The plants also give rise to infertile shoots that are usually shorter and have more rounded leaves, but do not produce flowers. After flowering, ovoid seed capsules containing small seeds form on the fertile stems, which die back. Infertile shoots persist through the winter, storing energy in their roots that fuels next spring's crop of fertile shoots.

Heal-all/Self-heal

Prunella vulgaris H 12–36 in (30–90 cm)

The two-lipped flowers of the Heal-all mature a few at a time, beginning at the bottom of the dense cylindrical flower head. This prolongs the bloom time of this member of the Mint family.

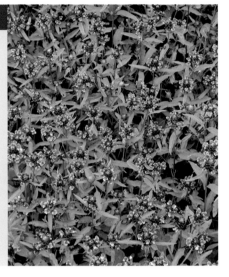

KEY FACTS

The small, 2-lipped purple- to blue-and-white flowers have hairy bracts that form a compact spike at the end of stems.

+ **habitat:** Fields, roadsides, and along woodland edges

+ **range:** Throughout much of the U.S. and Canada, except in far northern areas

+ **bloom period:** June–September

Heal-all flowers appear in profusion on erect flower heads. The leaves are lance shaped, untoothed or lightly toothed, and grow opposite each other; those near the top of the plant often lack stalks. The spikes continue to grow longer after flowering. Heal-all spreads from rhizomes that send up square, red-tinged stems. Introduced from Europe, with a native variety also present, Heal-all has a long tradition of use in treating wounds and afflictions of the throat; it has additional common names of Self-heal, Carpenter's Weed, and Woundwort. In traditional Chinese medicine, Heal-all was commonly used to treat liver and kidney ailments.

Virginia Bluebell

Mertensia virginica H 8–24 in (20–60 cm)

The Virginia Bluebell is beautiful in and of itself, with its nodding clusters of light blue, trumpet-shaped flowers, but to add to the delight, it tends to grow in local profusion.

KEY FACTS

The inch-long (2.5 cm), funnel-shaped flowers have 5 lobes, and the edges of the sepals are fringed.

+ **habitat:** Moist woods, stream banks, floodplains, and wet meadows

+ **range:** Throughout much of eastern U.S. and Canada; also cultivated

+ **bloom period:** March–June

The Virginia Bluebell's striking blue blossoms unfold from pink buds amid thick gray-green foliage, making a plant in transition a very colorful sight. The stem leaves are oval and have smooth edges, and those at the base are longer, up to 8 inches (20 cm). Although the species is particular about the moistness of its habitat, it rewards a suitable venue with an extensive display. *Mertensia* species are often called Lungwort for their similarity to a European species used to treat lung disease. In North America, it is also known as Virginia Cowslip. Butterflies favor this species and aid in its pollination.

Allegheny Monkeyflower
Mimulus ringens H 12–36 in (30–90 cm)

The Allegheny Monkeyflower's comical appearance inspired the genus name *Mimulus*, derived from the Latin *mimus*, referring to a mime, a comic actor, or a buffoon.

KEY FACTS

The light purple to blue flower has a 2-lobed top lip and a 3-lobed bottom lip; both lips are broad and wavy.

+ **habitat:** Wet meadows, stream banks, and pond edges

+ **range:** Throughout much of the U.S. and Canada, except southwestern U.S. and in far northern areas

+ **bloom period:** June–September

Leaves greatly outnumber flowers on the Allegheny Monkeyflower plant. They are lance shaped, toothed, and grow opposite each other, clasping the smooth, square, hollow stems. Sparse blossoms arise from slender stalks in the axils of upper leaves. The flower displays two bright yellow spots in its throat that may serve as a signal to bees. The species keeps small insects from its limited supply of nectar by closing off access to the throat unless the heftier weight of a pollinating bee on the blossom's lower lip creates an opening. A relative is the Common Monkeyflower (*M. guttatus*), described on page 19.

Fringeleaf Wild Petunia/Hairy Ruellia
Ruellia humilis H 12–24 in (30–60 cm)

The Fringeleaf Wild Petunia, an endangered native species, resembles a petunia but belongs to the Acanthus family, and the genus is named for 16th-century French herbalist Jean de la Ruelle.

KEY FACTS

The funnel-shaped flowers about 2.5 in (6 cm) long have 5 lobes and range in color from lavender to blue to purple.

+ **habitat:** Prairies, woodland edges and openings, and thickets

+ **range:** Most of the eastern U.S., except the Northeast

+ **bloom period:** May–October

The blossoms of the Fringeleaf Wild Petunia have long tubes and flared lobes often finely lined with purple to guide hummingbirds and other pollinators to the nectar inside. The flowers develop without stalks on angled, green, hairy stems. The stems bear opposite, broadly oval, smooth-edged and stalkless leaves, which are covered with white hairs above and below. The stalkless flowers and leaves give the plant a bushy appearance. All told, the species really earns the common name Hairy Ruellia. The flowers commonly open during the morning, only to fall off the plant by the end of the day.

Harebell
Campanula rotundifolia H 6–24 in (15–60 cm)

The Harebell is a hardy wildflower that belies the delicate appearance of its slender stems and drooping blossoms, thriving in such challenging habitats as high altitudes in the mountains.

KEY FACTS

The delicate 5-lobed blue-violet bell-shaped flowers nod on thread-like stems.

+ **habitat:** Meadows, woods, rocky slopes, sand shorelines

+ **range:** Throughout nearly all of the U.S. and Canada, except the southeastern U.S.

+ **bloom period:** June–September

The Harebell is a member of the Bellflower family, and the genus has many members with bell-shaped or conical flowers that characteristically nod from slender stalks. In the Harebell's case, the stalks are very slender and fragile looking; the flowers grow on them singly or in clusters. Slender, straight leaves on most of the plant stem give way to rounded ones at the base, the source of the species name *rotundifolia*. As the plant matures, the basal leaves may wither and disappear. This perennial Harebell forms small patches, not large stands. The plant's genus name is from the Latin, meaning "little bell."

Prairie Flax/Lewis's Flax/Wild Blue Flax
Linum lewisii H 12–24 in (30–60 cm)

A close relative of the cultivated flax species from which linen and linseed oil are made, Prairie Flax provides nutritious forage and seeds for livestock, birds, and other wildlife.

KEY FACTS

The pale blue flowers about 1.5 inches (4 cm) across with darker blue veins have 5 petals, 5 sepals, and 5 stamens; they grow on multiflowered stems.

+ **habitat:** Grasslands, meadows, open woods, and mountain brush

+ **range:** Throughout Canada and western U.S.; also a few populations in eastern U.S.

+ **bloom period:** March–September

Attractive and highly symmetrical Prairie Flax flowers grow in profusion in loose clusters. The plants have numerous long, narrow leaves that mostly grow alternate on the stem, but sometimes appear opposite one another. As the flower clusters develop and bloom, the stems remain green but begin to shed their leaves. The top-heavy flower stems rarely stand erect but instead lean. The Prairie Flax is used widely to control erosion, to beautify landscapes, and to restore native plant biodiversity. This species is able to withstand fires and resprout afterward. It is named for Meriwether Lewis, who collected a specimen in 1806.

Common Chicory
Cichorium intybus H 24–72 in (60–180 cm)

Roadsides and other disturbed places are replete with the erect-stemmed Common Chicory, punctuated at intervals with striking blue radiating flowers. It is a member of the Aster family.

KEY FACTS

The stalkless blue flower has numerous strap-shaped florets, the inner ones shorter, with toothed ends.

+ **habitat:** Roadsides, fields, and disturbed places

+ **range:** Naturalized throughout much of the U.S. and Canada, except in far northern areas

+ **bloom period:** June–October

Common Chicory's sky blue flower heads open several at a time on a stem and are ephemeral, lasting only one day. Small, inconspicuous oblong leaves clasp the stem; leaves at the base are jagged and dandelion-like. The plant exudes a milky white sap. The species has a long history as human food and animal fodder, and as a remedy for ailments of the liver, kidneys, and stomach. The French popularized the addition of the dried and ground root to coffee to enhance flavor and to stretch a quantity of the expensive beans. Today, prolific Common Chicory often is regarded as an undesirable weed.

Water Hyacinth
Eichhornia crassipes H 3–36 in (7.5–90 cm)

The floating Water Hyacinth, with its attractive flowers and dense foliage, forms immense rafts, clogging lakes and waterways and impeding water traffic.

KEY FACTS

Six-lobed purplish blue flowers have a distinctive yellow eye on the upper central lobe.

+ **habitat:** Fresh water, including marshes, bayous, lakes, streams, rivers, and ditches

+ **range:** Native to South America; naturalized mainly in southeastern U.S. and eastern Canada

+ **bloom period:** June–September or all year, depending on climate

Water Hyacinth blossoms, some 2 inches (5 cm) wide, appear on stalks amid shiny green, fleshy leaves. The leaf blades are round to kidney shaped and grow on stalks up to 3 feet (90 cm) tall from spongy, inflated bulbous bases. Roots dangle beneath the water's surface. At one time, the Water Hyacinth, a relative of the native Pickerelweed, was considered a desirable ornamental plant. But the species reproduces not only by seeds, but also rapidly by means of offsets—a method that can double the size of a plant mass in as few as six days. This has made it one of the most reviled invasive plants.

Greater Fringed Gentian

Gentianopsis crinita H 12–36 in (30–90 cm)

One of the last wildflowers to bloom in the year, the Greater Fringed Gentian's arresting deep blue flowers sometimes witness the first snowfalls in northern and mountainous parts of its range.

KEY FACTS

The bright blue corollas with 4 heavily fringed, flaring lobes above a tube grow at the end of an erect, often branching stem.

+ **habitat:** Moist meadows, forest edges, and stream banks

+ **range:** Mainly north-central and northeastern U.S. and Canada

+ **bloom period:** August–November

The Greater Fringed Gentian flower follows the sun: It opens in the morning on sunny days and closes in the evening, the buds appearing spirally twisted. Four folded sepals, two shorter than the others, clasp the corolla tube. Leaves are mostly lance shaped and rounded at the base. A rosette of leaves forms the first year; the flowering stalks appear in the second. The plant's common and scientific names tie it to King Gentius, a second-century B.C. king of Illyria on the Balkan Peninsula who championed its medicinal properties that centered on treating digestive ailments. The species' exquisite beauty has led to its decline through collecting and overpicking.

Virginia Iris/Southern Blue Flag

Iris virginica H 12–24 in (30–60 cm)

After the Virginia Iris flowers, its weak stems sometimes collapse, leaving a jumble of stems, leaves, and spent flowers on the ground. This species is a smaller version of *I. versicolor.*

KEY FACTS

The flower has 3 purple-blue petals and 3 petal-like sepals that are fused below; sepals have a greenish yellow nectar guide at the base.

+ **habitat:** Marshes, swamps, wet meadows, edges of ponds, lakes, streams, and ditches

+ **range:** Found mainly in southeastern and south-central U.S.

+ **bloom period:** May–June

The leaves of the Virginia Iris usually grow higher than the species' showy blooms on their sometimes branched stems. Long, narrow, and folded, the stems fan out from the plant base to form a V in cross section. The leaf shape accounts for the plant's common name; it derives from the Middle English *flagge*, which means "rush" or "reed." The conspicuous signal patches on the blossoms attract pollinators to the nectar. The species reproduces by rhizomes and by seeds. A variety of this species, Shreve's Iris (*I. virginica* var. *shrevei*), has more branches, and the stems usually do not collapse after flowering.

Blue-eyed Grass

Sisyrinchium angustifolium H 4–18 in (10–45 cm)

Examined closely, the Blue-eyed Grass's grasslike foliage looks more like a member of the Iris family than a grass. Unlike many other irises, these leaves may be shorter than the flower stalk.

KEY FACTS

The flowers have 3 each of pointed violet-blue petals and sepals, and yellow centers.

+ **habitat:** Damp meadows, low road-sides, prairies, old fields, damp fields, open woods, and ditches

+ **range:** Through-out much of central and eastern U.S. and Canada

+ **bloom period:** March–July

The star-shaped flowers of the Blue-eyed Grass grow at the top of long, flat, twisted, and often branched stems. The flower stalks appear amid a tight clump of bright green, linear leaf-like bracts that are about a quarter inch (0.6 cm) wide. Usually only one flower is in bloom on a stalk at a time. When the plant withers, its foliage turns black. The Cherokee, Iroquois, and other American Indian groups used the leaves and roots of the plant to treat both stomach and intestinal difficulties, including bouts of worms. A distinct yel-low "eye" gives the unassuming flower a bit of pizzazz. Although its leaves are very narrow, this is not a true grass.

Old-field Toadflax/Blue Toadflax

Nuttallanthus canadensis H 6–24 in (15–60 cm)

Old-field Toadflax is partial to sandy soil and disturbed areas, seeming to choose sites that present little competition from other plants. The species is closely related to Snapdragons.

KEY FACTS

Half-inch-long (1.3 cm) blue to blue-violet 2-lipped flowers have 2 white ridges on the lower lip; the flower base shows a slender, curved spur.

+ **habitat:** Open, dry, rocky sites, sandy fields, roadsides, and open woods

+ **range:** Most of central and eastern U.S. and Canada and U.S. West Coast

+ **bloom period:** March–September

The blossoms of the Old-field Toadflax grow from short stalks in an open spikelike cluster in the upper portion of a stem that rarely branches. The leaves of this species are more than an inch (2.5 cm) long and are fairly linear and smooth. They have two growing strategies: Very nar-row leaves appear alternately on erect plant stems, and others grow opposite each other on prostrate stems. Both the flowering stalks and the leaf stalks are green to reddish green and smooth. Only a few flowers bloom at one time, and a small colony of the plants may bloom continuously for two or three months.

American Brooklime

Veronica americana H 6–36 in (15–90 cm)

An exceptionally pretty member of the group known as Speedwells or Veronicas, the American Brooklime has stout but weak stems that take root at leaf joints when they touch the ground.

KEY FACTS

Small, deeply 4-lobed light blue or purple corollas have purple stripes, a white spot at the base of each lobe, and two conspicuous stamens.

+ **habitat:** Along streams and in ditches, swamps, and other moist areas

+ **range:** Much of U.S. and Canada, excluding Southeast and far north

+ **bloom period:** May–September

A delicate-looking plant, the American Brooklime bears loose, elongated 2- to 6-inch (5–15 cm) clusters of tiny flowers on branches rising upright from the leaf joints of its upper leaves. The leaves are oval to lance shaped and usually slightly toothed; those on the upper flowering stems clasp the stems at their base. This species is in a group called Speedwell for a reason: Much of its growth sprawls rapidly along the ground, punctuated at intervals by the smooth and hollow flower-bearing branches. The American Brooklime frequently grows partly in and partly out of the water.

One-flowered Cancer Root/One-flowered Broomrape

Orobanche uniflora H 3–10 in (7.5–25 cm)

Each stalk of One-flowered Cancer Root sends up one blossom. The parasitic species does not photosynthesize and obtains nourishment by tapping into the roots of its host plants.

KEY FACTS

The flowers are white to violet-blue; the fused petals form a long tube that bends and flares to 5 lobes, often with yellow stripes on the lower lobes.

+ **habitat:** Damp woods, thickets, and seeps

+ **range:** Throughout the U.S. and Canada, except in far northern areas

+ **bloom period:** April–August

One-flowered Cancer Root has scaly, short, and mostly subterranean stems that send up one to four flower stalks, each bearing a single flower. The stem may measure no more than 1.5 inches (4 cm). The plant does not have true leaves; instead, it has small brown scales at the base of the flower stalk. The species hijacks nutrients from a host plant and it parasitizes a wide variety of other plant species. A look around at the other plants surrounding a One-flowered Cancer Root creates a mystery as to which of the species might be hosting the boldly flourishing parasite.

Azure Bluet
Houstonia caerulea H 3–8 in (7.5–20 cm)

Perfection in miniature, slender stems, tiny leaves, and dainty, delicately tinted blossoms distinguish Azure Bluet, a member of the Madder family.

KEY FACTS

The pale blue, trumpet-shaped flowers have 4 spreading lobes and deep yellow centers, and grow singly on unbranched stems.

+ **habitat:** Fields, meadows, woods, and disturbed areas

+ **range:** Throughout much of eastern U.S. and Canada, excluding Florida

+ **bloom period:** April–July

Azure Bluet flowers are mostly less than a half inch (1.25 cm) wide, but they make a big impression, especially as they tend to grow in large patches. They are also named Quaker Ladies for their demure appearance. The plant's genus name refers William Houstoun, an early 18th-century Scottish botanist who collected plants mainly in the American tropics. Although the leaves on most Bluet stems are tiny and sparse, those at the base of the plant are larger and spatula shaped, and they grow in tufts. The Cherokee made an infusion from the plant, which they administered to prevent bed-wetting.

Woodland Forget-me-not
Myosotis sylvatica H 5–12 in (12.5–30 cm)

The Woodland Forget-me-not was certainly not forgotten: The introduced Eurasian native escaped from cultivation and now is naturalized in a number of areas of North America.

KEY FACTS

Small blue to white flowers have 5 spreading lobes that emerge from short tubes; the flowers grow in a rounded cluster.

+ **habitat:** Moist woods, stream banks, swamps, ditches, and roadsides

+ **range:** Naturalized in disparate areas mostly in northern and central U.S. and Canada

+ **bloom period:** April–September

The bractless flower clusters of the Woodland Forget-me-not grow on erect stems that branch from the base of the plant. The leaves near the base are longer and wider than those higher up the stem. The genus name *Myosotis* comes from the Greek for "mouse ear," referring to the various species' furry leaves. In the language of flowers, a presentation of Forget-me-nots signifies true love and remembrance through a legendary association with a tragic lost love. Despite this endearing allusion, the species perpetuates itself by aggressively self-seeding, which increases its potential as an undesirable invasive in some regions.

Common Arrowhead

Sagittaria latifolia H to 36 in (90 cm)

These aquatic perennials form large colonies in watery environments. Their genus name comes from the Latin for "arrow," a reference to the distinctive leaf shape.

KEY FACTS

The white flower is formed of 3 white to pinkish petals and 6 or more stamens; pistils are on separate flowers.

+ habitat: Aquatic habitats such as marshes, swamps, ponds, mud banks, wet sand, and ditches

+ range: Throughout most of U.S. and Canada, except far northern areas

+ bloom period: July–September

Leaves are all basal, with long stalks and large, mostly arrowhead-shaped green blades that vary in width and emerge from the water. Underwater, the leaves may be long and narrow. The emergent leaves are followed by taller stalks with whorls of three white flowers. Some variants display short hairs over much of the plant. The Common Arrowhead's starchy tubers, developing under the mud, are favored by ducks and muskrats and give the plant one of its alternate common names: Duck-potato. The tubers were also an important source of food and medicines for American Indians, who called the plant Wapato.

Soapweed Yucca

Yucca glauca H 24–48 in (60–120 cm)

Soapweed Yucca played many roles in traditional American Indian hygiene and health. Soap and dandruff shampoo were made from the roots, and a solution was used to kill lice in hair.

KEY FACTS

The flower is greenish white and bell shaped and has 6 pointed tepals (look-alike petals and sepals).

+ habitat: Prairies and waste areas at low to middle elevations

+ range: Central U.S. into Alberta; escapes from cultivation

+ bloom period: June–August

A member of the Asparagus family (previously in the Agave family), Soapweed Yucca often forms small colonies. First, a crowded clump of daggerlike, fibrous, evergreen leaves with hairy edges is established. Later, a fast-growing flower stalk to more than 4 feet (120 cm) tall rises from the center of the basal rosette. It is densely but loosely populated with pendent, bell-like blooms on short stalks. Each species of Yucca relies on a symbiotic partnership with its corresponding species of Yucca Moth for reproduction. In exchange for transfer of pollen, the Yucca flower and seedpods offer shelter and food for the moth's eggs and larvae.

Smooth Solomon's Seal
Polygonatum biflorum H 8–84 in (20–210 cm)

"Smooth" is an apt word for identifying this species, referring to its smooth leaves. The almost identical Hairy Solomon's Seal *(P. pubescens)* has minute hairs on the underside of its leaves.

KEY FACTS

The pendent 6-lobed tube, whitish to greenish yellow, is composed of 6 tepals, which are united at their bases.

+ habitat: Rich woods, old fields, thickets, and along roadsides

+ range: Central and eastern U.S. and Canada

+ bloom period: March–June

Flowers usually dangle in pairs from leaf axils on an erect or arching stem that has many smooth, stalkless, parallel-veined, lance-shaped leaves. Thick, fleshy, and gnarly rootstocks send up new stems each year. When these die back, they leave noticeable scars on the roots. These are the "seals," resembling the signet ring of King Solomon, and counting them gives a good estimate of a plant's age. A variety, *P. biflorum* var. *commutatum*, or Great Solomon's Seal, has proportionately larger flowers and leaves, and occasionally may grow to more than 6 feet (180 cm) tall.

Large-flowered White Trillium
Trillium grandiflorum H 8–18 in (20–45 cm)

This trillium boasts the largest flower of all *Trillium* species. It measures up to 4 inches (10 cm) wide and turns from white to light pink as it ages.

KEY FACTS

The flower has 3 large, oval, white petals with wavy edges; 3 green, lance-shaped sepals; and 6 yellow stamens.

+ habitat: Moist woods, swamps, flood-plains, and roadsides

+ range: Throughout much of eastern North America; also highly cultivated

+ bloom period: April–June

The showy blossom of the Large-flowered White Trillium, a member of the Bunchflower family (previously in the Lily family), rises on a central stalk above a whorl of three terminal leaves. The leaves—in reality, leaflike bracts—measure up to 6 inches (15 cm) long and have smooth margins and prominent parallel veins. There are variations of this species involving flower color or markings, including some that may be related to the presence of an infection by bacterial microorganisms. American Indians used the roots of the plant to treat a number of ailments. Trilliums of any species should be left undisturbed in their natural habitat.

Gunnison's Mariposa Lily/Gunnison's Sego Lily

Calochortus gunnisonii H 6–18 in (15–45 cm)

The genus name for this wildflower, *Calochortus,* derives from the Greek for "beautiful grass," referring to the plant's grasslike leaves, which wither as the flowers bloom.

KEY FACTS

The flower has 3 white, or sometimes purple, wide petals, each with a hairy, yellow glandular base bordered above by a purple line.

+ **habitat:** Open woods, prairies, dry gulches, and mountain meadows

+ **range:** Western U.S. in the Rocky Mountain and adjacent states

+ **bloom period:** May–July

The tulip-like Gunnison's Mariposa Lily, also known as Gunnison's Sego Lily, emerges from a bulb along with its linear basal leaves. One to three flowers may appear on each unbranched stem. The species occurs as scattered plants or in expansive colonies that make a dramatic impression. Meriwether Lewis collected the first *Calochortus* lily specimen for scientific study in Idaho in 1806. Various American Indian groups used the bulb of *C. gunnisonii* for food and medicine. The namesake of the species, Captain J. W. Gunnison, led an ill-fated surveying expedition to Utah in 1853, during which he and eight colleagues were murdered.

Common Chickweed

Stellaria media H 4–8 in (10–20 cm)

The common name of this Eurasian import refers to the fact that chickens favor its small, delicious flowers and leaves, which humans historically also have enjoyed.

KEY FACTS

The small white flower has 5 radiating petals, which are so deeply cleft that they often appear to be 10 petals.

+ **habitat:** Fields, open woods, lawns, and disturbed areas

+ **range:** Naturalized throughout most of the U.S. and Canada

+ **bloom period:** Nearly year-round under favorable conditions

The flowers of the Common Chickweed have five white petals arranged in a star shape, and five green, hairy sepals. The flowers form a cluster at the end of a stem or can emerge singly from a leaf axil. Leaves are smooth edged and oval, and grow opposite each other. The species forms low, sprawling, dense mats of stems that reach 16 inches (40 cm) long and have a single line of hairs on one side. It often germinates in fall and winter, and can be in bloom at any time of the year. Common Chickweed is often considered an undesirable weed in lawns and crop fields.

Wood Anemone
Anemone quinquefolia H 4–10 in (10–25 cm)

The delicate Wood Anemone, a member of the Buttercup family, graces woodland settings in early spring. Its habit of trembling in the breeze gives it another common name: Windflower.

KEY FACTS

The small white or sometimes pink flower typically has 5 petal-like sepals and a greenish center filled with 30–50 stamens.

+ habitat: Woods, woodland edges, and stream banks

+ range: Throughout much of central and eastern Canada and the eastern U.S.

+ bloom period: April–June

Wood Anemone tends to develop in large stands. Individual plants may start out with one long-stalked basal leaf with three to five coarsely toothed lobes. A single flower, about an inch (2.5 cm) wide, arises on a stalk above a whorl of three stem leaves. The flower may have four to nine petal-like sepals, but most commonly has five. Stem leaves are deeply divided into three, and sometimes five, leaflets. Even the leaves with three leaflets may appear to have five because the side leaflets are deeply cleft. By summer, the plant has died back and all but disappeared.

Early Meadow Rue
Thalictrum dioicum H 8–36 in (20–90 cm)

Male and female flowers of Early Meadow Rue occupy separate plants, giving rise to the species name, *dioicum*, derived from the Greek for "two households."

KEY FACTS

The male flower has elongated yellow stamens dangling from 4 to 5 greenish petal-like sepals; the female flower has a cluster of pistils.

+ habitat: Woods, slopes, and ravines

+ range: Throughout much of central and eastern North America

+ bloom period: April–June

Early Meadow Rue gets a head start on other plants and even on itself in the spring. The small, tasseled flowers of the perennial plants form large clusters from stem ends and leaf axils. The clusters start to bloom before the leaves themselves have fully developed. The leaves grow on long stems and have from 3 to as many as 12 roundish leaflets that may have rounded teeth. Flowers on male plants droop; those on female plants stand erect. Despite gender separation, flowering time is coordinated. The petal-like sepals often drop off, leaving the stamens and pistils.

Eastern Virgin's Bower

Clematis virginiana L 6–20 ft (2–6 m)

This native Virgin's Bower climbs over supporting structures with leaf stalks that act like tendrils, or it sprawls horizontally as a jumbled ground cover.

KEY FACTS

The flower has 4 white petal-like sepals and a center of long stamens in the male flower and long pistils in the female flower.

+ **habitat:** Moist woods, thickets, stream banks, pond edges, and fencerows

+ **range:** Much of central and eastern U.S. and Canada

+ **bloom period:** July–October

Eastern Virgin's Bower, a vigorous vining member of the Buttercup family, rapidly creates a mass of foliage and fragrant flowers that begin to blossom in late summer. The foliage takes the form of compound green leaves with three to five oval toothed leaflets. Male and female flowers grow in clusters from leaf axils on separate plants, relying on bees and butterflies to accomplish pollination. After fertilization, the female flower transforms into a seed head with a showy plume-like tail attached to each fruit. Repeated in large numbers, the plumes give the plant a grayish, bearded look.

Mayapple

Podophyllum peltatum H 12–18 in (30–45 cm)

Large colonies of green umbrella-like leaves shelter the nodding white flowers of the Mayapple, a shade-loving plant that often spreads abundantly across rich woodland soils in spring.

KEY FACTS

The flower has 6 to 9 white petals and a yellow center with double the number of stamens as petals.

+ **habitat:** Woods, shady fields, along roadsides, and stream banks

+ **range:** Central and eastern U.S. and Canada

+ **bloom period:** March–June

The apple blossom–like flower of the Mayapple dangles from its stalk at the junction of a pair of large, palmately lobed and deeply cleft leaves that are up to a foot (30 cm) around. The stalked leaves appear at the tip of a single stem that rises from spreading underground rhizomes. The plant also produces solitary leaves directly from the rhizomes that are not associated with flowers. The petals fall from the flower, and an egg-shaped yellowish fruit develops containing several seeds. All parts of the prolific Mayapple are poisonous, except for the ripe fruits.

Bloodroot

Sanguinaria canadensis H 4–8 in (10–20 cm)

American Indians used the reddish sap produced by the Bloodroot as a dye for clothing and crafts, as well as facial decoration.

KEY FACTS

The flower has 8–16 white petals surrounding a center with yellow-anthered stamens and a green pistil with a yellow stigma.

+ **habitat:** Woods and stream banks

+ **range:** Central and eastern U.S. and Canada

+ **bloom period:** March–April

In early spring, a large, grayish green leaf emerges from the ground that enfolds the stalk of a single flower. Before the leaf can fully unfold to reveal its multiple major and minor lobes and prominent veins, the flower has pushed past to unfurl like a small water lily. Both the flower and the leaf stalks continue to grow. The blossom opens during the day and closes at night for the few days of its existence. The foliage and rhizomes of the Bloodroot produce an acrid reddish juice, which gives the species its namesake "blood."

Twinleaf

Jeffersonia diphylla H 6–18 in (15–45 cm)

The genus name of this delicate and unusual plant is an honor bestowed on Thomas Jefferson, an avid horticulturalist, by American botanist William Bartram.

KEY FACTS

The bowl-shaped flower has 8 oval white petals and a center with 8 yellow stamens.

+ **habitat:** Moist woods and rocky slopes, often in limestone areas

+ **range:** Eastern U.S. and Canada, excluding the Northeast and parts of the Southeast

+ **bloom period:** March–May

The namesake basal leaves of the clump-forming Twinleaf, a member of the Barberry family, rise on long stalks and give a doubled appearance due to their distinct lobes. The cup-shaped white flower tops a separate leafless stalk. The plant bears a passing resemblance to the unrelated Bloodroot, a major difference being the latter's large, five- to nine-lobed leaf. Like the Bloodroot, the Twinleaf leaf and flower stalks continue to grow after the flower has blossomed and as the fruit ripens. The Twinleaf fruit is an unusual pear-shaped brownish pod with a hinged lid.

Annual Prickly Poppy
Argemone polyanthemos H 12–36 in (30–90 cm)

Annual Prickly Poppy is a tad less prickly than some others in its genus. Its name comes from *argema*, Greek for "cataract." *Argemone* species were once used to treat eye ailments.

KEY FACTS

The flower has 6 thin, white, crinkled petals and a mound of 20 or more yellow stamens and a maroon stigma in the middle.

+ **habitat:** Prairies, meadows, roadsides, and disturbed areas

+ **range:** Native to the central U.S.; introduced farther east and northwest

+ **bloom period:** April–October

Delicate large blossoms resembling crepe paper, up to 4 inches (10 cm) wide, top the branched stems of the Annual Prickly Poppy, a prominent annual or biennial wildflower species native to the central plains and prairies. The rest of the plant is covered with prickles (hence, the common name Prickly Poppy). The blue-green, waxy, and spiny lobed leaves are dense on the stiff stems, which are somewhat spiny. The pointed oval seed-pod, with its protruding stigma, is also prickly Disturbances such as highway construction probably aided the spread of this species, particularly to the West. All parts of the plant are poisonous to eat.

Carolina Horse Nettle
Solanum carolinense H 12–36 in (30–90 cm)

Not a true nettle but a member of the Nightshade family, the Carolina Horse Nettle sports prickly stems and leaves that discourage grazing animals.

KEY FACTS

The white or pale violet star-shaped flower is composed of 5 lobed, spreading corollas and 5 stamens with long yellow anthers erect in a group in the center.

+ **habitat:** Fields, prairies, roadsides, and disturbed areas

+ **range:** Much of western U.S. and eastern U.S. and Canada

+ **bloom period:** May–October

The Carolina Horse Nettle's attractive star-shaped flowers with their distinctive banana-like anthers share the plant with hairy and prickly leaves, stems, and stalks. The flowers grow in clusters at the top of erect stems that bear lance-shaped, angular, lobed leaves. The plant's prickly surfaces and Nightshade family toxicity are off-putting to mammalian herbivores; however, the mature yellow fruits that resemble little tomatoes are tolerated and eaten by birds, skunks, rodents, and other animals. For those who consider this species an undesirable weed, gloves should always be worn when attempting to pull it.

Dutchman's Breeches
Dicentra cucullaria H 4–12 in (10–30 cm)

The wide pantaloons Dutch men traditionally wore inspired the common name for this woodland perennial, a relative of Bleeding Hearts.

KEY FACTS

The flower has 2 outer white petals that form stout spurs and 2 inner yellow petals that curve up at the base.

+ **habitat:** Woods, ravines, ledges, and stream banks

+ **range:** Mainly in the eastern U.S. and Canada and the Pacific Northwest

+ **bloom period:** March–May

In the early spring, the double-spurred flowers of Dutchman's Breeches dangle upside down in pairs from a curved, leafless scape, or flower stem, that rises from the plant's rhizome. Their triangular nectar spurs, formed by the outer white petals, require the services of long-tongued bees for efficient pollination. The scape overhangs a profusion of deeply cut, fernlike, grayish green leaves that form a rosette on the forest floor. The compound leaves attach to the rhizome by a long, brownish stalk and have three leaflets, which are divided into three more leaflets that are further divided into linear lobes.

Wild Strawberry
Fragaria virginiana H 3–6 in (7.5–15 cm)

The large and becoming larger cultivated strawberries we enjoy had their start as a hybrid of the Wild Strawberry and a South American species.

KEY FACTS

The small flower is composed of 5 white petals around many bright yellow stamens and numerous separate pistils on a central cone.

+ **habitat:** Wood edges, fields, meadows, open ground, roadsides, and disturbed areas

+ **range:** Throughout the U.S. and Canada, except some far northern areas

+ **bloom period:** April–June

The native perennial Wild Strawberry, a member of the Rose family, hugs the ground, sending up new plants by means of sprawling, hairy runners. The compound leaves have three hairy, oval, toothed leaflets. The small, white flowers appear in clusters on short stems or emerge directly from the underground root. They turn into the small red, iconic strawberry with a multitude of seedlike fruits embedded in depressions on its surface. The plant sometimes goes dormant during the heat of the summer. This species feeds a wide range of animals, from insects to birds to mammals—including, of course, humans.

White Wild Indigo

Baptisia lactea/B. alba H up to 6 ft (180 cm)

This *Baptisia* species produces no blue dye, unlike some other members of the genus. When new growth pushes through the ground, it somewhat resembles asparagus.

KEY FACTS

The pealike white or cream flower is composed of 5 petals: a top banner petal, 2 side wing petals, and 2 lower ones that form a keel.

+ **habitat:** Open woods, marsh and lake edges, and prairies

+ **range:** Much of the central and eastern U.S.

+ **bloom period:** April–July

The White Wild Indigo's long, erect spikes of pealike flowers seem to go on forever at the top of stiff stems that form the bushy plant and can grow to 6 feet (180 cm) tall. At that height, the flowers frequently tower over nearby vegetation. The stems branch near the top and are covered with compound, three-parted green leaves. The leaves turn black in the fall, and the flowers are replaced by hairless, inflated seedpods, about 2 inches (5 cm) long that are green at first and then turn black. The pods often stay on the plant through the winter. Larvae of some butterflies use *Baptisia* as a food plant.

Flowering Spurge

Euphorbia corollata H 12–36 in (30–90 cm)

A relative of the Poinsettia, Flowering Spurge shares some features with the iconic holiday plant, such as bracts—modified leaves—that masquerade as petals.

KEY FACTS

What appears to be a flower incorporates 5 rounded, white glandular appendages that surround a small cup of tiny yellow flowers.

+ **habitat:** Open woods, old fields, prairies, sand dunes, roadsides, and disturbed areas

+ **range:** Central and eastern U.S. and Canada, except far northern areas

+ **bloom period:** June–October

The tiny real flowers of the Flowering Spurge are part of an elaborate architecture. The flowering forms appear in clusters in a multitiered branching structure that occurs at the top of a long, central stem. Linear to oblong leaves climb the stem alternately, ending in whorls of three just below the flower clusters. Typical of spurges, the species has milky sap in all tissues that can cause skin irritation in susceptible individuals. The name "spurge" comes from the Latin for "to purge," and the plant was used for that purpose, although it has other, mildly toxic properties.

Tufted Evening Primrose
Oenothera caespitosa H 6–12 in (15–30 cm)

The Tufted Evening Primrose displays its beauty by night, when the white blossom pops open and emits a sweet fragrance that lets pollinators know it is open for business.

KEY FACTS

The flower has 4 large white petals that are notched and appear heart shaped, and 8 yellow stamens.

+ **habitat:** Open woods, clearings, open desert, arroyos, dry slopes, and roadsides

+ **range:** Much of the western U.S. and Canada, except in far northern areas

+ **bloom period:** April–August

The Tufted Evening Primrose flower is stemless, rising from a substantial basal rosette, but the flowers have a long tube that holds the petals above the base. The rosette is formed of many long, narrowish, lance-shaped leaves that often have scalloped, toothed, or lobed edges. The 4-inch-wide (10 cm) bloom is a one-night wonder, fading to pink as the corolla withers. In warm climates, this perennial species often is evergreen, but it can be deciduous in colder regions. Despite the name, evening primroses are not members of the genus *Primula*, which is in the Primrose family.

Bunchberry Dogwood
Cornus canadensis H 3–9 in (7.5–22.5 cm)

Dogwood as ground cover, Bunchberry Dogwood looks much like its tree relatives and is a rare representative of that family in herb form.

KEY FACTS

The four petal-like white bracts surround a cluster of tiny, white flowers with purple centers.

+ **habitat:** Coniferous and mixed woods, thickets, and swamps, sometimes in mountainous areas

+ **range:** Northern U.S. and Canada and in the southern Rocky Mountain states

+ **bloom period:** May–September

The Bunchberry Dogwood's white bract-and-flower combination stands out in the midst of its lush, green foliage. This low-growing plant has pointed, oval, crowded pairs of opposite leaves that misleadingly appear to grow as whorls of four to six on slender stems. Whorls with four leaves typically do not flower, whereas those with six do. The plant forms extensive colonies by means of rhizomes, underground stems that put out roots and shoots, an important method of reproduction given that fruit production by the species is spotty. When it does fruit, clumps of dark red, two-seeded berrylike drupes replace the flowers.

Indian Pipe
Monotropa uniflora H 3–10 in (7.5–25 cm)

A ghostly denizen of dimly lit woods, the pale Indian Pipe contains no chlorophyll. Instead, it takes in nourishment by parasitizing soilborne fungi that are associated with the roots of trees.

KEY FACTS

The single white to pinkish to reddish flower is composed of 3 to 6 petals and about twice as many stamens.

+ **habitat:** Moist, shady woods and slopes

+ **range:** Throughout most of the U.S. and Canada, except in the central Rockies, Southwest, and far northern areas

+ **bloom period:** June–September

The Indian Pipe has no green foliage to distinguish flower from stem; the bend marks the nodding flower's beginning in the younger plant. The thick stem is fleshy, translucent, and covered in scales. As the flower and then the seed capsule mature, the entire plant begins to turn black, as it also does if it is picked and dried. After flowering, the flower base begins to unbend until the mature fruit is erect. Ripe seeds emerge from slits that open in the five-segmented capsule. This species parasitizes mycorrhizal fungi often associated with the roots of oaks and conifers.

Elliptical-leaved Shinleaf
Pyrola elliptica H 6–12 in (15–30 cm)

The leaves of this native wildflower traditionally were used as a poultice on wounds and bruises to reduce pain and inflammation—that is, a shin plaster—the source of the common name.

KEY FACTS

The white to greenish, cup-shaped flower has 5 petals, often green veined, and a curved, protruding style.

+ **habitat:** Dry or moist woods

+ **range:** Northern U.S. and Canada, except in far northern areas

+ **bloom period:** June–August

In May, a long scape, or leafless flowering stem, grows from the nearly basal cluster of bright green, broadly oval evergreen leaves of the Elliptical-leaved Shinleaf. It will grow up to 12 inches (30 cm) long and develops an elongated floral cluster. It bears up to about 20 fragrant, nodding blossoms. By the time the blooms of spring woodland wildflowers are long gone, the shinleaf flower comes to life in deep shade. The genus name is the Latin diminutive for "pear," referring to the shape of the leaves of some species.

Field Bindweed

Convolvulus arvensis L vining to 72 in (180 cm)

A member of the Morning Glory family, Field Bindweed is one of the more aggressive of the introduced Eurasian species, deep rooted and climbing or trailing its way across the landscape.

KEY FACTS

The funnel-shaped white or pink flower has a 5-petaled, shallowly lobed corolla.

+ **habitat:** Fields, open areas, roadsides, gardens, and disturbed sites

+ **range:** Introduced and naturalized throughout the U.S. and Canada

+ **bloom period:** April–October

The genus name of the Field Bindweed derives from the Latin for "to twine around," and that certainly is the method of operation for this perennial. Its stems are long and hairy, with triangular leaves arranged alternately along them. The flowers emerge singly or in pairs from the leaf axils. The plant adapts easily in many climatic zones and plant communities. Field Bindweed arrived here by the mid-18th century, probably first entering as a contaminant in crop or garden seeds. Many U.S. states and a number of Canadian provinces place this species on their noxious weeds lists.

Jimsonweed/Thorn Apple

Datura stramonium H 12–48 in (30–120 cm)

The name Jimsonweed is a corruption of Jamestown Weed. In 1679, soldiers in the Virginia colony allegedly ate the plant in a salad and suffered extreme psychological effects.

KEY FACTS

The white or purple funnel-shaped flower, up to 3 in (7.5 cm) long, has 5 lobes and a violet throat.

+ **habitat:** Sandy soils, fields, barnyards, and disturbed areas

+ **range:** Introduced and naturalized throughout the U.S. and Canada, except in far northern areas

+ **bloom period:** July–October

The large Jimsonweed blossoms with their pointed lobes appear in the axils of branch forks along the strong purple or green stems. Long, stalked leaves grow alternately from the stem and are usually jaggedly lobed. After blooming, the corolla falls off; in its place a green, oval capsule develops, about 2 inches (5 cm) long and covered in prickles, source of the alternate name Thorn Apple. Jimsonweed, introduced in temperate North America from the tropics, is a poisonous member of the Nightshade family. It contains alkaloid compounds with many toxic and psychoactive properties, making it harmful to humans and livestock.

American Ginseng
Panax quinquefolius H 6–24 in (15–60 cm)

This all-purpose healing plant has been overharvested due to worldwide demand for its roots, which began in the 1700s. Its genus name, *Panax*, derives from a Greek word meaning "cure-all."

KEY FACTS

The tiny, greenish white flowers form a small umbellate cluster on a separate stalk rising from a group of leaves.

+ habitat: Rich woods

+ range: Native to the eastern U.S. and Canada; widely and commercially cultivated

+ bloom period: June–July

The tiny, white umbel of the American Ginseng flower is dwarfed by the trio of palmately compound leaves that surround it. Three of the usually five, toothed leaflets are large; the other two are much smaller. Red berrylike fruits with two seeds each develop from the flowers in the fall. The common name of this species, Ginseng, derives from a Chinese term meaning "manlike," a reference to the root shape. The Chinese and others turned to the North American forms of this important medicinal plant when local species were overharvested.

Fragrant Water Lily
Nymphaea odorata H 2–6 in (5–15 cm)

The Fragrant Water Lily might appear to be tropical—and many people know it as a fish pond ornamental—but it is a widespread native perennial in temperate North America.

KEY FACTS

The white flower, up to 8 in (20 cm) across, has 4 sepals, 20–30 pointed white petals, and dozens of bright yellow stamens.

+ habitat: Aquatic habitats such as ponds, lakes, and slow-moving streams

+ range: Throughout U.S. and Canada, except some northern areas; highly cultivated

+ bloom period: March–October

Floating effortlessly on the surface of a pond or lake, the Fragrant Water Lily is strongly tethered to its rootstock in the muddy bottom. Long, separate stalks link the blossom and leaves to their source. The circular leaves develop first; up to a foot (30 cm) across, they are green on top, purplish below, and deeply cleft on one side. The flowers float lightly on their sepals, opening in the morning and closing at night. They last only a few days, and after withering are replaced by dark, round fruits that mature underwater to release their seeds.

Milk Vetch
Astragalus species H 4–48 in (10–122 cm)

Some milk vetches are the bane of ranchers and livestock farmers. Many species contain a powerful and addictive toxin that slowly debilitates and can eventually kill grazing animals.

KEY FACTS

The pealike flowers in colors from white to yellowish to pink to red to purple appear in loose clusters to dense spikes.

+ **habitat:** A variety, including prairies, desert, open woods, shores, and mountain slopes

+ **range:** Nearly 400 species occur in the U.S. and Canada

+ **bloom period:** March–August

Many of the hundreds of species of native milk vetches tend to sprawl. Some, such as the Canada Milk Vetch (*A. canadensis*), are more upright than others. Milk vetch leaves arise at the base of the plant and are pinnate, also alternating along the stem in many species. Milk vetches are closely related to Locoweeds (genus *Oxytropis*), which produce many of the same toxic effects. To confuse matters, milk vetches also are commonly called Locoweed, and the typical toxicity of most plant parts of species in both genera, along with overlapping habitats and range, tends to perpetuate the association.

White Sweet Clover
Melilotus albus H 12–96 in (30–240 cm)

Beekeepers often plant White Sweet Clover, a Eurasian import and member of the Pea family, as a fragrant and flavorful nectar source for their hives.

KEY FACTS

The small white flowers have 5 petals: a banner petal at top; two side petals, or wings; and 2 bottom petals joined as a keel.

+ **habitat:** Open areas, roadsides, and disturbed sites

+ **range:** Throughout most of the U.S. and Canada, except in some far northern areas

+ **bloom period:** May–September

The White Sweet Clover's pealike flowers grow as long, slender spikes from the leaf axils of erect stems. The compound leaves have three oblong leaflets and they emit a vanilla-like fragrance when crushed. The same chemical compound responsible for the fragrance is a source of the active ingredient in well-known anticoagulant medications that prevent blood clots. The species is frequently planted as a nitrogen-fixing cover crop to improve agricultural land. White and Yellow Sweet Clover (*M. officinalis*) are virtually indistinguishable when in flower or fruit.

Red Baneberry

Actaea rubra H 12–36 in (30–90 cm)

Beware of plants with "bane" in the name, including the Red Baneberry, a member of the Buttercup family with bright red fruits, which is highly poisonous.

KEY FACTS

The flowers have 4–10 spoon-shaped white petals and numerous long white stamens.

+ **habitat:** Moist woods, thickets, and stream banks

+ **range:** Throughout much of the northern and western U.S. and Canada, except in some far northern areas

+ **bloom period:** April–July

Red Baneberry flowers form dense clusters on the leafy stems of bushy plants. The large, divided leaves are composed of numerous oval leaflets with sharply toothed edges. The flower petals tend to drop off, leaving long stamens that give the clusters a feathery look. Red berries with black dots develop that attach to the clusters with slender stalks. A close relative the White Baneberry (*A. pachypoda*) has white berries with prominent black dots that suggest its other common name, Doll's Eyes. To confuse matters, the Red Baneberry, which differs from the white in flower stalk, flower, and fruit features, sometimes produces white berries.

Pokeweed

Phytolacca americana H to 10 ft (3 m)

An alternative common name for Pokeweed is Inkberry, which speaks to the practice of using the berry's juice as ink, as was commonly done by Civil War soldiers in letters home.

KEY FACTS

The tiny stalked flower, composed of 5 white petal-like sepals and up to 25 stamens, grows in an elongate cluster.

+ **habitat:** Open woods, thickets, old fields, roadsides, and disturbed areas

+ **range:** Much of the eastern U.S. and Canada, southern U.S., and in parts of the West

+ **bloom period:** July–October

Pokeweed flower clusters are terminal on the stem branches. The plant's leaves are very long, lance shaped, and heavily veined. As the berries form, the heavy axis supporting them starts to droop. Berries start out green and eventually become a blackish purple. Although the species is considered mildly toxic, parts of the Pokeweed have long been used in traditional medicine, and the developing greens are cooked as a spring green in the Southeast. Its documented antiviral properties are being studied as a potential treatment for HIV. "Poke" may derive from "puccoon," an American Indian term for a plant that yields dye.

Solomon's Plume/Feathery False Solomon's Seal
Maianthemum racemosum H 12–36 in (30–90 cm)

Erect or arched stems with branched clusters of small white flowers hold their own against imposing foliage in the Solomon's Plume, a native perennial.

KEY FACTS

The tiny white flowers are composed of 6 tepals (look-alike sepals and petals) that are shorter than the blossoms' 6 stamens.

+ **habitat:** Moist woods, floodplains, and stream banks

+ **range:** Throughout most of the U.S. and Canada, except in some far northern areas

+ **bloom period:** April–July

The flowers of the Solomon's Plume appear on short stalks at even intervals on short branches. The branches form a pyramid-shaped terminal cluster on an unbranching stem. The species name, *racemosum*, references the raceme, which is the term for a kind of flower cluster. Large, oval, veiny green leaves up to 8 inches (20 cm) long attach alternately to the stem. After blooming, the flowers are replaced at first by round green berries with coppery or purple spots that mature to a deep red. Western plants of this species tend to have erect stems; eastern plants often have arched ones.

American Elderberry
Sambucus canadensis H 5–12 ft (1.5–4 m)

Growing often into a stately, multibranched shrub, the American Elderberry has a long history of medicinal and other practical uses. It has immune system–stimulating properties.

KEY FACTS

The small, 5-lobed white flowers form large clusters, up to 10 in (25 cm) across.

+ **habitat:** Stream banks, woods, field edges, thickets, and roadsides

+ **range:** Native to the eastern U.S. and Canada; cultivated and escaped elsewhere

+ **bloom period:** May–July

The foliage and flower-laden American Elderberry often arches, its blossoms emitting a musty smell. The leaves are compound, with lance-shaped, toothed leaflets up to 7 inches (17.5 cm) long. After flowering, purplish black berrylike fruits appear, a favorite of wildlife. Both the flowers and the berries have many medicinal uses, and they are used to make cordials and wine; the juice also is used to flavor commercial beverages. The genus name derives from the Greek for a musical instrument and relates to the soft pith in the twigs, which are fashioned into flutes and whistles.

Field Pussytoes
Antennaria neglecta H 2–12 in (5–30 cm)

Small clusters of fluffy flowers produce a flower head with the shape that gives the Field Pussytoes its name. The genus *Antennaria* includes more than 30 native species and many variations.

KEY FACTS

The rounded white disk flowers grow in small, dense, fuzzy clusters at the top of the stems.

+ **habitat:** Prairies, fields, pastures, and roadsides

+ **range:** Throughout the U.S. and Canada

+ **bloom period:** April–July

Multiple fuzzy and fluffy parts reinforce the feline imagery of the Field Pussytoes. The fluffy, rayless flower heads grow atop fuzzy stems from a fanned-out base of spoon-shaped woolly leaves, each with a single prominent vein. The flowers morph into even fluffier seed heads. Stem leaves are smaller and lance shaped. Some species of the genus *Antennaria* do not have to rely on fertilization to produce seed; the female flower can manufacture fertile seed alone. The plant can also create clones by means of runners. These two reproductive methods lead to many variations in cloned colonies, making identification of Pussytoes species a challenge.

Wild Carrot/Queen Anne's Lace
Daucus carota H 12–48 in (30–120 cm)

The Wild Carrot, or Queen Anne's Lace, a Eurasian import, over time was selectively bred into a subspecies we know as a delicious and nutritious orange vegetable.

KEY FACTS

The flat, lacy white umbels (umbrella-like clusters) grow on hairy stems with deeply dissected leaves.

+ **habitat:** Fields, roadsides, and in disturbed areas

+ **range:** Introduced from Eurasia and naturalized throughout the U.S. and Canada

+ **bloom period:** May–October

A single purplish floret rests in the center of each of the Wild Carrot's exquisite open, umbrella-shaped flower clusters, which in turn are formed of smaller white umbrellas. The mature flower cluster takes a curved, hollowed-out shape that suggests a bird's nest, which is another of its common names. The fernlike leaves give off a distinct "carroty" odor when crushed. The long, white taproot of the Wild Carrot is not something to be munched raw, as is done with its cultivated cousin, but it can become edible through cooking. The species has long been used as an herbal remedy, especially in the treatment of bladder and kidney ailments.

Cow Parsnip
Heracleum maximum H 2–10 ft (1–3 cm)

The Cow Parsnip, largest native member of the Carrot family in North America, has a flower head similar to that of the Wild Carrot, but its leaves and stems differ greatly.

KEY FACTS

The small 5-petaled white or yellowish flowers form small umbels (umbrella-like clusters) that are part of large compound umbels.

+ habitat: Meadows, fields, pastures, woods, and marshes

+ range: Throughout U.S. and Canada, except far northern areas and the extreme South

+ bloom period: February–September

Characteristic of the Carrot family, the Cow Parsnip displays flat-topped umbels composed of small white flowers. This native plant generally has large leaves—up to 20 inches (50 cm) wide with an inflated basal sheath and a blade divided into three maple-like leaflets with serrated and lobed margins. The terminal leaflet is often the largest. The leaves appear alternately on the ribbed, woolly, and hollow stem. Flower buds in waiting form large growths on the stem, often the size of an orange. The genus name may refer to the great size of the plant or suggest that consumption of the Cow Parsnip aided Hercules in his tasks.

Oxeye Daisy
Leucanthemum vulgare H 6–30 in (15–75 cm)

A Eurasian import that is better known than many native wildflowers, the Oxeye Daisy was first planted here as a garden ornamental and now is naturalized nearly everywhere.

KEY FACTS

The flower head has 15–25 white ray florets and an indented center of many yellow disk florets.

+ habitat: Fields, pastures, and disturbed areas

+ range: Introduced and naturalized throughout the U.S. and Canada, except in some far northern areas

+ bloom period: June–August

The Oxeye's iconic daisy flower, up to 2 inches (5 cm) wide, tops an erect, often hairy stem that rises from a base of dark green leaves on short, slender stalks. Leaves farther up the stem are smaller, oblong, and attach directly to the stem. This perennial species is so hardy that its rootstock can survive bulldozing and removal, only to resurrect in another location. The word "daisy" is a corruption of the Old English "day's eye," a reference to the flower head's habit of opening in the morning and closing in the evening. A number of states consider the Oxeye a noxious weed.

White Wood Aster
Eurybia divaricata H 12–36 in (30–90 cm)

The flower heads of the White Wooc Aster grow in such profusion on zigzagging stems that they make a big impact wherever they grow.

KEY FACTS

The flower head is composed of 5 to 10 narrow, white ray florets and a dozen or so yellow disk florets.

+ **habitat:** Open woods, woodland edges, and roadsides

+ **range:** Throughout much of the eastern U.S. and Canada

+ **bloom period:** August–October

White Wood Aster flowers grow on flat-topped clusters at the top of very leafy stems. The leaves are heart shaped to oval to lance shaped, pointed, and toothed. The leaves near the base of the plant start to die back as the flowers bloom in late summer and early fall, and the flower heads' yellow centers turn purplish as they age. This native perennial species tolerates dry, shady conditions and spreads prolifically by means of underground rhizomes and seeds. The genus *Eurybia* is named for a Greek sea goddess who was the mother of Astraeus, Titan god of the stars and planets.

Philadelphia Fleabane
Erigeron philadelphicus H 6–36 in (15–90 cm)

The Philadelphia Fleabane is a little daisy on steroids, boasting up to 300 ray flowers a head. The genus name derives from the Greek for "old man," which may refer to the stem's beardlike down.

KEY FACTS

A small flower head has up to 300 white or sometimes pinkish threadlike ray flowers, surrounding dense yellow disk flowers.

+ **habitat:** Fields, meadows, open woods, lake edges, roadsides, and disturbed areas

+ **range:** Throughout U.S. and Canada, except parts of Southwest and far north

+ **bloom period:** March–June

The stem of Philadelphia Fleabane bears a stalked cluster of the small, composite flower heads with their distinguishing abundance of rays. Additional clusters often grow from the axils of the upper leaves, which are sparse near the flower. The leaves are generally lance shaped or oval, may be slightly toothed, and grow alternately, clasping the stem. They are larger and more numerous at the bottom of the plant. This early-flowering, basically biennial species has more or less played out by midsummer, another feature that may contribute to the "old man" name.

White Crownbeard/Frostweed
Verbesina virginica H 36–84 in (91–213 cm)

During frigid weather, fleshy green stems of the White Crownbeard split open and exude sap that freezes into often elaborate ribbonlike shapes called frost flowers.

KEY FACTS

The composite flower head typically has 5 white rays and a center of 5-lobed white disk flowers with white stamens and purple anthers.

+ habitat: Open woods, fields, stream banks, and roadsides

+ range: Central and southern portions of the eastern U.S. and Canada

+ bloom period: August–November

Flower heads of the White Crownbeard appear in large terminal clusters; frequently, individual heads may lack many or most of their rays. Large, oval to lance-shaped, pointed leaves alternate on the plant stems. The bases of the leaves extend downward on the stem, contributing to its distinctive longitudinal flanges, or wings. Below-freezing temperatures cause water and water vapor inside the stems to freeze and split the stems, leading to the formation of the icy shapes. White Crownbeard is an important nectar source for the Monarch Butterfly. American Indians used it to treat a number of medical ailments, including gastrointestinal problems.

Common Yarrow
Achillea millefolium H 12–36 in (30–91 cm)

The Common Yarrow often survives along sides of heavily trafficked roads where dust and pollution have killed off other plants.

KEY FACTS

The compact clusters of small flower heads have yellowish disk flowers that are surrounded by 5 rounded white rays.

+ habitat: Roadsides, old fields, and scrublands; also highly cultivated

+ range: Throughout the U.S. and Canada

+ bloom period: May–September

Common Yarrow flower heads grow on the ends of tough, hairy, gray-green stems. As the blossoms age, the florets at the center of the head turn brown or grayish. Recognized as a valuable medicinal plant for millennia, Common Yarrow contains more than 100 biologically active compounds, including some with strong anti-inflammatory properties. One explanation attributes its genus name, *Achillea*, to the legend that Achilles used the herb to stanch the bleeding of soldiers' wounds during the Trojan War. Another version links the name to the plant's discoverer, also called Achilles. Handling Common Yarrow can sometimes cause a rash.

Green False Hellebore
Veratrum viride H 18–96 in (45–240 cm)

A highly toxic plant, the Green False Hellebore sends up large, ribbed, bright green leaves and long flower clusters in wet meadows, posing a threat to grazing animals and their offspring.

KEY FACTS

The small, hairy, green, star-shaped flowers grow on long spikes.

+ habitat: Swamps, wet forests, wet meadows, seeps, and other wet areas, often in montane areas

+ range: Separate populations in the western and eastern U.S. and Canada

+ bloom period: May–July

A member of the Bunchflower family (previously in the Lily family), the Green False Hellebore emerges in early spring, developing large leaves that are up to a foot long (30 cm) and 6 inches (15 cm) wide, are heavily ribbed, and clasp the stem. The plant occurs in western and eastern North America but is absent from the continent's center, likely the result of glacial advance. Nevertheless, the two populations share common features, and many botanists consider them a single species. Green False Hellebore contains strong alkaloids that are used homeopathically in small doses to treat hypertension and heart disease, and have been studied in mainstream medical trials.

Common Ragweed
Ambrosia artemisiifolia H 12–60 in (30–150 cm)

Despite its genus name, *Ambrosia,* the Common Ragweed is a plain and much maligned wildflower, being a major instigator of the seasonal allergy known as hay fever.

KEY FACTS

The long central flower spikes have stalked drooping yellow-green male flower heads above stalkless, smaller green female flower heads.

+ habitat: Fields, roadsides, and disturbed areas

+ range: Throughout the U.S. and Canada, except in far northern areas

+ bloom period: July–October

The Common Ragweed gives rise to numerous small flower heads, source of the fine yellow pollen that spreads misery to allergy sufferers in late summer and fall. Its highly dissected leaves are more conspicuous; up to 6 inches (15 cm) long, they grow on green to pinkish hairy stems and create a profusion of foliage. The female flowers mature to produce hardy, burr-like fruits that can stay viable for more than five years and take hold in disturbed and nutrient-depleted soils. Although honeybees collect pollen from ragweed, the plant relies on wind pollination. Seed-eating bird species are attracted to the oil-rich seeds.

Eastern Poison Ivy

Toxicodendron radicans H ground cover or shrub to 3 ft (1 m); L vining to 150 ft (50 m)

Learning all the different appearances of Eastern Poison Ivy through the seasons is a good idea. It can save the discomfort of an inflamed rash, caused by a toxic resinous substance, urushiol.

KEY FACTS

The small, yellowish white or yellowish green flowers grow in elongated clusters from leaf axils.

+ **habitat:** Woods, thickets, roadsides, and disturbed areas

+ **range:** Throughout mainly central U.S. and Canada

+ **bloom period:** April–June

The genus name *Toxicodendron,* meaning "poison tree," explains it all. Urushiol, Poison Ivy's irritating oil, is toxic in all seasons and can remain potent on unwashed clothing. The old adage "Leaves of three, let it be" is your best guide, because species recognition is tricky. Poison Ivy leaves are compound, with three leaflets that can be small or large, flat green, shiny, or reddish. Notably, the central leaflet tapers to a stalk. The species can take the form of a trailing ground cover, an upright shrub, or a woody climber. Its stems can be slender or thick, smooth or hairy. Another adage—"Only a dope swings from a hairy rope"—is good advice.

Atlantic Poison Oak

Toxicodendron pubescens H to 10 ft (3 m)

With three-part leaves like its close relative Eastern Poison Ivy, Atlantic Poison Oak most often takes the form of a shrub. The urushiol-laden plant lacks Poison Ivy's climbing capabilities.

KEY FACTS

The small yellowish flowers grow in elongated clusters from the leaf axils.

+ **habitat:** Woods, thickets, glades, old fields, and sand hills

+ **range:** South-central and southeastern U.S.

+ **bloom period:** March–June

Atlantic Poison Oak has the potential to grow to 10 feet (3 m), but more often it stays in the 2- to 4-foot (60 to 120 cm) range as a smallish upright shrub. Its leaves are compound, with three oak-shaped leaflets that have downy stalks. After blooming, the inconspicuous flowers turn into small, tan, velvety, pumpkin-like fruits, and the foliage turns an alluring reddish brown in the fall. The similar western species, Pacific Poison Oak (*T. diversilobum*), grows as a shrub or vine, and its leaves can have three or five leaflets that are toothed or lobed.

Poison Sumac
Toxicodendron vernix H 6–30 ft (2–9 m)

Poison Sumac prefers wetter habitats than Poison Ivy, so the two rarely appear together. It does share some habitats with nonpoisonous sumacs (*Rhus* species).

KEY FACTS

The 5-petaled green-ish flowers grow in loose clusters from leaf axils.

+ habitat: Wet areas, including bogs, fens, swamps, marshes, and stream banks

+ range: Eastern U.S. and Canada, including Texas

+ bloom period: June–July

Poison Sumac is a flowering shrub with a proper trunk, branches, and twigs. The twigs bear alternate, compound leaves composed of an odd number of pinnate oval leaflets, usually between 7 and 13, attached to their stems by reddish stalks. The leaves turn orange or red in the fall. The bark of the trunk is gray and mostly smooth, except for the small bumps of the lenticels, pores allowing gas exchange. Urushiol-containing Poison Sumac resembles *Rhus* species (p. 118), but has smooth, not toothed, leaf margins and its drupes (fruits) are smooth and white, instead of red and hairy.

Multiflora Rose
Rosa multiflora H 6–15 ft (1.8–4.5 m)

An East Asian import, Multiflora Rose grows prolifically as a many-blossomed shrub or climber that is known for its ability to form an impenetrable hedge.

KEY FACTS

The flowers have 5 lance-shaped sepals and 5 white or pink petals with many stamens and pistils.

+ habitat: Fields, roadsides, woodland edges, prairies, and some wetlands

+ range: Introduced but naturalized in the eastern U.S. and Canada and also along the west coast of U.S. and Canada

+ bloom period: May–July

Large clusters of small white flowers grow on long, arching stems full of thorns. Leaves are divided into 5 to 11 toothed leaflets and have fringed stipules—paired, winglike appendages. The fragrant flowers produce bright red hips, or fruits, that are a favorite of birds. The plant reproduces by seeds or by developing new plants where stems root on touching the ground. The species was introduced as a rootstock for ornamental roses and then deployed for erosion control and as barriers. It invades pastures and crowds out native flora, especially in the woodland understory; cultivation is prohibited in numerous states.

Porcelain Berry
Ampelopsis brevipedunculata L vine to 15 ft (4.5 m)

Looking similar to native grape species, the Porcelain Berry vine sends tendril-bearing branches up, out, and over other plants. Its dense foliage robs light and space from native species.

KEY FACTS

Inconspicuous greenish white flowers occur in flat-topped to dome-shaped, large, stalked clusters opposite the shallowly to deeply trilobed leaves.

+ habitat: Forest edges, thickets, and disturbed land; along ponds and streams

+ range: Introduced; now naturalized in the eastern and midwestern U.S.

+ bloom period: June–August

The perennial Porcelain Berry is known more for its multicolored fruits than for its small flowers. Attractive clusters of berries ranging from green to lavender to porcelain blue often occur together on the same cluster. These berries and the lush, quite variable foliage made it a popular ornamental plant when brought here from Asia in the 1900s. Birds and other animals feed on the berries and disperse the seeds, although the species self-sows efficiently. Porcelain Berry proliferates very rapidly and is difficult to eradicate once it has taken hold. The plant resists pests, and it tolerates poor soil if there is enough light.

English Ivy
Hedera helix L vine to 100 ft (30 m)

Brought to North America by nostalgic European settlers hoping to create a familiar presence, English Ivy wasted no time climbing buildings and trees, and crawling across landscapes.

KEY FACTS

Tiny, inconspicuous, yellow-green flowers grow in clusters at ends of mature climbing plants that have reached open sunlight.

+ habitat: Woodlands, forest edges, fields, and coastal areas with not too much moisture

+ range: Native to Eurasia; widely naturalized in western, central, and eastern U.S.

+ bloom period: September–October

Most people never see English Ivy in bloom. The flowers appear in the fall only near the tops of mature vines. In its creeping and early climbing phase, English Ivy produces evergreen, three- to five-lobed leaves on woody stems that attach to surfaces by small rootlets aided by a sticky substance. The leaves of mature flowering stems are unlobed and oval. The vines block light needed for photosynthesis—often killing the tree from the bottom and leaving only a "broccoli top"— and potentially causing it to uproot or break during storms. English Ivy along the ground also crowds out native species, creating a monoculture.

Japanese Honeysuckle
Lonicera japonica L vine to 30 ft (9 m)

For many, summer nostalgia includes plucking and sipping the sweet nectar from a fragrant Japanese Honeysuckle blossom. Its luxuriant vines climb over shrubs and small trees.

KEY FACTS

A strongly 2-lipped, tubular flower with long, projecting stamens; ovate, hairy, untoothed leaves grow opposite each other.

+ **habitat:** Woods, wood edges, thickets, and roadsides

+ **range:** Naturalized in central, eastern, and southwestern U.S., and in California

+ **bloom period:** April–September

The blossom of the Japanese Honeysuckle starts out white, as one of a pair at leaf axils on a hairy, woody stem that climbs over bushes or trails on the ground; over time, the flower ages into a golden yellow. Often overlooked are the black berries that form later. One of several introduced Asian honeysuckles, Japanese Honeysuckle can easily overtake and choke out woodland trees. As with many exotic species, there is a less aggressive native equivalent that can provide a similar look in the garden. Among the possibilities, Trumpet Honeysuckle (*L. sempervirens*) has long, red blossoms and produces nectar favored by hummingbirds.

Japanese Wisteria
Wisteria floribunda L vine to 70 ft (21 m)

In some areas, Japanese Wisteria is a common sight, draped over a porch front or arbor. But this aggressive, woody vine escapes cultivation and spreads rapidly, thwarting native flora.

KEY FACTS

Violet-blue, lilac, white, or pink pealike flowers grow in large, pendulous clusters from 6–18 in (15–45 cm) long; compound, pinnate leaves have 7–19 leaflets.

+ **habitat:** Various in cultivation; escapees thrive in forests.

+ **range:** Native to Japan; escaped cultivars widely established in central and eastern U.S. and Canada

+ **bloom period:** April–June

The strong, climbing vines and fragrant, showy flowers of the Japanese Wisteria, a member of the Pea family, were introduced to North America in the mid-19th century. Seeds form in slender, fuzzy, flattened pods about 6 inches (15 cm) long. Escaped Japanese Wisteria, often from abandoned homesteads and nurseries, infiltrates forests, where it can smother trees and crowd out native species. Gardeners who want the same look should seek out cultivars of American Wisteria (*W. frutescens*), a native species, with narrower and more delicate flower clusters. Its native range is the southeastern and midwestern United States.

Morning sun silhouettes the towering trees of
Redwood National Park, California.

Trees & Shrubs

The Living Landscape

What is a tree? What is a shrub? Most of us would say we knew, and that we could tell you the difference. But because this chapter deals with 160 species of trees and shrubs likely to be encountered in the United States and Canada, it would be worth a few minutes to have a closer look at this issue. Debating how to define all species as either a shrub or a tree, especially considering how variable many species are, seems less valuable than having a look at how these plants are able to exist in the first place. The word is *wood*.

Herbaceous annuals, biennials, and perennials grow mostly at their tip, or apex, which contains the apical meristem, where the growing cells divide. So do trees and shrubs, but they also grow outward, laterally. Lateral growth occurs in lateral meristems. Growth patterns and aging result in several kinds of tissues that in concert allow these sometimes behemoth plants to arise. Woody plants include small trees, bushes, shrubs, and many vines.

■ Anatomy of a Tree Trunk

Let us imagine that we are looking at the stump of a just-felled tree that had been evolved just so naturalists could learn the basic anatomy of wood. We are looking at a cross section of the tree, taken near its base, in which we can see perfectly its bark and other layers. Let us consider those layers, from the outside in, remembering that the same layers may be found on a smaller scale on the tree's branches, twigs, and roots. There are three main layers—the bark, the cambium, and the wood, some of them divided further.

Bark, of course, is a tree's external covering. This layer of dead cork-like cells protects the interior from injury and disease and conserves water.

The **inner bark,** or **phloem,** is a transport system that moves food and other substances made during photosynthesis (usually taking place in the leaves) to the rest of the tree. As new bark is produced, it pushes older cells outward. Eventually the older bark cannot retain all this newer bark, and its smooth complexion cracks, flakes, or peels.

Between the inner bark and the wood lies the **cambium,** the lateral meristems, where the cells that make the tree grow wider multiply. New cells form to its outside and its inside.

Toward the center of the trunk, inside the cambium, is the **wood,** or **xylem.** Wood contains strengthening fibers and other conducting vessels. There are two kinds of wood. The sapwood, adjacent to the cambium, transports water and dissolved minerals and nutrients from the soil (taken up by roots) to the rest of the tree.

When the cambium creates new wood cells, their size and shape depend on the conditions in the area at the time, so in the generally wetter spring, new cells are bigger, and in the drier summer, they are smaller. These differences make visible annual rings (each actually representing a layer of wood), which of course are an index of a tree's age.

With time, some xylem layers closer to the center of the trunk die, no longer needed for transport. This older wood, called *heartwood,* fills with substances, like resins, that protect it from rot. It is often darker than the surrounding wood.

Without the development of this miraculous system, we would not have trees and shrubs. The wood varies with the kind of tree or shrub, and properties of the different parts of the trunk also vary. People have exploited these differences for millennia, using different species in making canoes, clothing, medicines, food, shelter, charcoal, decorations, or fire.

■ Chapter Organization

The chapter is organized in several ways. The overarching scheme is the most major. First discussed are the gymnosperms. The word means "naked seed," and for our purposes, it is another way of saying conifers. The conifers include the cedars, the cypresses, the pines, and the spruces, whose seeds are bare in the cone and not surrounded by a fruit. (The Ginkgo, which does make fruit, and a nasty one at that, is considered a conifer.) The other major group is the angiosperms, whose seeds are enclosed by fruits (the mature ovary). The angiosperms are divided into the dicots (which have two cotyledons, or seed leaves), which include oaks, the Mountain Laurel, birches, locusts, and the Jumping Cholla, and the monocots (which have but one cotyledon), which include the Joshua Tree, the palms, and the palmettos.

KEY IDENTIFYING FEATURES

Conifers have four types of leaf arrangement:

❶ **Fascicles** are bunches of needles growing from one bud.

❷ **Pectinate rows** are arranged in ranks along opposite sides of the shoot.

❸ **Awl-shaped leaves** are set radially around the shoot.

❹ **Scalelike leaves** are set densely around the shoot.

Broadleaf trees have more varied foliage than the conifers. Some examples of types of foliage shape include:

❶ **Pinnately lobed:** many lobes radiating from the leaf's midrib

❷ **Ovate:** egg-shaped with broad end at the base

❸ **Pinnately compound:** many small leaflets arranged in pairs on either side of a central midrib

❹ **Deltoid:** triangular with stem attached to the middle of the base

❺ **Palmately lobed:** hand-shaped structure with more than three lobes branching from a single point at the base

❻ **Lanceolate:** general shape of a lance—longer than wide and a pointed apex

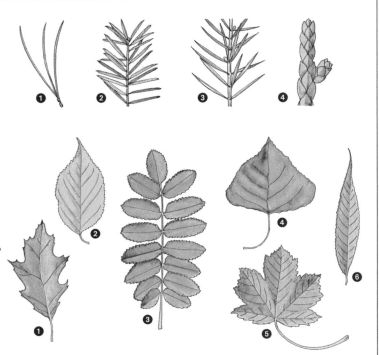

Snow dusts aspen and spruce trees in Colorado's Rocky Mountain National Park.

■ Principles Used in the Entries

As you use the entries, keep in mind a few principles that are followed. *General* is probably the most important trait for these descriptions. Multivolume sets have been published on wood morphology, reproductive strategies, taxonomy, ranges, and ecology of trees and shrubs. Detailed, torturous keys get to the nitty-gritty of identification. All these books are available at libraries and bookstores. This chapter is meant to whet readers' interest and to truly help them identify trees and shrubs that they may see on field trips and travels, and to tell them something about the species' ecology, value to wildlife and humans, and status—including whether or not they are native.

Geographic ranges are general and refer to a plant's native range. It is not noted that conifers are evergreen, which most are; some, however, are deciduous (the Baldcypress, for example), and this is

Why are there trees
I never walk under
but large and melodious thoughts
descend upon me?
—WALT WHITMAN

pointed out. Likewise, it is not stated when an angiosperm tree is deciduous, because most of them are; but some are evergreen or mostly so (the Wax Myrtle, for example), and this too is pointed out.

Though some entries mention culinary uses of various tree or shrub parts, readers should not eat any part of any plant unless they are 100 percent certain that it is 100 percent safe. Other species are described as being toxic; if a species is not so described, it should not be deduced that any part is safe to eat.

The descriptions of the trees and shrubs are designed to convey the species' unique features and those that are most helpful in identification, most important the height (following the scientific name) and the information under Key Facts; the photographs and illustrations complement each other. Features of twigs are rarely described. Floras often describe twigs, as do books that teach users to identify deciduous trees in winter. Depending on latitude, altitude, soil characteristics, or sunlight, two individuals of a species can look quite different. The species entries take into account much of the variation by a descriptive phrase, or a range of the numbers.

■ Trees & You

Not only will this chapter help you get to know and more deeply appreciate this starter set of 160 species, it will help you see the forest in addition to the trees—and something larger than that, albeit not definable or quantifiable. The spirit of North America, and of its peoples, indigenous and not, is inseparable from these amazing woody plants.

OFFICIAL TREES OF THE U.S. STATES AND CANADIAN PROVINCES

+ **Alabama:** Longleaf Pine (*Pinus palustris*)
+ **Alaska:** Sitka Spruce (*Picea sitchensis*)
+ **Arizona:** Palo Verde (*Parkinsonia florida*)
+ **Arkansas:** Pine Tree (*Pinus*)
+ **California:** Coast Redwood (*Sequoia sempervirens*), Giant Sequoia (*Sequoiadendron giganteum*)
+ **Colorado:** Blue Spruce (*Picea pungens*)
+ **Connecticut:** White Oak (*Quercus alba*)
+ **Delaware:** American Holly (*Ilex opaca*)
+ **Florida:** Cabbage Palmetto (*Sabal palmetto*)
+ **Georgia:** Live Oak (*Quercus virginiana*)
+ **Hawaii:** Kukui or Candlenut (*Aleurites moluccana*)
+ **Idaho:** Western White Pine (*Pinus monticola*)
+ **Illinois:** White Oak (*Quercus alba*)
+ **Indiana:** Tuliptree (*Liriodendron tulipifera*)
+ **Iowa:** Oak (*Quercus*)
+ **Kansas:** Eastern Cottonwood (*Populus deltoides ssp. deltoides*)
+ **Kentucky:** Tuliptree (*Liriodendron tulipifera*)
+ **Louisiana:** Baldcypress (*Taxodium distichum*)
+ **Maine:** Eastern White Pine (*Pinus strobus*)
+ **Maryland:** White Oak (*Quercus alba*)
+ **Massachusetts:** American Elm (*Ulmus americana*)
+ **Michigan:** Eastern White Pine (*Pinus strobus*)
+ **Minnesota:** Red Pine (*Pinus resinosa*)
+ **Mississippi:** Magnolia (*Magnolia*)
+ **Missouri:** Flowering Dogwood (*Cornus florida*)
+ **Montana:** Ponderosa Pine (*Pinus ponderosa*)
+ **Nebraska:** Eastern Cottonwood (*Populus deltoides ssp. deltoides*)
+ **Nevada:** Singleleaf Piñon (*Pinus monophylla*)
+ **New Hampshire:** Paper Birch (*Betula papyrifera*)
+ **New Jersey:** Northern Red Oak (*Quercus rubra*)
+ **New Mexico:** Piñon Pine (*Pinus edulis*)

+ **New York:** Sugar Maple (*Acer saccharum*)
+ **North Carolina:** Longleaf Pine (*Pinus palustris*)
+ **North Dakota:** American Elm (*Ulmus americana*)
+ **Ohio:** Ohio Buckeye (*Aesculus glabra*)
+ **Oklahoma:** Eastern Redbud (*Cercis canadensis*)
+ **Oregon:** Douglas-fir (*Pseudotsuga menziesii*)
+ **Pennsylvania:** Eastern Hemlock (*Tsuga canadensis*)
+ **Rhode Island:** Red Maple (*Acer rubrum*)
+ **South Carolina:** Cabbage Palmetto (*Sabal palmetto*)
+ **South Dakota:** Black Hills Spruce (*Picea glauca var. densata*)
+ **Tennessee:** Tuliptree (*Liriodendron tulipifera*)
+ **Texas:** Pecan (*Carya illinoensis*)
+ **Utah:** Blue Spruce (*Picea pungens*)
+ **Vermont:** Sugar Maple (*Acer saccharum*)
+ **Virginia:** Flowering Dogwood (*Cornus florida*)
+ **Washington:** Western Hemlock (*Tsuga heterophylla*)
+ **West Virginia:** Sugar Maple (*Acer saccharum*)
+ **Wisconsin:** Sugar Maple (*Acer saccharum*)
+ **Wyoming:** Plains Cottonwood (*Populus deltoides ssp. monilifera*)

Canadian Provinces and Territories
+ **Alberta:** Lodgepole Pine (*Pinus contorta*)
+ **British Columbia:** Western Redcedar (*Thuja plicata*)
+ **Manitoba:** White Spruce (*Picea glauca*)
+ **New Brunswick:** Balsam Fir (*Abies balsamea*)
+ **Newfoundland and Labrador:** Black Spruce (*Picea mariana*)
+ **Northwest Territories:** Jack Pine (*Pinus banksiana*)
+ **Nova Scotia:** Red Spruce (*Picea rubens*)
+ **Nunavut Territory:** none
+ **Ontario:** White Pine (*Pinus strobus*)
+ **Prince Edward Island:** Northern Red Oak (*Quercus rubra*)
+ **Quebec:** Yellow Birch (*Betula alleghaniensis*)
+ **Saskatchewan:** White Birch (*Betula papyrifera*)
+ **Yukon Territory:** Subalpine Fir (*Abies lasiocarpa*)

TREES & SHRUBS

Ginkgo/Maidenhair Tree
Ginkgo biloba H to 130 ft (40 m)

The Ginkgo is considered a conifer, though it does not produce cones. It is deciduous, quickly losing its distinctive leaves after a bright yellow autumn phase.

KEY FACTS

This tree fares well in gardens and cities; *biloba* refers to its leaves, usually 2-lobed.

+ **leaves:** Fan-shaped, recalling maidenhair fern

+ **flowers/fruits:** Sexes on different trees, male in catkin-like clusters, female 1 or 2 on short stalk

+ **range:** Survives mainly in cultivation

Ginkgo biloba is the only species in a genus that millions of years ago had nearly worldwide distribution. The native of China is now prized around the world for its beauty and its ability to withstand pollution and other adverse conditions—including atomic bomb blasts. (A group of trees just one mile/1.6 km from ground zero at Hiroshima survived, regained their health, and still grow there.) The fruit mature on the female flower stalks and recall small plums—until they fall to the ground and decay, earning the tree another alias, "stinkbomb tree."

Alaska Cedar/Yellow Cypress/Nootka Cypress
Chamaecyparis nootkatensis H 49–125 ft (15–38 m), but shrub-like at highest altitudes

Typical of false cypresses (often incorrectly called cedars), the Alaska Cedar is aromatic and resinous. Its cones mature in two years (one year in its southern extent, like other false cypresses).

KEY FACTS

Branches droop in flat sprays.

+ **leaves:** Scales pressed together, bluish green; unpleasant odor

+ **cones:** Male and female on same tree at branchlet tips; mature seed cones of 4–6 short-pointed scales, reddish brown

+ **range:** Coastal mountains, southeastern Alaska to northern California

Prized for its durable, fine-grained wood, the Alaska Cedar is slow growing and long lived, so that a tree can be much older than the diameter of its trunk would suggest. One individual is 3,500 years old. The tree was important to indigenous peoples of its range, including the Nuu-chah-nulth (formerly called Nootka). It is used in landscaping in the coastal Northwest. This species was originally assigned to the genus *Cupressus* (the true cypresses), then to *Chamaecyparis*, but it may be assigned to a new genus, *Xanthocyparis*, because it is felt to differ strongly from the other two genera.

Atlantic White Cedar

Chamaecyparis thyoides H to 82 ft (25 m)

A medium to large tree reminiscent of the junipers, the Atlantic White Cedar grows in pure stands or, less often, with wet-woods trees like Baldcypress and Blackgum.

KEY FACTS

Branches are in fan-shaped sprays; bark is not shredded.

+ **leaves:** Scales tightly overlapping

+ **cones:** Male and female on same tree but separate branches; mature cones of 6 pointed, woody scales

+ **range:** Maine to Florida to Mississippi, mainly in swamps and wet sands of Coastal Plain

The wood of the Atlantic White Cedar is so appealing that the species has been overharvested—since the time of the American Revolution. Its stands are much reduced, though it is still commercially important in parts of its range. Its wood is light, strong, and easily worked, ideal for use in posts and telephone poles, barrels, shipbuilding, and decoy carving. Aromatic, it repels insects, retards decay, and resists disease, such as cedar-apple rust (unlike the Eastern Redcedar, for which the tree can be mistaken if not in fruit). Shade tolerant and attractive, the tree is often planted in yards and gardens.

Eastern Redcedar

Juniperus virginiana H to 100 ft (30 m)

The Eastern Redcedar is our most widespread conifer and one of the most common, found in fields, fencerows (thanks to seed dispersal by birds on fences), dry, open woods, and along roads.

KEY FACTS

Branchlets often droop.

+ **leaves:** Mature leaves scalelike, overlapping; foliage needle-like

+ **cones:** Male and female cones are mostly on separate trees; female cones mature into bluish berries.

+ **range:** East of Great Plains, from southeastern Canada to the Gulf of Mexico

The aromatic wood of this slow-growing juniper (it is not a true cedar) is perhaps best known for its use in lining cedar chests and closets; its aroma repels moths and other insects. It is also used for posts, paneling, and carving. Its reddish brown bark peels off in strips. Many cultivars have been developed for landscaping. The tree is the primary host to the fungus that causes cedar-apple rust, which produces masses of orange, gelatinous spore tubes among its needles. The fungus's secondary hosts are members of the rose family, such as pear and apple trees, where it is more damaging, affecting fruit crops.

Redwood/California Redwood

Sequoia sempervirens H to 375 ft (115 m)

The giant Redwood is locally common in protected areas along rivers experiencing extensive rainfall and constant fogs off the ocean.

KEY FACTS

The world's tallest tree, named for its thick, reddish bark, can live longer than 2,000 years.

+ needles: Dark green

+ cones: Male and female in separate clusters on same tree; seed cones oblong, red-brown

+ range: Pacific coast, southwestern Oregon to central California

Redwoods have limbs to a height of 50–100 feet (16–30 m), above which the trunk has many vertical stems. This unique habitat accumulates water, develops soils, and supports vertebrates, insects, and epiphytic plants. Not only is the tree's habitat characterized by fog, the tree induces fog drip, condensing water and supplying organisms in its crown and the ground below. Its wood is smooth, straight grained, and strong, with great appeal for building and woodworking. Since Spanish settlement, more than 90 percent of the largest Redwoods have been logged. The tree is now well managed.

Giant Sequoia

Sequoiadendron giganteum H 250–290 ft (76–88 m)

This massive tree was named for Sequoyah, a 19th-century Cherokee silversmith, who devised a character set for writing in the tribe's language and published papers and books in Cherokee.

KEY FACTS

Columnar trunk is fluted; bark is reddish brown, as thick as 2 ft (0.6 m) at the base.

+ leaves: Scalelike, pointed, blue-green

+ cones: Egg-shaped, woody seed cones, similar to a Redwood's, but larger

+ range: Restricted to 75 groves in California, on western slopes of the Sierra Nevada

The sole surviving *Sequoiadendron* species, the Giant Sequoia includes some of the world's largest and oldest organisms. The crown can develop secondary trunks and hollow spaces, providing cover, nesting, and foraging opportunities for wildlife. The wood resists decay, but it is brittle, and harvest difficulties have also limited its commercial importance. Fire suppression threatens the Sequoia, as species naturally limited by fire can thrive in the understory. The Sequoia's thick bark can normally resist fires, but the understory trees act as fire ladders, allowing the flames to reach the giants' highly flammable crowns.

Baldcypress

Taxodium distichum H 65–130 ft (20–40 m)

A majestic, deciduous conifer of wet areas, the Baldcypress in the deeper South is commonly home to the epiphytic, grayish bromeliad Spanish moss, hanging from its limbs.

KEY FACTS

The trunk base is enlarged, fluted, and surrounded by knobby, woody "knees."

+ leaves: Flat, 2-ranked, feather-like, turning russet in autumn

+ cones: Rounded, wrinkled seed cones

+ range: Southeast; wet areas in Coastal Plain, inland to Texas and up the Mississippi River to Illinois and Indiana

An array of cultivars of the Baldcypress have been developed, and the tree is grown coast to coast, and it can survive in areas that are drier and decidedly less gothic than its familiar swampy habitat—including on city streets. The species' most menacing disease is "pecky cypress," a sometimes deadly brown pocket rot caused by a fungus that attacks the heartwood. Insects damage the leaves, cone, and bark. Humans, too, pose a threat, by draining wetlands and overharvesting for timber. The attractive wood is easily worked and decay resistant, used in shingles, flooring, cabinetry, trim work, beams, and barrels.

Arborvitae/Eastern White Cedar

Thuja occidentalis H to 65 ft (20 m)

The Eastern White Cedar is at home in moist or wet soil, where it outcompetes other trees. It is most populous in the Great Lakes region and farther north and is somewhat shade tolerant.

KEY FACTS

Short branches to the ground form flat sprays; reddish, shredding bark.

+ leaves: Scales sometimes long-pointed

+ cones: Male and female cones on same tree; female greenish, turning brown, with 8–12 pointed scales

+ range: Manitoba to Hudson Bay, south to Iowa and South Carolina

This native evergreen can live more than 800 years, and individuals growing in Ontario are the oldest trees in eastern North America. The Ojibwe discovered its value in construction and medicine. After they taught 16th-century French explorer Jacques Cartier to use its vitamin C–rich foliage to treat scurvy, it earned the name Arbor-vitae ("tree of life" in Latin). The decay-resistant, fragrant wood is used as posts and in boats, cabins, and shingles. Its oils find purpose in disinfectants and insecticides. The Eastern White Cedar is popular as an ornamental, and numerous cultivars have been developed.

Pacific Silver Fir

Abies amabilis H to 151 ft (46 m)

The Pacific Silver Fir is commonly found in moist coastal coniferous forests, to the tree line in the mountains but to sea level from Vancouver Island, British Columbia, northward.

KEY FACTS

Spire-like crown rounds with age.

+ needles: Attached spirally but twisted at base, lying flat on and above the stem; dark green above, silvery below

+ cones: Male and female on same tree; cylindrical, upright seed cones, not stalked, green, becoming purple to brown

+ range: Alaska to northwestern California

Soft and weak, the Pacific Silver Fir's wood is an important source of pulp used in plywood, crates, and poles. But the tree is attractive, and cultivars have been developed, though they are planted mostly in its native range and in parts of New Zealand and Scotland that can supply the requisite cool, humid summers. Indigenous peoples in its range used its sap as chewing gum, made bedding of the branches, and burned the wood. The trees are associated with hemlocks and other firs, among others, assemblages that provide food and habitat for sooty and spruce grouse, squirrels, and a bird named Clark's Nutcracker.

Balsam Fir

Abies balsamea H to 66 ft (20 m)

The Balsam Fir, whose shape is a slim pyramid, is common in moist places and swamps, as well as on well-drained hillsides. Growth is strongest and fastest in full sun.

KEY FACTS

Seed bracts extend slightly if at all beyond cone scales.

+ needles: Two-ranked, flattened, grooved above

+ cones: Erect, rounded cylinders; purple, browning with age

+ range: From Alberta to Labrador south to the Great Lakes and northeastern states, south to Virginia in higher mountains

The Balsam Fir is common, often dominant in its moist habitats, occurring with White Spruce, Black Spruce, Trembling Aspen, or Paper Birch. It has been heavily damaged by the balsam woolly adelgid, an invasive insect introduced from Europe. The foliage and seeds provide food and cover for wildlife, including birds, squirrels, moose, and porcupines. In addition to providing pulpwood, this handsome tree is planted as an ornamental and favored as a Christmas tree. Its fragrant resin (balsam) is used in incense, rodent repellent, and to make Canada balsam, which has served as a cement for eyeglass lenses and been used to treat colds.

Fraser Fir/She-balsam

Abies fraseri H mostly to 80 ft (25 m)

The population of the Fraser Fir has been reduced as much as 95 percent by the Balsam Woolly Adelgid, an insect introduced into North America from Europe in the early 1900s.

KEY FACTS

The seed bracts protrude between cone scales.

+ **needles:** As in Balsam Fir, with a pine-like scent

+ **cones:** Male and female on same tree; seed cones purple, browning with age

+ **range:** Rare in high Appalachian forests of southwestern Virginia and adjacent North Carolina and Tennessee

In the wild, the Fraser Fir grows in pure stands or with other evergreens and with birches and other hardwood. It is popular as a Christmas tree, for which it is farmed, but it is intolerant of warm climate, so its culture is being reconsidered in some areas where temperatures are increasing. For better or for worse, it is sometimes called She-balsam because resin can be "milked" from its bark blisters (not true of the balsam-bearing Red Spruce *Picea rubens*, called He-balsam). "Ba sam" describes fragrant oily or resinous substances found in a number of trees and shrubs that are not necessarily close botanical relatives.

Grand Fir

Abies grandis H to 260 ft (80 m)

The majestic Grand Fir reaches its greatest size in the rain forests of Washington's Olympic Peninsula and inhabits moist forests from sea level to 4,900 feet (1,500 m).

KEY FACTS

Branches are short and drooping.

+ **needles:** Shorter at stem tips; 2-ranked, dark green above, 2 white bands beneath

+ **cones:** Seed cones barrel-shaped, upright on crown branches; seed bracts not visible

+ **range:** Southern British Columbia to northern California

This stately, long-lived fir is common in its range, growing in mixed coniferous and hardwood stands, often with Douglas-fir and Larch. The Grand Fir's resin has been used in wood finishing, and its wood is valuable in the paper and building industries. The tree has also been used medicinally. Commercially, it is most important in Idaho, although in general it has not proved to be a good species for landscaping outside its native range. It is an important source of food and cover for small mammals, seed-eating birds, and game birds, and it is often a host to the parasitic fir Dwarf Mistletoe.

California Red Fir

Abies magnifica H to 120 ft (37 m)

A large evergreen conifer that often dominates the montane forests of California and Oregon, the California Red Fir is distinguished by its blue-gray needles and reddish brown bark.

KEY FACTS

Short, horizontal limbs make the tree slim.

+ **needles:** Sharp, curving atop shoot; with whitish coating when young, aging to bluish green, with white bands beneath

+ **cones:** Seed cones rounded, cylindrical, borne at crown; bracts not visible beyond scales

+ **range:** Mountains of Oregon and California

At home in altitudes from 5,250 to 9,350 feet (1,600 to 2,850 m), the long-lived California Red Fir often grows in dense, pure stands, but is sometimes associated with other conifers when near the tree line. Its branches have a camphor-like aroma. As in many firs, the fatty seeds provide energy-rich food for rodents and other small mammals, and the shoots are fodder for deer. Its wood is stronger than that of other firs, and it is an important source of wood for the pulp industry and construction and as fuel and Christmas trees. The tree is plagued by heart rot and Dwarf Mistletoe.

Tamarack/Eastern Larch

Larix laricina H to 80 ft (25 m)

A slow-growing, conic, deciduous conifer, the Tamarack is heat intolerant, which explains its greatly northern distribution and why it is seldom sought in landscaping far from its native range.

KEY FACTS

Bark is scaly, pinkish to red-brown.

+ **needles:** Soft, 3-sided, in tufts on spurs, turning yellow before falling

+ **cones:** Male and female in separate clusters on same plant; seed cones upright, unstalked, egg-shaped, persisting

+ **range:** All Canadian provinces, Great Lakes and northeastern states

A common tree of swampy forests and moist uplands in its boreal zone, the Tamarack is one of our most northerly trees and has a large natural range. It grows in pure stands or is found in conjunction with Balsam Firs and other conifers. Browsed but not seriously harmed by deer, a tree can fall victim to porcupines that girdle it while feeding on its bark. The larch sawfly can damage or even kill Tamaracks by defoliating them. The wood is hard and resinous and provides lumber used for railroad ties, construction, poles, and pulp. Wild animals use its seeds.

White Spruce

Picea glauca H to 131 ft (40 m)

Our northernmost tree, the White Spruce is widespread in bogs, on bodies of water, and on rocky hills, often found with such trees as Balsam Fir, Eastern Hemlock, and Red Maple.

KEY FACTS

This species is dense, with a conical crown, becoming cylindrical.

+ **needles:** Four-sided, individual, on peg-like stalk

+ **cones:** Unisexual, on same tree; seed cones pendulous, slender, maturing to brown; scales flexible

+ **range:** Alaska through all Canadian provinces, Great Lakes states, to New England

The White Spruce is valued in Canada for lumber and paper pulp, and it is popular for Christmas trees, landscaping, and as a windbreak. Its needles are aromatic—but they are not exactly mountain fresh and have earned it the labels "skunk spruce" and "cat spruce." *Picea* is Latin for "pitch," a reference to resins in the bark. Spruce Beetles have destroyed 2.3 million acres in Alaska; it is susceptible to Spruce Budworm and Spruce Sawfly, and rusts make it shed its needles early. South Dakota's state tree is a variety known as the Black Hills Spruce (*P. glauca* var. *densata*), but not all authorities recognize the variety.

Black Spruce

Picea mariana H to 98 ft (30 m)

A widely distributed, small, slow-growing conifer of the continent's coldest regions, the Black Spruce inhabits moist flatlands and lake margins and, in its southern extent, sphagnum bogs.

KEY FACTS

This spruce is spire-like; its branches droop with ends upturned.

+ **needles:** Four-sided, blue-green above, paler, powdery below

+ **cones:** Unisexual, on same tree; seed cones clustered in crown

+ **range:** Alaska through all Canadian provinces to Great Lakes states and New England

In addition to reproducing by seeds, the Black Spruce can propagate asexually; the lower limbs can touch the ground, often under the weight of snow, and sometimes take root, creating a circle of smaller trees around the main trunk of the parent. The tree is similar to the Red Spruce but is usually found in more extreme conditions and is more northern. Its form varies; for example, at the tree line, it is often prostrate. Because the tree is small, the wood is useful for little more than pulp and fuel. It is of low value for wildlife food but provides cover for small mammals and important nesting habitat for birds.

Blue Spruce/Colorado Spruce

Picea pungens H 65–82 ft (20–25 m)

The slow-growing Blue Spruce is a medium-size evergreen probably best known from its widespread planting in Canada, the U.S., and Europe. At least 38 cultivars have been developed.

KEY FACTS

In aging, this spruce becomes more layered.

+ **needles:** Bluish green, sharp, 4-sided, on all sides of stem

+ **cones:** Unisexual, on same tree; seed cones slender, cylindrical, hanging from upper branches

+ **range:** High in Rocky Mountains from Idaho and Wyoming across Utah and Colorado to Arizona and New Mexico

The Blue Spruce has been called the "most beautiful species of conifer" for its stately form and "vibrant" blue to silvery hue (though this is more common in cultivars than in natural trees). Its primary value is in its looks, which horticulturists have exploited. It usually has a sporadic distribution or is a component of mixed-conifer forests, with Douglas-fir, Engelmann Spruce, and others. Not a favorite forage for wildlife, it does provide seeds for birds and cover for deer. It is plagued by the Spruce Bark Beetle, Spruce Gall Aphid, Spruce Budworm, and Spruce Spider Mite.

Engelmann Spruce

Picea engelmannii H to 164 ft (50 m)

With the Subalpine Fir *(Abies lasiocarpa),* Engelmann Spruce forms one of the most frequent forest types in the Rocky Mountains. At high elevations, it is one of the largest trees.

KEY FACTS

A spire-like tree, its trunk has few small limbs between branches.

+ **needles:** Linear, 4-sided, blue-green

+ **cones:** Unisexual on same tree; seed cones yellow to purple-brown; bracts hidden by thin scales notched at tip

+ **range:** Yukon, British Columbia, and Alberta south to Arizona and New Mexico

Soft and knotty, Engelmann Spruce wood is not prized for lumber, restricted to framing and other unseen construction uses and to paper manufacturing. Instead, it enters the limelight in construction of guitars and piano soundboards. The bark provides tannin, and cord has been made from its branches and roots. Young male cones are eaten raw or cooked, used as flavoring, and added to breads and cereals; the seeds are small but edible. A vitamin C–rich infusion is made from the young shoots. Another, made from the bark, is used to treat respiratory ailments. Various resins have been used to treat eczema.

Red Spruce/He-balsam

Picea rubens H to 148 ft (45 m)

A handsome, medium-size evergreen, the Red Spruce remains one of the most important forest trees in the Northeast, despite having been overexploited for myriad purposes.

KEY FACTS

Crown is conical but broader than in other eastern spruces.

+ **needles:** Four-sided, curved, on all sides of twig, bright green

+ **cones:** Male and female on same tree; seed cones long, egg-shaped, red-brown; scales untoothed

+ **range:** Cape Breton to Ontario, south in mountains to North Carolina

Red Spruce grows in pure stands or with other conifers such as Eastern White Pine, Balsam Fir, and Black Spruce. Light, straight-grained, and resilient, its wood is used in construction and paper-making; its resonance makes it sought after for building guitars, violins, piano soundboards, even organ pipes. Its resin was used into the 20th century in chewing gum (replaced now by a substance from a tropical plant). The buds, foliage, and seeds provide food for small mammals and birds (up to half the diet of a White-winged Crossbill). The Spruce Bud-worm is especially damaging where this spruce grows with Balsam Fir.

Sitka Spruce

Picea sitchensis H to 197 ft (60 m)

The largest spruce, the tall, grand Sitka, can be found in its long band of Pacific habitat, from 980 feet (300 m) to sea level, its range inland determined by that of ocean fogs.

KEY FACTS

The trunk base is buttressed.

+ **needles:** Yellow- to blue-green, sharp, flat, with 2 white bands beneath

+ **cones:** Male and female on same plant; seed cones cylindrical, hanging from upper shoots; scales thin, irregularly toothed

+ **range:** Narrow strip along coast from south-central Alaska to northern California

Different parts of the Sitka Spruce have many uses in food, varnish, medicine, and for making rope and cord. The wood is exceptionally strong, thanks to its long, straight grain, especially when its light weight is considered. That combination caught the imagination of airplane designers in both world wars. So did its cost, then less than that of steel or aluminum. It is still used in aerobatic craft. (The largest plane ever built was an immense seaplane, made entirely of wood, called the "Spruce Goose"—even though it was made of birch.) Sitka Spruce is a food source for birds and small mammals. It is rarely cultivated.

Jack Pine
Pinus banksiana H to 89 ft (27 m)

Our most northerly ranging pine, the Jack Pine is usually a small to medium-size tree that soon loses the pyramidal form of its youth to assume a more gnarled look.

KEY FACTS

Fire is often needed to open cones.

+ **needles:** In 2s, olive to gray-green, stiff, curved to slightly twisted; margins rough

+ **cones:** Crooked toward twig tip; scales thick, stiff, with fragile spine; resinous

+ **range:** Much of Canada, Great Lakes states, New England, south to Missouri, West Virginia

Large stands of Jack Pine are requisite as breeding habitat for the endangered Kirtland's Warbler (*Setophaga kirtlandii*), which nests almost exclusively in the Lower Peninsula of Michigan, near the southern extent of the tree's range. Historically, fires maintained this open habitat, but now it must be maintained by controlled burning. The tree supplies wood for use as pulp, posts, and firewood. Young trees are a host for the sweet fern blister rust, which causes orange cankers to grow on the trunks, and galls to form on the lower branches. Mature trees are defoliated by the Jack Pine Budworm.

Lodgepole Pine
Pinus contorta H to 98 ft (30 m)

Often found in expansive pure stands, the slender Lodgepole Pine is expected to decline drastically as a result of warming in its cool range, drought, and epidemic levels of Pine Sawyers.

KEY FACTS

Inland, this pine is tall, slender, and straight; in coastal or wet areas, it is shrubby and twisted.

+ **needles:** In 2s, curved, thick, stiff

+ **cones:** Broadly oval; scales with curved spine

+ **range:** Alaska to Baja California; most abundant in northern Rockies and Pacific coast area

American Indians used the slim, flexible limbs to build tepees (thus its common name), the soft inner bark for food, and the sap in medicines. The wood is used in framing, posts, railroad cross ties, and paneling, and to make pulp. It is a close relative of the Jack Pine, with which it hybridizes where the two occur together. Like its relative, it is plagued by Pine Sawyers; the insects spread fungi, and, girdling the tree, they kill it. But in an odd turn of events, the dead trees provide fuel for fires, enhancing germination of Lodgepole seeds. Seedlings outperform those of other species.

Shortleaf Pine
Pinus echinata H 80–100 ft (24–30 m)

The Shortleaf Pine is second in commercial importance (behind Loblolly) in the Southeast, where it is used for turpentine production, plywood, flooring, beams, and (taproot included!) pulp.

KEY FACTS

The crown is pyramidal to rounded; the trunk is clear of scrubby limbs.

+ **needles:** In 2s and 3s, long, slender, dark blue-green

+ **cones:** Cylindrical, conical, or egg-shaped; scales thin, prickle-tipped

+ **range:** Southeast; also New York and New Jersey to Missouri, Oklahoma, and Texas

The most wide-ranging yellow pine of the Southeast, the Shortleaf has needles that could be deemed short only in comparison with those of other native southern pines. The branches are irregular, compared with the straighter ones of the Loblolly, with which it hybridizes where their ranges overlap (as it also does with the Pitch Pine). Its seeds provide cover and forage for birds, squirrels, chipmunks, and mice. The Shortleaf's plated bark has resin ducts, appearing as small holes. In addition to Pine Sawyers, it is threatened by littleleaf disease, caused by a root fungus in poorly drained soils.

Piñon Pine/Two-needle Piñon
Pinus edulis H to 20 ft (6 m)

A handsome, small, gnarled tree, the Piñon has a dense, rounded, conic crown. Increased droughts have recently killed trees whose habitat is the less dry portion of the species' range.

KEY FACTS

Branches persist almost to the trunk base; the limbless portion is taller, and the silhouette is more irregular with age.

+ **needles:** In 2s, sometimes 1s or 3s, dark green, curving

+ **cones:** Spherical; scales lack prickle

+ **range:** Utah, Arizona, Colorado, New Mexico; small numbers in adjacent states

The seeds of the slow-growing Piñons do not fall when the cone opens. Their dispersal depends on the foraging of Pinyon Jays (named for the tree), which pluck the seeds from the cone scales and then stash them for future eating. Some seeds inevitably go forgotten and eventually germinate. The jays have mammalian competitors. Humans also collect the seeds but do not lose track of them so often. Indigenous peoples have harvesting rights to these traditional foods—called pine nuts in English, *pignoli* in Italian—and there is a commercial harvest. The wood is used as fuel, for posts, and in incenses.

Slash Pine

Pinus elliottii H 59–100 ft (18–30 m)

Usually inhabiting slashes, or overgrown swampland, Slash Pines grow fast and so are favorites in reforestation programs.

KEY FACTS

These have a rounded crown and reddish brown bark, becoming plated.

+ **needles:** In 2s or 3s; long, straight, stout

+ **cones:** Long, conical to ovoid, glossy, stalked; scales with outcurved prickle

+ **range:** Coastal Plain of South Carolina to Florida Keys and Louisiana; naturalized in nearby states

A tree of warm, humid flatwoods in poor, sandy Coastal Plain soils, the Slash Pine is grown in suitable areas worldwide for its wood, which is heavy, strong, durable, and compares favorably with that of the Longleaf Pine; its resin is used to make turpentine and rosin. (The species has become invasive in Hawaii and Australia and is under scrutiny in the entire Pacific Rim.) It is like the related Loblolly, except the Slash Pine often has needles in twos (mostly threes in the Loblolly) and larger cones. Large stands are especially hard-hit by the fusiform rust fungus, the most serious disease of pines in the Southeast.

Sugar Pine

Pinus lambertiana H 98–164 ft (30–50 m)

Our largest, most majestic member of the genus *Pinus*, the long-lived Sugar Pine inhabits mainly mixed-conifer forests. It can take 100 years or more to begin producing seeds.

KEY FACTS

This pine grows mostly on damper northern slopes. It has heavy seed cones (largest of any tree), which can weigh down branch tips.

+ **needles:** In 5s, blue-green, straight, slender

+ **cones:** Long, brown (purple when young), long-stalked, hanging

+ **range:** Mountains of Oregon and California

The Sugar Pine takes its name from the sweet-smelling and sweet-tasting resin that oozes from injured wood and forms candy-like beads. It is said to taste better than maple sugar, a trait native peoples took advantage of—though in moderation, because it is a laxative. They also used the resin to affix points and feathers to arrow shafts. The timber is appreciated for its workability, lightness, and strength and is used in framing and molding. Despite the tree's rapid growth rate, it cannot keep pace with harvests. Although it is planted in its native range, it has not been successful in culture elsewhere.

Singleleaf Piñon

Pinus monophylla H 16–66 ft (5–20 m)

The Singleleaf Piñon is often the dominant tree, a key element in the piñon–juniper woodlands of the mountains. As do other piñons, it depends on the Pinyon Jay for dispersal of its seeds.

KEY FACTS

Bark is gray and scaly, developing reddish furrows; the trunk branches to the base.

+ **needles:** Gray-green to blue-green, mostly solitary; stout, curved, sharp-pointed

+ **cones:** Broadly egg-shaped; a few woody scales

+ **range:** Idaho south to California, Nevada, Arizona, and New Mexico

Humans have long eaten the seeds of the Singleleaf Piñon. The seeds are known as pine nuts, but American Indians still roast the cones and grind the seeds to make traditional soups and cakes. This practice was disrupted in the late 1800s, when the resin-rich trees were harvested for use as a fuel in smelting silver. More recently, deemed low-quality forage for livestock, the trees have been destroyed in favor of other species. Other uses are as fence posts and Christmas trees. The parasitic Dwarf Mistletoe affects the Singleleaf, and a number of insects and fungal diseases especially harm piñons.

Western White Pine

Pinus monticola H 45–160 ft (14–49 m)

Closely resembling its relative the Eastern White Pine, this species grows fast and is quick to become established following a disturbance, such as fire.

KEY FACTS

This tall and narrow pine's trunk is clear of scrubby branches.

+ **needles:** In 5s, long, sometimes twisted, blue-green, whitish waxy wash

+ **cones:** Long, narrow cylinders hang from a long stalk; scales thin, curved back

+ **range:** British Columbia to California; Alberta to Nevada and Utah

Primarily a mountain tree, this species ranges to sea level at its northern extent. The wood is light, soft, straight-grained, and nonresinous. Easily worked and excellent for finer products than that of many conifers, it is used in interior details, trims, floors, and for matches and toothpicks. White Pine blister rust, caused by a fungus native to Asia and introduced into North America via Europe around 1900, attacks the five-needle pines and is especially grave in this species, having killed as much as 90 percent of trees in some regions. Bears, tearing into trunks for the sweet sapwood, also cause irreparable harm.

Longleaf Pine
Pinus palustris H 80–130 ft (24–40 m)

The grand Longleaf Pine was once widespread, the keystone of unique habitats of the southeastern Coastal Plain.

KEY FACTS

The needles are longest of our pines; the branches are sparse, relatively short, and stout; the crown is open and asymmetrical.

+ needles: In 3s, slender, flexible, usually hanging

+ cones: Largest of eastern pines; woody, hanging

+ range: Coastal Plain sandhills and flats, Virginia to Texas

As a seedling, the Longleaf Pine passes through a "grass stage" during which it grows thicker instead of taller, and its taproot becomes established; the long needles at this unbranching stage recall tufts of bunchgrasses. Historically, fires created a savanna characterized by the enormous pines and other fire-adapted species. Overharvest (for lumber, pulp, turpentine, and resin, especially for naval purposes), elimination of fires, and management for other species wiped out up to 95 percent of this habitat, but the species and the savannas are being reestablished in much of its former range by extensive plantings maintained by prescribed burning.

Ponderosa Pine
Pinus ponderosa H 60–140 ft (18–39 m)

Our most widely distributed and most abundant pine, the stately Ponderosa gets its name from the Spanish adjective meaning heavy, or ponderous.

KEY FACTS

Mature bark is cinnamon and plated.

+ needles: Mostly in 3s, with tiny teeth on edges, sharp, with turpentine odor

+ cones: Male and female on same tree; seed cones reddish, egg-shaped, not stalked; scales with stout prickles

+ range: Pacific coastal mountains and Rockies

The Ponderosa Pine is found in pure stands and in mixed-conifer forests, where it towers over other trees. One of the most important sources of timber in the West, its wood is used for construction and cabinetmaking, as well as for pulp and firewood. Though sometimes used for Christmas trees, it is not farmed. Larvae of the moth *Chionodes retiniella* feed exclusively on the Ponderosa's needles. Government fire suppression since the early 1900s threatened the Ponderosa's native parklike habitat, because its competitors were no longer killed by the flames. It is now often managed with the aid of controlled burning.

Red Pine/Norway Pine

Pinus resinosa H 50–100 ft (15–30 m)

Norway Pine is a misnomer for this North American native. It was probably so labeled by European explorers and settlers for whom it recalled the Norway Spruce.

KEY FACTS

This pine of cold regions usually towers over other trees; the crown is dense and rounded.

+ **needles:** In 2s, sometimes twisted, snapping when bent

+ **cones:** Seed cones near branch tips, small, egg-shaped, not prickly

+ **range:** Cape Breton Island to Manitoba, around the Great Lakes, south to Virginia

Formerly one of the main timber trees in much of its range, the Red Pine has light, close-grained wood, which is well suited for construction and the manufacture of pulp. As in many pines, the resin is made into turpentine. Vanillin, used in flavorings, is isolated during pulp processing. The tree is often cultivated as an ornamental or as a shade tree, and several cultivars, including a dwarf version, have been developed. Fire is needed to create the condition the seeds need for germination, namely bare mineral soil, and when there are no fires, other species will eventually replace it.

Eastern White Pine

Pinus strobus H 50–220 ft (15–67 m)

The Iroquois tree of peace, the Eastern White Pine once had extensive stands in the Northeast but has been so exploited that only one percent of those stands survive.

KEY FACTS

Branches are tiered, long, and horizontal.

+ **needles:** In 5s, light green to bluish, edges rough

+ **cones:** Seed cones slim, hanging, stalked; resinous, and somewhat curved

+ **range:** Newfoundland to Manitoba, around Great Lakes; New England to West Virginia; mountains south to Georgia

The White Pine was already being cut by colonists by the 1650s, and aggressive logging for the next two centuries did irreparable damage. The tree provided resins, pitch, and turpentine for use on ships, and its straight, strong trunks were ideal for ships' masts. Today, the wood is used in construction and for trim, furniture, and pulpwood, the demand met by cultivation. Many cultivars are grown, including dwarf and weeping varieties. In the United Kingdom, it is called the Weymouth Pine, for British explorer George Weymouth, who returned in 1605 with seeds he had collected in what is now Maine.

Loblolly Pine

Pinus taeda H 60–140 ft (18–43 m)

Cultivated varieties of the large, fragrant, and resin-rich Loblolly Pine include a 20-foot version often planted in windscreens; the wild type is not suitable because it sheds its lower branches.

KEY FACTS

This fragrant tree is pyramid-shaped when young, and eventually drops lower branches.

+ **needles:** In 3s, slender, stiff, sometimes slightly twisted

+ **cones:** Seed cones woody; scales thin with short prickle

+ **range:** Mainly in Coastal Plain and Piedmont region from southern New Jersey to eastern Texas

The Loblolly's rapid growth habit is a mixed blessing. The tree is one of the first to be naturally established in abandoned fields and is a good soil stabilizer. It is the most commercially important southern pine, but often at a price. Raised in dense monocultures, it is productive in lumber and forest products and renewable for those purposes, but these tree farms never will revert to the mixed-hardwood stands that they often replaced and that otherwise would develop. Mixed forests sustain greater biodiversity, and their wholesale substitution by depauperate Loblolly stands is receiving attention in restoration projects.

Virginia Pine

Pinus virginiana H 25–49 ft (10–15 m)

Common in the Piedmont and foothills of the Appalachians, the Virginia Pine has been called the Oldfield Pine because it is one of the first trees to grow on abandoned agricultural fields.

KEY FACTS

This smallish pine's limbs grow irregularly from the trunk, which often bears stubs of old branches.

+ **needles:** In 2s, short, pointed, twisted, spreading, serrate

+ **cones:** Seed cones small, conical to egg-shaped, woody; scales with slim, curved prickle

+ **range:** New York to Ohio, south to Mississippi

Thanks to its scrubby look, the Virginia Pine is also known as Scrub Pine, but its low branches remain on the tree, which helps make it the most popular species in the South for use as a Christmas tree. It grows in pure stands or mixed with various other trees, including hardwoods, which, in nature, will eventually shade it out. The Virginia Pine pioneers on old fields and is planted on abandoned fields and cutover lands. The wood is often used for pulp or rough lumber. In older trees, woodpeckers nest in the trunk, creating cavities where fungi have softened the wood.

Douglas-fir
Pseudotsuga menziesii H 80–150 ft (24–46 m)

The grand, long-lived Douglas-fir is among the most important of our timber trees. Unlike the upright cones of the true firs, those of the Douglas-fir hang from the branches.

KEY FACTS

Pyramidal when young, the tree loses its lower branches later for a cylindrical appearance.

+ **needles:** Flattened, on slim twigs

+ **cones:** Seed cones with 3-pointed seed bracts projecting beyond the rounded scales

+ **range:** Coast to mountains from British Columbia to California, and to tree line in Rockies

One of the largest trees in the world (unusually can reach 300 feet/92 m), the Douglas-fir produces a great volume of timber, which is used in support beams, interior woodworking, and plywood veneer. The species is used in landscaping, and numerous cultivars have been developed. In fact, the tree is planted as an ornamental in cool regions worldwide, and can be invasive outside its natural range. A variety called the Rocky Mountain Douglas-fir (*P. menziesii* var. *glauca*) is smaller, its foliage with a bluish or grayish cast (i.e., glaucous). Its seed bracts curve back on themselves, pointing toward the cone base.

Eastern Hemlock
Tsuga canadensis H 66–100 ft (20–30 m)

In the southern reaches of its range, the Eastern Hemlock is under attack by the Hemlock Woolly Adelgid, an insect from Asia that feeds on a tree's sap, killing the tree in only a few years.

KEY FACTS

Branches are drooping, with pubescent twigs and fissured bark.

+ **needles:** Flattened, in 2 ranks

+ **cones:** Seed cones to 0.8 in (2 cm), egg-shaped, stalked, borne at branch tips

+ **range:** Ontario to Nova Scotia, Minnesota through northeastern states, to mountains of Georgia and Alabama

The wood is of poor quality, sometimes used in general construction and as pulp. Many cultivars have been developed, including dwarf, shrubby, and weeping forms. The related but geographically distinct Western Hemlock—*T. heterophylla* (H 130–230 ft/40–70 m)—is the largest species of the genus, common and widespread in its range from Alaska to California (especially Oregon and Washington), as well as Montana. Its seed cones are 0.75–1 inch (2–2.5 cm) long, unstalked, its twigs finely pubescent. The hard, durable, and light wood makes it commercially superior to that of other hemlocks. It is rarely cultivated.

Pacific Yew

Taxus brevifolia H 15–50 ft (5–15 m)

A slow-growing evergreen, the Pacific Yew has lithe branches that often appear drooping or weeping.

KEY FACTS

The Yew is nonresinous, with thin, reddish brown bark.

+ **leaves:** Flat, pointed; borne spirally but appear in 2 ranks

+ **cones:** Berrylike, one seed surrounded by red, fleshy aril

+ **range:** Alaska to central California in mountains; isolated population from southeastern British Columbia to Idaho

A chemical compound isolated from the Pacific Yew (and found in other yews) is used in chemotherapy for cancer. The compound, known as the generic drug paclitaxel, from the name of the genus, is now manufactured with semisynthetic and cell culture procedures that do not harm the trees. The attractive wood is hard, heavy, strong yet flexible, fine grained, and easily worked. But the supply is limited, and the small size of the trunk and limbs limits their use to such things as canoe paddles, carvings, musical instruments, harpoon and spear handles, archery bows, and firewood. Grown ornamentally, the tree is especially good in hedges.

Common Sumac/Smooth Sumac

Rhus glabra H to 23 ft (7 m)

Common Sumac is a familiar shrub or small tree along roadsides, field edges, and train tracks. In the fall, red leaves and, on female plants, clusters of dark red berries, are its trademark.

KEY FACTS

This fast-growing sumac is nonpoisonous, but is related to Poison Ivy.

+ **leaves:** Pinnately compound; 7–31 narrow, toothed leaflets

+ **flowers/fruits:** Male and female inflorescences are on separate plants, at branch ends.

+ **range:** All 48 contiguous U.S. states and across Canada; most common in East

Common Sumac is popular in gardens for its striking red autumn foliage—giving it another name, Scarlet Sumac. But it can be weedy, spreading via underground stems. It may be confused with its look-alike cousin Poison Sumac (p. 91, *Toxicodendron vernix*), but the latter is rarely encountered, restricted to moist, even wet areas. Also, in *T. vernix,* male and female flowers grow on the same plant, the leaves have fewer leaflets, and the fruit clusters grow in the axils where two branches diverge. Sumac seeds provide food for birds and insects into the winter, and those of Common Sumac can be made into a lemonade-like beverage.

Pawpaw

Asimina triloba H to 40 ft (12 m)

The Pawpaw has few insect pests, possibly because its leaves contain unsavory chemical compounds. Compounds in the seeds show promise for use in cancer chemotherapy.

KEY FACTS

Colonies are often created by suckering.

+ **leaves:** Alternate, to 1 ft (30 cm) long; oval, narrowing basally

+ **flowers/fruits:** Flowers bell-shaped, 6-petaled, purple; berries fleshy, large, roughly cylindrical, ripening to yellow, then brown

+ **range:** Southern Ontario, eastern states west to Nebraska, Texas

The Pawpaw is a northern representative of the tropical custard apple family. Pawpaws were once harvested, but their supply has dwindled as forests have been cut. The fruit is delicious, but fierce competition from wildlife makes it a prize indeed. Contact with the fruit and leaves can occasionally cause dermatitis. The species is cultivated in its native range, but its wood has no value. The flowers' putrid scent attracts small flies and beetles, which effect pollination. *Asimina* comes from an American Indian name for the tree; the common name probably echoes the Spanish *papayo*, a word of Arawak origin in the West Indies describing a similar but unrelated fruit, the papaya.

American Holly

Ilex opaca H 20–49 ft (6–15 m)

An evergreen broadleaf, the American Holly is a common shrub or small tree of the understory, most common in the humid southeastern United States.

KEY FACTS

This holly sometimes has multiple trunks; bark is smooth and gray.

+ **leaves:** Alternate, oval, stiff, sparsely but strongly toothed; tooth tips sharp

+ **flowers/fruits:** Flowers unisexual, on separate trees, tiny, inconspicuous; fruits orange or red berrylike drupes, often dense

+ **range:** Mainly eastern and southern U.S.

The American Holly is the favorite of the evergreen hollies, with its handsome foliage, pyramidal form, and beautiful berries, ubiquitous in holiday and winter arrangements wherever it grows or is grown. More then 300 cultivars have been developed (but a female tree is needed, with a male nearby, if berries are expected). The berries also provide food for many birds and small mammals, which disperse the seeds. The wood is white and not important as a source of lumber. It is used in veneers, as pulpwood, and in creation of small wooden objects. The slow-growing tree tolerates a range of soil types.

Common Winterberry
Ilex verticillata H 8–15 ft (2.4–4.5 m)

The deciduous Common Winterberry is a popular ornamental and is striking in winter with its red berries borne on bare branches with dark, smooth bark.

KEY FACTS

This small tree is multi-trunked.

+ **leaves:** Alternate, elliptic, pointed; margin toothed, lacking spines

+ **flowers/fruits:** Flowers unisexual, on separate trees; greenish white; fruits bright-red berrylike drupes, often dense

+ **range:** From Quebec to Newfoundland south to Florida and Louisiana

The Common Winterberry is most abundant in wet areas, along streams, and in moist wooded habitats of the Coastal Plain. The red berries lend color to the winter landscape until the birds make off with them. A number of cultivars have been developed, and male and female cultivars must be paired up to ensure that they flower at the same time. The bark of the Winterberry was listed in the U.S. Pharmacopeia for most of the 19th century. Concoctions made from it have been used as an astringent, a bitter tonic, or a febrifuge (thus another of its common names, Feverbush).

Yaupon
Ilex vomitoria H 12–25 ft (3.7–7.6 m)

As in other hollies, the fruits of the Yaupon are an important winter food for birds and small mammals, which distribute the seeds.

KEY FACTS

This evergreen shrub branches heavily inside; can form dense thickets.

+ **leaves:** Alternate, oval, shiny, thick, small

+ **flowers/fruits:** Flowers unisexual, on separate plants, inconspicuous, greenish white; fruits bright-red drupes, numerous

+ **range:** Virginia to Florida, west to Oklahoma and Texas

The Yaupon's species label, *vomitoria*, is a misnomer. Certain American Indian tribes drank a ceremonial infusion that caused purging, and Yaupon was erroneously thought to be the emetic agent. Its twigs and leaves contain caffeine and are used to make a tealike drink not unlike maté, a beverage made from another holly, *I. paraguariensis*, and enjoyed in parts of South America. The name "Yaupon" is from a Catawban word related to the word for tree. This holly, too, is a popular ornamental, used for winter color in the landscape and in holiday decorations; cultivars include dwarf and weeping forms.

Devil's Walkingstick
Aralia spinosa H 15–23 ft (4.6–7 m)

The Devil's Walkingstick is in the Ginseng family, along with the invasive English Ivy, the *Panax* ginsengs, and the popular houseplant Schefflera.

KEY FACTS

This plant bears short, strong spines.

+ **leaves:** Alternate, to 32 in (81 cm) long, compound to doubly so; leaflets opposite, oval, pointed

+ **flowers/fruits:** Flowers unisexual, on same plant; tiny, white, in broad clusters; fruits fleshy, black berries

+ **range:** New Jersey to Florida, west to Texas and Missouri

Though its description may sound menacing, the Devil's Walkingstick is appreciated and even grown ornamentally for its intricate foliage, appealing inflorescences and fruits, and fall color. Its leaves turn bright yellow, and the black fruit are borne in clusters on bright pink stalks. The berries were eaten by the Iroquois, and also provide food for birds and small mammals, as well as black bear. The plant, a shrub or small tree of the understory, often forms dense clusters and has an open, irregular, flat crown. It is aromatic and deciduous, and its bark, roots, and berries have been used medicinally.

Big Sagebrush
Artemisia tridentata H 2–13 ft (0.6–4 m)

This evergreen member of the Aster family is one of the most wide-ranging shrubs in western North America, a dominant species to the tree line in much of the Great Basin.

KEY FACTS

The shrub is covered with fine gray hairs.

+ **leaves:** Alternate, wedge-shaped, 3-lobed at tip

+ **flowers/fruits:** Flowers perfect, tiny, tubular, yellow, in heads of 3–12; fruits seedlike, dry, hard, flat, broadest toward tip

+ **range:** British Columbia to Baja California, east to the Dakotas

The coarse, aromatic Big Sagebrush is a plant of dry, rocky soils that can live 100 years or longer. While not related to the sages, which are in the mint family, it has an aroma and a growth form that are not unlike those of some sages. Protein-rich (more nutritious than Alfalfa), it serves as forage for many large herbivores, especially in winter. Today it is used mainly as firewood and in smudges and incenses—its aromatic oils burn strongly and fragrantly. Native peoples, too, used it for those purposes but also made ropes and baskets from it and used it medicinally to treat a range of symptoms, as a tea and as a disinfectant.

Red Alder

Alnus rubra H to 60 ft (18 m)

Alnus rubra is our tallest alder and, unlike the others, a commercially viable source of timber. In fact, it is the Pacific Northwest's most important hardwood tree.

KEY FACTS

Inner bark and heart-wood are red, thus the common name.

+ **leaves:** Alternate, oval, pointed, with rounded teeth

+ **flowers/fruits:** Catkins unisexual, on same plant, in clumps, male drooping, female erect; fruits small cones

+ **range:** southeastern Alaska to southern California, within 125 miles of Pacific

Native peoples made a red dye from the inner bark and used it to color their fishing nets, which made them less visible to fish. The tree also had medicinal uses, possibly because it contains, as do the willows, aspirin-like chemical compounds that show antitumor properties. Its rapid growth makes it a good pioneer on disturbed lands, which it stabilizes. It is short-lived because when other trees invade, it cannot survive in the resulting shade. Associated with its roots are nitrogen-fixing bacteria, which provide the trees nitrogenous nutrients. The wood is used in furniture, pallets, plywood, spools, and boxes, and in pulp and as firewood.

Hazel Alder

Alnus serrulata H 12–20 ft (3.6–6 m)

Primarily an Atlantic coastal species, the Hazel Alder is a common large shrub or small tree of moist lowlands and stream banks, where it is mostly found in mixed stands.

KEY FACTS

This alder can be spindly with crooked trunks.

+ **leaves:** Alternate, oval, edge wavy, finely serrate

+ **flowers/fruits:** Catkins unisexual, on same plant, male pendent, female erect; fruits small, woody cones, lasting into winter

+ **range:** Nova Scotia and New Brunswick to Florida, west to Oklahoma and Texas

The Hazel Alder's growth habit makes the tree a superior colonizer and ideal in restoring wetlands, stabilizing stream banks, and mitigating storm-water runoff. When it is sold, it is generally for such purposes. The plant has an extensive root system and forms broad thickets by suckering. The roots are extensive and have symbiotic bacteria that fix nitrogen, providing the tree with nitrogenous nutrients. It is an important component of the American Woodcock's habitat and provides food and shelter for many other birds and small mammals. White-tailed Deer browse on the plants. The tree has had a range of medicinal uses, including as a pain reliever during childbirth.

Yellow Birch

Betula alleghaniensis H 60–75 ft (18–23 m)

The largest and most important birch, this species provides 75 percent of the birch wood used in the United States, and half of the species' entire stock is in Quebec.

KEY FACTS

Yellow-bronze bark (hence the common name) has horizontal lenticels.

+ **leaves:** Alternate, sharp-tipped, toothed

+ **flowers/fruits:** Catkins unisexual, on same tree, male pendent, female upright; fruits cone-like

+ **range:** Southeastern Canada; northeastern U.S., Great Lakes states, south in mountains to Georgia

A species of lower elevations, the Yellow Birch is a slow-growing, long-lived, single-trunked tree. When solitary, it develops a broad, open shape; when crowded, the trees grow tall and slender. The leaves turn a brilliant yellow in the fall. Deer and moose browse on the plants, and birds eat the seeds. A syrup is sometimes derived from the sap, and a tea is made from the twigs or inner bark, which have the aroma and flavor of wintergreen. The dark reddish brown to creamy-white wood is durable and heavy and is used widely in interior finishes, veneers, flooring, furniture, and cabinets.

River Birch

Betula nigra H 50–70 ft (15–21 m)

The fast-growing, shade-intolerant River Birch is a vigorous pioneer species that stabilizes soil on stream banks. The largest specimens grow in the Mississippi Valley.

KEY FACTS

The bark of young trees sheds in papery coils.

+ **leaves:** Alternate, broad, strongly serrate, pointed, wedge-shaped at base

+ **flowers/fruits:** Catkins unisexual, on same tree, male pendent, female upright; fruits cone-like

+ **range:** New Hampshire west to Minnesota, south to northern Florida and west to Texas

The lithe River Birch is the only spring-fruiting birch and the only birch whose native range includes the southeastern Coastal Plain. When young, it has a pyramidal shape, which becomes rounder with age. It often has multiple and highly branching trunks. Most common in floodplains and wet areas, it also grows in drier locations. Though the wood is close grained and sturdy, it is too knotty to be of much commercial value. It is used for furniture, baskets, small objects, and pulpwood. But it is an attractive tree, used as an ornamental; a number of cultivars have been developed. The sap has been concentrated for use as a sweetener.

Water Birch
Betula occidentalis H 20–35 ft (6–11 m)

The Water Birch is a shrub or small tree of wet, wooded sites, often growing near waterways or on banks, occurring sporadically throughout its range.

KEY FACTS

Bark is not exfoliating.

+ leaves: Alternate, rounded to wedge-shaped, with small marginal serrations on the larger teeth

+ flowers/fruits: Catkins unisexual, on same tree, male pendent, female erect; fruits are cones

+ range: In the west, especially in the Rocky Mountains, east to northwestern Ontario

The crown of the Water Birch is irregular and open; the branches are slender and drooping; and it usually has multiple trunks, forming thickets. The tree is too small to have much timber value, but its hard and heavy wood does find utility as posts and firewood. It is seldom cultivated but is planted as a stabilizer and buffer along stream banks; it is rather shade tolerant. Water Birch is an important component in wildlife habitat, and large mammals browse on the plants. Native peoples used a tea prepared from the tree as a diuretic and as a treatment for kidney stones.

Paper Birch/White Birch
Betula papyrifera H 50–70 ft (15–21 m)

The bark of the Paper Birch eventually becomes chalky white and papery, with prominent black lenticels, peeling off in horizontal strips to reveal orange beneath.

KEY FACTS

Pyramidal when young, but shape becomes less regular with age.

+ leaves: Alternate, oval, pointed; margins doubly toothed

+ flowers/fruits: Catkins unisexual, on same tree; fruits are cones

+ range: Alaska to Labrador, south into Rockies, Plains states, and Pennsylvania to North Carolina in mountains

For most of us, the Paper Birch is the tree that probably comes to mind when we think of birch trees, because of its beauty and because it is the most widespread of our birches. It can have one or multiple trunks. A number of cultivars have been developed. The wood is soft but moderately heavy and so is used for fuel, as well as in veneers, plywood, cabinets, furniture, and pulp. The tree's buds, catkins, and seeds provide food for many small mammals and birds, and a range of larger mammals find food in the bark and stems, including moose and porcupines.

Virginia Roundleaf Birch/Ashe's Birch

Betula uber H to 40 ft (12 m)

The first tree species listed as endangered by the U.S., the Virginia Roundleaf Birch is now listed federally as threatened (though still deemed endangered in Virginia, where it is endemic).

KEY FACTS

When crushed, the twigs smell of win-tergreen; the bark eventually splits into ragged plates.

+ **leaves:** Alternate, round to oval, with heart-shaped base, serrate

+ **flowers/fruits:** Catkins unisexual, on same tree, male pendent, female erect; cones erect

+ **range:** Endemic to a site in Smyth County, Virginia

The majestic Virginia Roundleaf Birch was first described in 1918, but it was not seen again and was believed extirpated until 1975, when a stand of 41 mature trees was found on a creek bank in Smyth County, Virginia. Thanks to propa-gation programs, in 2006, the count stood at 961, including 8 trees from the original 1975 group. In 1918, the species was described as a botanical variety of the Sweet Birch (*B. lenta*) but was elevated to its own species in 1945, a decision that is still debated. The 1975 rediscovery went mainstream with a feature in the *New Yorker* the fol-lowing January.

Southern Catalpa/Indian Bean Tree

Catalpa bignonioides H 25–45 ft (7–14 m)

One of the South's trademark trees, the Southern Catalpa is the host plant of a Sphinx Moth larva, the Catawba Worm, a popular bait for bream; the tree is also called Catawba and Fish Bait tree.

KEY FACTS

This tree can be wider than tall.

+ **leaves:** Oppo-site or whorled, large, rounded or heart-shaped

+ **flowers/fruits:** Flowers perfect, tubu-lar, 2-lipped, 5-lobed, white, in showy clus-ters, throat with yel-low and purple; fruits slender, beanlike pods

+ **range:** Georgia to Florida, west to Mississippi

The Southern Catalpa has been planted beyond its natural range for shade, beauty, and interest almost since its initial discovery by colonists, but drawbacks to some people are its falling flowers, unpleasant smell, large leaves, seedpods, and the tendency to sucker. It has become naturalized in those areas. The Northern Catalpa, *C. speciosa*, is the most north-ranging member of its family, with a limited natural range that includes southwest-ern Indiana and southern Illinois, Tennessee, and Arkansas. It is a larger tree, 75–100 feet (23–30 m), with larger leaves; smaller, fewer, and somewhat less attractive flowers; and thicker and longer pods.

Saguaro

Carnegiea gigantea H 40–60 ft (12–18 m)

The unforgettable shape of this tree-size cactus instantly brings to mind the Old West, an association that is exploited in movies and advertisements.

KEY FACTS

The trunk and branches are ribbed, with dense clusters of spines.

+ **leaves:** None

+ **flowers/fruits:** Flowers open at night, crown of white petals with yellow center, 3 in (7.5 cm) across, near branch ends; fruits oval, red, sweet; up to 4,000 seeds

+ **range:** Sonoran Desert in Arizona and into California

Our largest cactus does not develop its trademark upcurved arms until its 75th year; until then, the columnar plant is called a "spear." It can live 200 years. When it rains, the plant absorbs water and physically expands, then uses the water as needed. The cactus grows faster in areas with more rainfall. Its waxy coating helps conserve water. The Saguaro offers nesting habitat to birds including the Gila Woodpecker, Cactus Wren, and Pygmy Owl, and its seeds and fruit provide food. Arizona law makes it a crime to harm a Saguaro, and a permit is required to move one in order to build.

Jumping Cholla

Cylindropuntia fulgida H to 12 ft (3.6 m)

The name "Jumping Cholla" reflects the ease with which the spiny joints separate from the plant when bumped, sticking to one's clothes or skin, seemingly having jumped from the plant.

KEY FACTS

This shrubby cactus has one low-branching trunk.

+ **leaves:** Reduced to spines

+ **flowers/fruits:** Flowers pink to magenta, at branch and old fruit tips; 5–8 petals; fruits green, pear-shaped, many-seeded berries

+ **range:** Sonoran Desert in California and Arizona into New Mexico, Nevada, and Utah

The drooping branches of the Jumping Cholla are jointed, with cylindrical segments. When a fruit persists into the following year, a flower and another fruit can form at its tip, creating a chain of 25 years' worth of end-to-end fruits. This habit gives the plant another name, the Hanging Chain Cholla. The segments are armed with sharp, strong spines in crowded tufts. Because the segments disconnect easily, the ground at the base of the plant is often littered with plant parts, which can take root and form new plants. The fruits and seeds are food sources, especially for rodents, and the flesh of the plant provides water for desert animals, especially during droughts.

Roughleaf Dogwood
Cornus drummondii H 15–25 ft (4.6–7.6 m)

The Roughleaf is a dogwood whose flower clusters are not surrounded by large showy bracts, which may surprise those of us who think of the bracts as the symbol of dogwoods.

KEY FACTS

This dogwood can be a small tree or a multiple-trunked shrub.

+ **leaves:** Opposite, simple, entire; rough above

+ **flowers/fruits:** Flowers perfect, with 4 short, off-white petals, in clusters at branch tips; fruits hard berrylike drupes, white, in clusters

+ **range:** Great Plains, Midwest, Mississippi Valley

The Roughleaf Dogwood is a common component in the understory of rich woodlands. Though not showy, it is nonetheless attractive and has found uses in hedges and borders, around patios and in foundation plantings, and in parking lots and medians. As with many of its relatives, its fruits are eaten and its seeds are dispersed by many birds. And its suckering habit, resulting in those multiple trunks, is helpful in controlling erosion and in buffers. In the leaves, the veins come off the midrib, then arc toward the tip, and eventually nearly parallel with the margin, a dogwood trait.

Flowering Dogwood
Cornus florida H 15–35 ft (4.5–11 m)

In 2012, the U.S. sent 3,000 saplings of the Flowering Dogwood to Japan to mark the 100th anniversary of the famous cherry trees in Washington, D.C., a gift from Japan in 1912.

KEY FACTS

The branching is often tiered, with limbs to the ground.

+ **leaves:** Opposite, red in fall

+ **flowers/fruits:** Flowers perfect, tiny, in heads surrounded by 4 large white or pink petal-like bracts notched at the tip; fruits red, ovoid drupes

+ **range:** Mainly the eastern half of the U.S.

An eye-catching tree of the understory, the Flowering Dogwood is one of our most beautiful trees. But dogwood anthracnose, a fungal disease discovered in the U.S. in 1978, has spread throughout its range, causing significant losses of wild and planted trees, especially at higher altitudes and in shadier or moister sites. It first attacks leaves and stems, which die back, then it causes cankers, and the trees die. Guidelines are available for mediating the disease, and the many cultivars include a number of anthracnose-resistant varieties. The hard, dense wood has been used for golf club heads, mallets, tool handles, jewelry boxes, and spools.

Pacific Dogwood
Cornus nuttallii H 30–45 ft (9–14 m)

The drupes of the Pacific Dogwood are eaten and their seeds dispersed by birds and small mammals, as in other dogwoods. Large herbivores browse on the young plants.

KEY FACTS

This dogwood can bloom a second time, in early fall.

+ leaves: Opposite, simple, oval, hairy

+ flowers/fruits: Perfect, tiny flowers, in crowded heads with 4–8 showy, white to pinkish, petal-like bracts; fruits berrylike drupes, pink to red or orange

+ range: Pacific coast, British Columbia to California

The graceful Pacific Dogwood is a striking tree of the Pacific coast, where it is found inland to about 200 miles. It is much like the Flowering Dogwood, with its "flowers" of a small inflorescence and showy bracts. The Pacific Dogwood's flowers are larger, their bracts lack a terminal notch, and the two species' ranges are widely distinct. Anthracnose plagues the Pacific Dogwood; low air circulation and too much water worsen it. Dogwoods are sometimes called cornels, a name referring to their genus. The name is probably derived from the Greek *kerasos*, or "cherry tree," presumably because of the color of the drupes. A few cultivars are available.

Common Persimmon
Diospyros virginiana H 30–40 ft (9–12 m)

When fully ripe (following a solid frost, or if aged beyond technically ripe), a wild persimmon is delicious, tasting somewhat like a date. Earlier, it is so bitter and astringent as to be inedible.

KEY FACTS

Bark forms plates recalling charcoal briquettes.

+ leaves: Alternate, simple, entire, oval

+ flowers/fruits: Male and female flowers on separate trees; female, white, bell-shaped; male clustered, tubular, smaller; fruits fleshy, round, orange to purplish

+ range: East; southern, but found to Connecticut

Persimmons persist on the tree long after the leaves have fallen, which makes them one of the latest available wild fruits. This was nicely timed with the harvest feasts of American Indians and European settlers alike, so persimmons were served. The fruits are eaten fresh and are made into cakes, puddings, breads, candies, pies, jams, and beverages. Humans must vie with raccoons, other mammals, and birds to get the prized fruits. A member of the ebony family, the Persimmon has dark, heavy, close-grained wood, which is used in heads for golf club woods, in lathe work, and in carving. Several cultivars have been developed.

Mountain Laurel

Kalmia latifolia H 3-23 ft (1-7 m)

The blight that decimated the American Chestnut in the early 1900s benefited the Mountain Laurel, which prospered as sunlight gained entry to the once denser reaches of the forest.

KEY FACTS

This evergreen broad-leaf shrub is dense and gnarled.

+ **leaves:** Alternate, elliptic, leathery

+ **flowers/fruits:** Flowers perfect, white to pink clusters; 5 petals, fused into saucer shape; fruits small, round, capsules with withered pistil

+ **range:** East of a line from New Brunswick to Louisiana

The Mountain Laurel grows mostly in rounded stands sprawling on the forest floor. It is one of our most magnificent native shrubs (occasionally reaching tree size), common in the wild, and is planted for its showy beauty in yards and parks. Many cultivars have been developed. The stamens are held under tension by the petals, springing loose when tripped by a bee, peppering the insect with pollen, which it transfers to other flowers. All parts of the plant are toxic. It was used among native peoples as a means of suicide, as well as a source of a yellow-brown dye.

Sourwood

Oxydendrum arboreum H 40-60 ft (12-18 m)

The Sourwood finds many uses in landscaping, thanks especially to its leaves, whose vivid fall color has few rivals and which persist longer than most other leaves.

KEY FACTS

The branches often extend to the ground and droop.

+ **leaves:** Alternate, vivid orange or red in fall

+ **flowers/fruits:** Flowers cup-shaped, small, white, in spikes along one side of stem, showy; fruits are capsules, silver-gray in fall

+ **range:** Southeast; west to Kentucky, south to the Gulf and northern Florida

This narrow, small to medium-size tree of the understory has a graceful, open, and irregular shape. It was being cultivated as early as the mid-1700s, and cultivars have been developed. In nature, it is found in mixed stands on drier, upland, wooded sites, often with other heaths (as members of its family are known), such as *Rhododendron* species. It is sensitive to air pollution. The Cherokee and Catawba used the young trees to make arrows, and though the Sourwood's dense, close-grained wood has been used for paneling, fuel, and pulp, it has not had great commercial success.

Sparkleberry/Farkleberry
Vaccinium arboreum H 12–28 ft (3.6–8.5 m)

The Sparkleberry is our only tree-high *Vaccinium*, though some highbush blueberries come close. It is more tolerant of more alkaline soils than are its relatives.

KEY FACTS

The tree is spindly with twisted branches and is mostly evergreen.

+ **leaves:** Mostly evergreen, alternate, rounded, leathery, bright red in fall

+ **flowers/fruits:** Flowers white, bell-shaped, 5-lobed, in drooping clusters; fruits black, shiny

+ **range:** Southeast from Virginia to Missouri, to Florida and Texas

The name "Sparkleberry" probably refers to the fruit's shininess; "Farkleberry" may be a play on that name. The fruits are dry, bitter, and mostly ignored. The name has been tied to two Arkansas governors: Orval Eugene Faubus was nicknamed "Farkleberry" by a cartoonist who found it apt for satirizing Faubus and the state's politics. Frank White was dubbed "Governor Farkleberry" after saying that his family had been so poor they had to eat Farkleberries. This attractive plant provides habitat and food for wildlife, and the bark, peeling in red, gray, and brown patches and often splotched with lichen, adds visual interest.

Blueberries
Vaccinium species H various, to 16 ft (5 m)

Though these plants have provided food for humans and wildlife for centuries, cultivars have been developed that provide plumper, juicier, and sweeter berries.

KEY FACTS

Blueberries are mostly upright, small shrubs.

+ **leaves:** Deciduous or evergreen, oval to tapering at ends

+ **flowers/fruits:** Flowers bell-shaped, 5-lobed, white, pink, or red, in clusters at end of branches; fruits are blue berries

+ **range:** Arctic Circle south, especially in cooler areas

The genus *Vaccinium* includes 450 to 500 species of shrubs, vines, and small trees mainly in cooler areas of the Americas, Europe, and Asia, but also in southern Africa, Madagascar, and Hawaii. The "typical" blueberries are endemic to North America. Their classification is difficult. Genetic and molecular studies are incomplete, there are different schools of thought, and hybridization is a further complication. The genus includes the important Highbush (*V. corymbosum*), Lowbush (*V. angustifolium*), Velvetleaf (*V. myrtilloides*), and Rabbiteye (*V. virgatum*) blueberries and at least nine others, including the Cranberry (*V. macrocarpon*), a low-growing evergreen in acidic bogs in the Northeast.

Mimosa/Silktree

Albizia julibrissin H 20–50 ft (6–15 m)

A native of Asia, from Iran to Japan, the exotic Mimosa was introduced here as an ornamental in the 1700s. It remains popular in gardens, yet it is invasive on roadsides and in disturbed areas.

KEY FACTS

Flowers are borne at branch ends; trunks are often multiple.

+ **leaves:** Opposite, large, twice compound, feather-like; leaflets opposite

+ **flowers/fruits:** Flowers pink filamentous pompons, fragrant, clustered; fruits are thin, flat pods

+ **range:** Nonnative; widely naturalized, invasive, especially in Southeast

The Mimosa is vase-shaped, with a crown of large, feathery leaves. It has bacteria in its roots that can change inert atmospheric nitrogen into forms that can serve as plant nutrients. Wildlife eats its fruit and disperses its many seeds. The roots or an old stump can send up sprouts. As a result, and because it is nonnative, it has no natural enemies to keep it in check, and it is often invasive. The University of Florida Center for Aquatic and Invasive Plants suggests native alternatives, such as Serviceberry (*Amelanchier arborea*), Redbud (*Cercis canadensis*), and Fringe Tree (*Chionanthus virginicus*), but people still plant the Mimosa, and there are many cultivars.

Yellow Paloverde

Parkinsonia microphylla H to 25 ft (7.6 m)

The Yellow Paloverde is a trademark shrub or small tree of the Arizona desert. Its branches are photosynthetic, which gives it a yellow-green cast and its common name.

KEY FACTS

The branches are erect, not flowing.

+ **leaves:** Alternate, twice compound; 2 leaflets; secondary leaflets tiny, oval, soon dropping

+ **flowers/fruits:** Flowers perfect, in small clusters; 5 petals—4 yellow, 1 white; fruits are pods constricted between seeds

+ **range:** Sonoran Desert in California and Arizona

Like its relative the Jerusalem Thorn, the Yellow Paloverde is drought deciduous: It loses its leaves in the hottest and driest part of the year, which lessens water loss. It will even drop some branches if weather is especially severe. It is a "nurse plant" to the Saguaro; the cactus's seeds germinate in its shade, and it protects the seedlings from wind and trampling. Yellow Paloverde wood is used mainly as fuel, and it is cultivated in other arid regions. Buffelgrass (*Cenchrus ciliaris*) was introduced from Africa to grow as food for livestock, but it has become invasive and harms the Yellow Paloverde by robbing it of scarce water from the soil.

Eastern Redbud

Cercis canadensis H 15–28 ft (4.6–8.5 m)

The Eastern Redbud stages a striking display in spring, its pink flowers appearing before most trees have leafed out. Flowers are borne on all parts of the tree, even the trunk.

KEY FACTS

The flower, as in many legumes, recalls that of the familiar garden pea.

+ **leaves:** Alternate, simple, entire, heart-shaped, papery, pendent; yellow in fall

+ **flowers/fruits:** Flowers perfect, pink, stalked, in clusters of 4–8; fruits flat, hanging pods

+ **range:** Eastern half of the U.S.

The Eastern Redbud is not native to what is now Canada; *canadensis* in the scientific name refers to a rather different Canada, a French colony that extended into what is now the U.S., including areas where the Redbud is native. This graceful, often multistemmed shrub or small tree grows on a range of soils and in different light regimes, which, with its beauty at flowering time, have made it a successful ornamental. Cultivars include white-flowered varieties. The tree is too small to be a commercial lumber source. Like many leguminous plants, the Redbud can fix nitrogen.

Honeylocust

Gleditsia triacanthos H 30–80 ft (9–24 m)

The large, forked thorns on the trunk and lower branches of the Honeylocust have been used as pins or even nails. Thornless cultivars are popular in landscaping.

KEY FACTS

The bark becomes ridged, its edges curling.

+ **leaves:** Once or twice compound; leaflets oval

+ **flowers/fruits:** Flowers in hanging unisexual clusters on same tree (some flowers perfect), yellow-green, fragrant; fruits twisting, leathery pods, turning brownish

+ **range:** Midwestern and south-central U.S.

The wood of the Honeylocust is of good quality and durable when in contact with the soil. It is used to make furniture, interior finishings, and utilitarian products such as posts and firewood. The seedpods provide a sweet pulp (thus the common name) that was eaten by native peoples and has been fermented into a beer. The inconspicuous flowers are a source of pollen and nectar for honey. Pharmacognosists have isolated chemical compounds from the tree that have anticancer properties or that show promise in treating rheumatoid arthritis. The pods, bark, and young shoots are often eaten by mammals, the seeds by birds.

Kentucky Coffeetree
Gymnocladus dioicus H 60–80 ft (18–24 m)

The stout branches of the Kentucky Coffeetree lose their leaves early and so can go leafless for half the year. *Gymnocladus* means "naked branch."

KEY FACTS

The bark is fissured with scaly ridges.

+ leaves: Huge, twice compound; leaflets oval, pointed

+ flowers/fruits: Flowers fragrant, in upright clusters; 5 petals and calyx lobes, alternating, greenish white, sexes on same tree; fruits broad, leathery

+ range: Midwest to Appalachians, south to Louisiana

Though the Kentucky Coffeetree is widespread and tolerates a range of conditions, it is not common. Animals that might have dispersed its large seeds—Mastodon, Mammoth, and Giant Sloth—are long extinct, which may have curbed the tree's population. Colonists roasted and milled the seeds to make a coffee-like beverage, but the unroasted seeds are toxic. The drink contains no caffeine and was forgotten when coffee beans became available. In landscaping, male trees are usually used because females drop one part after another, which must be cleaned up. The males, however, lack the interesting pods in winter.

Honey Mesquite/Glandular Mesquite
Prosopis glandulosa H 20–30 ft (6–9 m)

Though only a large shrub or small tree (at its biggest in areas of highest moisture), the Honey Mesquite is still the largest tree in much of its range.

KEY FACTS

It has feathery foliage and long, paired spines.

+ leaves: Alternate, twice compound, smooth, with 2 leaflets, each with small leaflets

+ flowers/fruits: Flowers perfect, white to yellow, fragrant, in spikes; fruits are long cylindrical pods, constricted between seeds

+ range: Central Texas to California

Before Europeans introduced livestock, mesquites may have had more clearly defined ranges. Animals that eat the pods disperse the seed, and have thus blurred the boundaries. In addition to seeds, the Honey Mesquite reproduces by sprouting from the roots. It has found many uses and is being eyed as a promising species for the world's food supply (as well as for animal feed and building materials). Native peoples used the pods for food and to make alcoholic beverages and flour. Mesquite honey is prized for its flavor. Yet where stock is grazed, mesquites are considered weeds and are often destroyed.

Velvet Mesquite

Prosopis velutina H to 30 ft (9 m)

Overgrazing removed range plants that served as fuels in natural fires. Coupled with dispersal of its seeds by livestock, mesquite expanded into grazing lands, where it is considered a pest.

KEY FACTS

Larger than other mesquites, it is spiny, with feathery foliage.

+ **leaves:** Alternate, twice compound, velvety, 2 leaflets, each with smaller leaflets

+ **flowers/fruits:** Flowers perfect, creamy, fragrant, in spikes; fruits are long cylindrical pods, constricted between seeds

+ **range:** Central and southern Arizona

Mesquites such as the Velvet were essential to native peoples of the Southwest. The trees often grew in bosques (wooded assemblages that extended for miles along rivers), allowing the fruit to be gathered almost as if it had been planted. The beans were dried and ground into flour that was used to make bread, a dietary staple; the pods are about 25 percent sugar, and the seeds are high in protein. The trees were also important for articles of everyday life, such as tools, rope, baskets, and weapons. The original extent of the mesquite bosques is diminishing as the mesquites are cut for wood and cleared for agricultural and other development.

Black Locust

Robinia pseudoacacia H 30–60 ft (9–18 m)

The roots of the Black Locust have nodules that contain bacteria capable of fixing nitrogen, that is, of transforming inert nitrogen gas in the air into chemical forms that plants can use.

KEY FACTS

Most leaves have a pair of thorns at the base.

+ **leaves:** Alternate, compound, feathery; entire, large, oval leaflets

+ **flowers/fruits:** White, fragrant flowers, typical pea flower, in large, showy, hanging clusters; fruits are flat pods, sometimes persisting

+ **range:** Central Appalachians and Ozarks

The Black Locust's many seeds and ability to sprout from its roots enable it to establish quickly, even on poor soils—ideal in reclamation projects and erosion control, but not so much in a garden. Nevertheless, it has been widely planted and is naturalized in the 48 contiguous states and much of Canada. It grows (and can be invasive) in many parts of the world, sometimes farmed as an alternative to taking tropical woods. It produces a strong, hard wood used in boats (instead of teak), furniture, paneling, and flooring, in addition to more utilitarian uses including the manufacture of charcoal.

American Chestnut
Castanea dentata H to 100 ft (30 m) before the blight

"Not only was baby's crib likely made of chestnut, but chances were, so was the old man's coffin," wrote George H. Hepting in "Death of the American Chestnut" (*Journal of Forest History*, 1974).

KEY FACTS

The tree is spreading and dense.

+ **leaves:** Alternate, 5–8 in (13–20 cm), pointed; base rounded; margin sharp-toothed

+ **flowers/fruits:** Flowers unisexual, both sexes on same plant—male in catkins, female near catkin bases; fruits spiny capsules with 3 nuts

+ **range:** Eastern United States, Ontario

The American Chestnut has been called the ideal tree. Majestic, massive, it made up 25 to 50 percent of the hardwoods in its range. Its wood was unequalled, it was a key source of tannins used in tanning leather, and the nuts had a ready market in the East. But a fungus, brought from Asia around 1900, caused a blight that destroyed nearly all mature chestnuts in 40 years. The species survives in unaffected areas to which it had been transplanted. Also, the root systems did not succumb and send up sprouts. Studies are under way to exploit what natural resistance exists, and to explore hybridization with a resistant chestnut from China.

American Beech
Fagus grandifolia H to 66 ft (20 m)

No matter how remote the tree, it seems initials will be found carved into the trunk of an American Beech, the carving sealed by the bark and remaining for posterity.

KEY FACTS

The smooth bark is light gray.

+ **leaves:** Alternate, to 5.5 in (14 cm), pointed; margin toothed

+ **flowers/fruits:** Flowers unisexual, male in round heads, female 2–4 in spikes, on same tree; fruits spiny husks with 2 3-faced, winged nuts

+ **range:** New Brunswick to Florida to Mississippi Valley

The American Beech is a large, imposing, slow-growing, and long-lived tree usually found in mixed hardwood forests, where it is the dominant or a codominant tree, sturdy and with many branches. The tree is deciduous, but many leaves persist through the winter, their light reddish brown somehow almost showy against the gray bark. The nuts are delicious, but competitors are many, including squirrels, chipmunks, mice, raccoons, porcupines, opossums, rabbits, deer, black bear, and foxes, not to mention ruffed grouse, turkeys, bobwhite, and pheasants. Beech wood is used in furniture, veneers, and flooring, and for fence posts and fuel.

White Oak

Quercus alba H to 82 ft (25 m)

The White Oak is possibly our most abundant tree, usually dominant when present, and our most commercially important oak. There are 58 species of oak in North America north of Mexico.

KEY FACTS

The leaves are red-purple in fall.

+ **leaves:** Alternate, to 7 in (18 cm); 7–10 rounded lobes

+ **flowers/fruits:** Flowers in unisexual catkins on same tree; male lax, green-yellow; female reddish, in leaf axils; acorn caps cover a third of nut

+ **range:** Southern Ontario and Quebec to eastern and central U.S.

The White Oak is usually found in mixed hardwood forests. Half of U.S. hardwood production is White Oak, because of its quality and abundance. It is used in furnishings, interiors, shipbuilding, and wine and whiskey barrels. This tree is the flagship of the White Oak group, whose acorns mature the first year on new branches. The inside of the acorn shell is not hairy, and the cap's scales are brown and flat. The meat is sweet to somewhat bitter. The leaf lobes are rounded and do not have bristles at the tip. The bark is often pale gray and blocky. Species of the White Oak group will hybridize with one another.

Arizona White Oak

Quercus arizonica H to 60 ft (18 m)

The Arizona White Oak is found in arid and semiarid areas in oak and piñon woodlands, growing larger in areas with more moisture, such as canyons.

KEY FACTS

The trunk is short, and the branches are twisted.

+ **leaves:** Evergreen or nearly so, alternate, to 3.5 in (9 cm) thick

+ **flowers/fruits:** Catkins typically oak; acorns oblong, to 1 in (2.5 cm); caps cover a third of nut

+ **range:** Central Arizona to southwestern New Mexico into Texas

The long-lived Arizona White Oak (White Oak group) is a shrub to a medium-size tree, yet one of the largest of the southwestern oaks. It has stout, spreading branches and an irregular crown, and its light gray bark can be an inch (2.5 cm) thick on a mature tree. The acorns are eaten by cattle and deer, but they are not the preferred acorn. Birds forage in the stands, and deer browse on the sprouts. The wood is neither straight enough nor large enough to provide commercial timber, and, though difficult to cut and work, it is used for fuel and sometimes in furniture.

Scarlet Oak

Quercus coccinea H to 100 ft (30 m)

The Scarlet Oak is widely planted in the U.S., Canada, and Europe as an ornamental and a shade tree, and cultivars have been developed. The leaves turn brilliant red in autumn.

KEY FACTS

The trunk of this oak swells at the base.

+ **leaves:** Alternate, to 6 in (15 cm); 5- to 9-lobed, each with a bristled tip; sinuses are deep and C-shaped

+ **flowers/fruits:** Catkins typically oak; acorns to 1.2 in (3 cm); caps cover half of nut

+ **range:** Eastern and central U.S.

The Scarlet Oak (a member of the Red Oak group; see Northern Red Oak, p. 143) is a large, fast-growing tree with stout, upright, spreading branches and an open, rounded crown, the bark thin and gray-black. It is common in dry upland forests, where it grows in large pure stands and mixed stands. The wood is reddish brown, coarse-grained, and strong, inferior to that of the Northern Red Oak but, as wood of many other species of the Red Oak group, is marketed as Red Oak. It is used as lumber. Small mammals and birds use the acorns and seedlings as food, and the trees provide nesting sites, including cavities.

Coastal Sage Scrub Oak

Quercus dumosa H 3–7 ft (1–2 m)

The Coastal Sage Scrub Oak usually grows within sight of the Pacific Ocean—that is, on prime real estate. The tree, which has never been common, is at great risk from human encroachment.

KEY FACTS

The tree has a scraggly look.

+ **leaves:** Evergreen or nearly so, to 1 in (2.5 cm), irregularly toothed or shallowly lobed, with erect curly hairs on underside

+ **flowers/fruits:** Catkins typically oak; acorns narrow, pointed; reddish caps cover a third of the nut

+ **range:** Southern California

Most Coastal Sage Scrub Oaks of the White Oak group (see White Oak) have at one time been considered *Quercus dumosa*, but the species concept is now much narrower. It refers to a plant growing in a limited part of southern California and restricted to Coastal Sage Scrub habitats. Most trees once considered this species are now included as the more widespread and common California Scrub Oak, *Q. berberidifolia*, which ranges to the high north coast mountains and to the foothills of the Sierra Nevada. Its acorns are more rounded, and its leaves lack hairs on the underside. Its range does not overlap with that of *Q. dumosa*.

Southern Red Oak

Quercus falcata H to 100 ft (30 m)

A medium-size to large tree, the Southern Red Oak is also known as Spanish Oak, probably because it grows in some of the former Spanish colonies.

KEY FACTS

Its trunk is straight, its crown rounded.

+ **leaves:** Alternate, to 9 in (23 cm); 3–7 lobes, terminal longest, sharp, often curved; sinuses deep, U-shaped

+ **flowers/fruits:** Catkins typically oak; acorns round, orangish, 0.5 in (1.3 cm); caps cover a third of nut

+ **range:** Southeast to southern Missouri, eastern Texas

The Southern Red Oak (a member of the Red Oak group; see Northern Red Oak, p. 143) is a handsome tree that is rather fast growing and fairly long lived. It grows on dry, poor, sandy, upland soils. The bark is dark and thick with deep furrows between broad, scaly ridges. It is an important timber tree, providing light-red wood that is coarse grained, hard, and strong. It is used in furniture, cabinets, veneer, and for pulp and fuel. Native peoples used oaks for myriad medicinal purposes. The Southern Red Oak was used to treat such conditions as indigestion, chapped skin, and fever, and as an antiseptic and tonic.

Gambel Oak

Quercus gambelii H 10–30 ft (3–9 m)

For the caterpillar (larva) of the Colorado Hairstreak Butterfly to undergo metamorphosis, it must feed on the leaves of the Gambel Oak.

KEY FACTS

The tree has crooked branches.

+ **leaves:** Alternate, 3–6 in (8–15 cm), yellow-green, broadest beyond middle to uniformly wide; 5–9 lobes; deep sinuses

+ **flowers/fruits:** Catkins typically oak; acorns to 1 in (2.5 cm), mostly round; caps cover to a third of nut

+ **range:** Utah, Wyoming, Arizona, New Mexico

A tall shrub or small tree, the Gambel Oak (White Oak group) is the most common deciduous oak in most of the Rocky Mountains, widespread and abundant in the foothills and lower elevations on dry slopes and in canyons. The slopes can be covered in clonal thickets of the tree, the result of sprouting from tuber-like parts of the root system. Farther south, it exists in mixed stands. The tree's small size prevents the fine wood from being commercially significant, but it is used for fires and to make fence posts. Deer and livestock browse the foliage, and acorns provide food for turkeys and squirrels, as well as domestic stock.

Oregon White Oak/Garry Oak

Quercus garryana H to 50 ft (15 m)

The only oak native to British Columbia, Washington, and northern Oregon, the Oregon White Oak is the most commercially important oak in the West.

KEY FACTS

The tree often bears mistletoe.

+ leaves: Alternate, 3–4 in (8–10 cm), broadest beyond middle to uniformly wide; 5–9 lobes; teeth variable, sinuses deep

+ flowers/fruits: Catkins typically oak; acorns to 1.5 in (3.8 cm); caps cover to a third of nut

+ range: Central California to southwestern British Columbia

The Oregon White Oak is a shrub to medium-size tree with a dense, spreading crown. It exists in pure stands or mixed with other hardwoods or conifers. Its ecological foes include fire suppression. Natural grass fires did not harm the oaks but burned out young Douglas-firs beneath them. In the absence of fires, the Douglas-firs become established, and the oaks cannot survive in their shade. Other threats are development and invasive nonnatives. In Oregon White Oak woods in British Columbia, more than 80 percent of the understory is nonnative, and in Oak Bay on Vancouver Island, a permit is required to cut even a branch from established trees.

Shingle Oak

Quercus imbricaria H to 65 ft (20 m)

A stately, medium-size tree, the Shingle Oak is planted for shade or as a windbreak, a hedge, an ornamental, or a street tree. It is our most cold-hardy oak.

KEY FACTS

The lower branches may droop.

+ leaves: Alternate, 4–6 in (10–15 cm), oval, unlobed, broadly pointed, with one bristle; margins somewhat wavy; yellow-brown to russet in fall

+ flowers/fruits: Catkins typically oak; acorns to 0.7 in (1.8 cm); caps cover about half of nut

+ range: Midwest and upper South

The Shingle Oak (a member of the Red Oak group; see Northern Red Oak, p. 143) is never a dominant species in a mixed woods, probably because of its intolerance of shade. It is not a commercially important timber tree, although one of its main practical uses gives it its common name. The wood is pale reddish brown, heavy, hard, and coarse grained, and is used to make split shingles. This use is probably related as well to its scientific name, *imbricaria*, derived from the Latin word for "tile." It also may refer to the overlapping scales of the winter bud. The acorns are especially bitter.

Valley Oak/California White Oak
Quercus lobata H to 82 ft (25 m)

The largest oak tree in the U.S. and Canada is a Valley Oak in California's Round Valley called the Henley Oak, which is 151 feet (46 m) tall and more than 500 years old.

KEY FACTS

The twigs often weep; salt in the air makes the tree scrubby to 4 mi (6.4 km) inland.

+ **leaves:** Alternate, to 4 in (10 cm); 9–11 deep, rounded lobes

+ **flowers/fruits:** Catkins typically oak; acorns 1.2–2.5 in (3–5.6 cm), narrow, acutely conical or bullet-shaped; caps covering nut base

+ **range:** Endemic to California

A California icon, the Valley Oak (White Oak group) is a tree of valleys in the state's inner and middle coastal ranges. When English explorer George Vancouver visited the Santa Clara Valley in 1792, the expanses of Valley Oak savanna reminded him of a closely planted stand with the understory removed, and he called the trees the "stately lords of the forest." More than 90 percent of the Valley Oak stands were gone by World War II—victims of orchard and vineyard agriculture. The San Francisco Estuary Institute is investigating "re-oaking," hoping to reintegrate Valley Oaks and other natives into California's highly developed landscape.

Overcup Oak
Quercus lyrata H to 65 ft (20 m)

The Overcup Oak's scientific name refers to the overall shape of the leaf, which is said to recall the outline of a lyre.

KEY FACTS

The tree has a short trunk.

+ **leaves:** Alternate, 6–10 in (15–25 cm); 5–9 lobes, irregular, outer pair often making cross shape; deep sinuses

+ **flowers/fruits:** Catkins typically oak; acorns about 1 in (2.5 cm), round, often almost covered by cap; very long stalk

+ **range:** Deep South and Mississippi Valley

The Overcup Oak (White Oak group) is a slow-growing, small to medium-size tree of the warm, humid Southeast. Also called Swamp Post Oak, it grows among the Baldcypress and Water Tupelo common in many southern swamps. Its wood warps easily and is generally inferior to that of most of the White Oak group; nonetheless, it is marketed as White Oak and used for general construction and barrels. The Overcup Oak is planted to enhance land for wildlife, and its acorns are eaten by deer, turkey, and squirrels; ducks also eat the acorns, though because these nuts are so large, they are less useful than those of other oaks.

Bur Oak/Mossycup Oak

Quercus macrocarpa H to 98 ft (30 m)

The Bur Oak has the largest acorn of all our oaks, and its scientific name reflects that: *Macrocarpa* means "big seed."

KEY FACTS

The bark forms rectangular blocks.

+ leaves: Alternate, to 12 in (30 cm); widest beyond middle; 5–9 lobes; deep sinuses

+ flowers/fruits: Catkins typically oak; acorns to 2 in (5 cm); caps cover three-quarters of nut, with mossy or bur-like fringe

+ range: Midwest and eastern U.S., south-central Canada

A medium-size to large, spreading tree, the Bur Oak (White Oak group) is cold tolerant and, because of its long taproot, drought resistant. Its large acorns are important forage for wildlife, including black bears. It is planted as an ornamental and is the most forgiving White Oak of the urban environment. The Bur Oak savannas of the eastern prairie were vital to settlers, providing wood and grazing land. In the early 1900s, these savannas covered 32 million acres. By 1985, as a result of development, including agriculture, and fire suppression, high-quality savanna had declined to 6,400 acres (2,592 ha)—a loss of more than 99 percent.

Cherrybark Oak/Swamp Red Oak

Quercus pagoda H to 130 ft (40 m)

The acorns of the Cherrybark Oak provide food for domestic hogs and for wildlife, including larger birds and a number of mammals.

KEY FACTS

The leaf shape recalls a pagoda (thus its scientific name).

+ leaves: Alternate, 5–8 in (13–20 cm), broadest at base; 7–11 lobes, regularly shaped

+ flowers/fruits: Catkins typically oak; acorns 0.6 in (1.6 cm); caps cover a third to half of nut

+ range: Southeastern United States, Mississippi Valley

One of the largest oaks in the South, the Cherrybark Oak (Red Oak group; see Northern Red Oak, p. 143) is similar to the Southern Red Oak, of which it was formerly considered a botanical variety. It prefers moist areas, in riverbanks and in floodplains, especially in the Coastal Plain. The bark recalls that of the Black Cherry, smooth, with flaky ridges, and with a red tint. The trunk is straight, with relatively few limbs, and the wood is strong, making this a good source of timber. The light red-brown wood is heavy, hard, and coarse grained and is used in fine work, including interiors, furniture, cabinets, and floors.

Pin Oak

Quercus palustris H to 82 ft (25 m)

"Pin Oak" is said to refer to persisting dead branches that resemble pins driven into the tree's trunk; another theory is that its branches were used as dowels, or "pins," in barn construction.

KEY FACTS

The tree's interior is dense.

+ **leaves:** Alternate, to 6 in (15 cm), broadest beyond middle; 5–9 lobes, bristle-tipped; very deep sinuses

+ **flowers/fruits:** Catkins typically oak; acorns round, with short beak at tip; caps cover only nut base

+ **range:** Mid-Atlantic and central states and extreme southern Ontario

The fast-growing Pin Oak (Red Oak group; see Northern Red Oak) thrives in wet, soggy floodplains and flatlands and tolerates some flooding, but it also inhabits some well-drained and upland areas. The branches spread in age, and the lower ones often droop toward the ground. The wood is weaker than that of the Red Oak and is used in general building, as posts, and for fuel. The Pin Oak is planted extensively as an ornamental and shade tree (even in parts of Australia and Argentina), and cultivars have been developed. Birds that eat its acorns include mallards, wood ducks, blue jays, turkeys, and woodpeckers.

Willow Oak

Quercus phellos H to 98 ft (30 m)

The Willow Oak is so named because its leaves, narrow and tapering at both base and tip, recall the leaves of willows. It is not, however, related to the willows.

KEY FACTS

The leaves are borne stiffly on the twigs.

+ **leaves:** Alternate, to 5 in (13 cm), tipped by tiny awn; unlobed

+ **flowers/fruits:** Catkins typically oak; broadly rounded acorns, to 0.5 in (1.3 cm); caps cover a third of thin, saucer-like nut

+ **range:** Atlantic and Gulf Coastal Plains, Mississippi Valley

The Willow Oak (Red Oak group; see Northern Red Oak) inhabits bottomlands and drier areas and is often planted in landscapes and on streets. It consistently has good acorn crops and is important to wildlife. While acorns feed larger animals, the trees host thousands of insect species, few of them pests. For many butterflies and moths, an egg-laying stopover on an oak (for some, a particular species) is required. The larvae need oak leaves to eat, or they cannot metamorphose into adults. Leafhoppers, wasps, and beetles also use oaks, and all these insects provide food for birds and other animals.

Northern Red Oak

Quercus rubra H to 98 ft (30 m)

One of our top timber trees, the Northern Red Oak produces hard, strong, coarse-grained wood, often used in interiors. It is widely planted in Europe for such purposes.

KEY FACTS

The dark gray-brown bark is ridged.

+ **leaves:** Alternate, 4–9 in (10–23 cm), often widest beyond middle; 7–9 lobes; sinuses rounded halfway to midrib

+ **flowers/fruits:** Catkins typically oak; acorns to 1 in (2.5 cm); caps cover a quarter of saucerlike nut

+ **range:** Extreme southeastern Canada, eastern U.S.

The attractive and moderately fast-growing Northern Red Oak is the banner tree of the Red Oak group, which also includes Black Oaks. These oaks' acorns mature in two years, on the previous year's branches. The husk is thin, clinging, and papery, and woolly on the inside. The acorn meat is typically very bitter due to the high levels of tannins, which had to be leached out before native peoples could use them as food. The leaf lobes are often toothed and bear bristles at their tip. The bark is dark, and the texture varies. Species of the Red Oak group will hybridize with one another.

Shumard Oak

Quercus shumardii H to 115 ft (35 m)

A large, stately southern oak, the Shumard Oak is long lived, the crown opening with age, and the trunk buttressed in older trees.

KEY FACTS

This deep-rooted oak is often planted near paved areas.

+ **leaves:** Alternate, to 8 in (20 cm), broad; 5–9 paired lobes; tips toothed, bristled; sinuses deep

+ **flowers/fruits:** Catkins typically oak; acorns broadly ovate, to 1 in (2.5 cm); caps cover a third of nut

+ **range:** Eastern, central United States, southwestern Ontario

Acorns of the Shumard Oak (Red Oak group) feed many wildlife species. Humans, too, ate acorns, and there is now renewed interest in the nuts' dietary potential. But acorns, especially in the Red Oak group, contain bitter tannins, which make unprocessed acorns inedible. The Menominee in Wisconsin removed the tannins in a process that was probably not unique to them. They toasted the acorns and removed the husks. They then boiled the acorns in water, which was discarded. They boiled them again, adding alkaline wood ash to accelerate leaching. A final boil removed the ash. The processed acorns were dried and ground into meal or stored.

Post Oak

Quercus stellata H to 65 ft (20 m)

Not a very important timber species, the Post Oak does produce lumber, sold as White Oak, that is used in railroad ties and flooring and, of course, for fence posts.

KEY FACTS

The trunk is gray to light red-brown, becoming ridged.

+ **leaves:** Alternate, oblong, to 6 in (15 cm); 5 lobes, the center 2 largest, lending a cross-like shape

+ **flowers/fruits:** Catkins typically oak; acorns to 1 in (2.5 cm); caps cover half of nut, not fringed

+ **range:** Eastern and central U.S.

Identification of the Post Oak (White Oak group) is difficult because it varies in plant habit, leaf shape, and bark; aberrant populations have been incorrectly designated as a separate species. A character sometimes deemed certain for identification is its cross-shaped leaf, created by two main lobes, which are square and large and project opposite each other from the leaf axis. But other species can have the same arrangement. A better way, though a bit technical, involves use of a hand lens to examine the twigs and the underside of the leaves for stellate hairs, which resemble tufts branching in a starlike (stellate) pattern.

Black Oak

Quercus velutina H to 82 ft (25 m)

The bark of the Black Oak is so rich in tannins that their extraction was once a commercial pursuit. Tannins, used to tan leather, are now synthesized.

KEY FACTS

The bark is gray-black, in vertical plates.

+ **leaves:** Alternate, to 10 in (25 cm); 5–9 lobes, pointed, bristle-tipped; sinuses deep, U-shaped

+ **flowers/fruits:** Catkins typically oak; acorns 0.5 to 0.75 in (1–2 cm); caps cover half of nut

+ **range:** Extreme southwestern Ontario, eastern and central U.S.

The Black Oak (Red Oak group) is common in dry, upland deciduous forests, as well as in savannas, where the eastern forests cede to the prairie. Oak flowers are similar among species, unisexual, and borne on separate catkins, which appear on the same tree before or with the leaves. They usually go unnoticed until they fall onto sidewalks and cars. The male catkins are yellow to greenish, long, lax, and pendulous, bearing many flowers that make pollen. The stiffer, smaller female catkins bear one to several cupules, each of which contains a flower and will become an acorn cap. Oaks, like most trees bearing catkins, are wind pollinated.

Live Oak
Quercus virginiana H to 115 ft (35 m)

The young U.S. Navy bought extensive Live Oak stands for the use of its shipbuilders. The strong, curved limbs were ideal for fashioning the ships' ribs and other supports.

KEY FACTS

The trunk becomes buttressed with age.

+ leaves: Nearly evergreen, alternate, to 5 in (12 cm); oblong, widest beyond middle, stiff, waxy; sometimes toothed, sometimes curled under

+ flowers/fruits: Catkins typically oak; acorns long, slim; caps scaly, deep

+ range: Coastal Plain from Virginia to Texas

The long-lived and fast-growing Live Oak (White Oak group) has been described, without exaggeration, as majestic, noble, and picturesque. The tree is wide spreading with an extensive, rounded crown. Its massive branches can sprawl so far from the trunk that they bend to touch the ground, sometimes even curving up again. The dark gray bark recalls an alligator's skin. Spanish moss, which often graces its limbs, completes the picture of this icon of the Deep South. The Live Oak can also be shrubby, and it sprouts vigorously from stumps. Native peoples extracted an oil similar to olive oil from its s im acorns.

Ocotillo/Coach Whip
Fouquieria splendens H 25 ft (7.6 m)

The Ocotillo is unique and unmistakable. As many as 100 unbranching, whiplike canes grow from the root crown, all angling out slightly for a narrow, vaselike outline.

KEY FACTS

Spiny canes arise from the root crown.

+ leaves: Alternate, 2 in (5 cm), oval, fleshy, in bunches

+ flowers/fruits: Flowers perfect, scarlet, tubular, clustered at stem tips; fruit capsules with winged seeds

+ range: Southern tip of Nevada through Mojave and Sonoran deserts from California to Arizona

The Ocotillo is called drought deciduous, but it is fairer to say that it leafs out after a rain several times a year. When leafless, the tortuous, gray-green canes handle the photosynthesis. The shrub is even more striking when flowering, the tips of the stems bearing bright red flowers that attract hummingbirds and bees, which pollinate them. *Ocotillo* means "little ocoto," the ocoto being the Montezuma Pine, *Pinus montezumae*, of Mexico and Central America. But the plants don't look alike; they are both just extremely resinous. The stems of the Ocotillo are sometimes removed at the root, and then sunk n the ground to make a living fence.

Witch Hazel

Hamamelis virginiana H to 20 ft (6 m)

Cutting a Y-shaped twig from a Witch Hazel is the first step in making a dowsing rod (several other species are also favored), which "water witches" use to locate, or dowse, water.

KEY FACTS

The Witch Hazel flowers in the fall.

+ **leaves:** Alternate; somewhat scalloped; base rounded to wedge shaped; yellow in fall

+ **flowers/fruits:** Flowers perfect, small but showy, 4 petals, narrow, curly, spreading ribbonlike; fruit capsules 2-beaked, becoming woody

+ **range:** Eastern U.S., Nova Scotia

The Witch Hazel is a shrub or small tree, common in the understory of many eastern woods. The biggest specimens are seen in the mountains of the Carolinas. It has upright branches and an irregular crown, often leaning strongly. Its wood and fruits are of minor value, but it has been used medicinally for centuries. The native peoples boiled many parts of the tree, then used the astringent liquid to treat inflammations. Settlers followed suit, and commercial production eventually began and continues today. Witch Hazel is used as an aftershave lotion, as an eyewash, to treat insect bites, and to soothe hemorrhoids, and may act as a UV protectant.

Sweetgum

Liquidambar styraciflua H to 150 ft (47 m)

Many people may decry the spiny gumballs the Sweetgum produces, but it is widely cultivated for its attractively shaped leaves and its brilliant red, yellow, and orange fall foliage.

KEY FACTS

Maple-like, but the leaves are alternate (opposite in maples) and star shaped.

+ **leaves:** 5 lobes, pointed, deep, wide

+ **flowers/fruits:** Flowers in unisexual clusters on same tree; fruits spiky balls, to 1.5 in (3.8 cm)

+ **range:** Southeast, to Mississippi and Ohio valleys and on the coast to New York

A medium-size to large tree, the Sweetgum is one of the most common hardwoods in the eastern United States, and one of the most commercially important, second only to the oaks in production. The wood, marketed as satin walnut, takes a finish well and is used in furniture, plywood, cigar boxes, barrels, and pulp. Many cultivars have been developed. The genus name refers to a juice that oozes from the bark, a balsam similar to turpentine, which is extracted from the bark by boiling and made into resin. The juice also will dry on the tree, and dried bits were picked off and used as chewing gum by Native Americans.

Black Walnut

Juglans nigra H to 125 ft (38 m)

The Black Walnut is allelopathic, that is, most plants cannot grow beneath it. It produces a compound, juglone, that inhibits metabolism in many plant species.

KEY FACTS

The trunk is often straight for half its height.

+ **leaves:** Alternate, compound, to 2 ft (0.6 m); 15–23 leaflets.

+ **flowers/fruits:** Male flowers in catkins, female on new growth, on same tree; fruits single or paired; husk thick, green; nut black, grooved

+ **range:** Eastern U.S., southern Ontario

The Black Walnut was once so common that furniture was solid walnut, a beautifully grained, brownish wood. The large walnuts are gone, and the wood is used in gun stocks and some furniture and cabinets. Most walnut furniture is only walnut veneer. Walnuts for ice cream, candy, and cakes support a commercial harvest. These nuts have a stronger, richer flavor than English walnuts, but extracting the meat is real work. The hull is thick and hard, and the interior is a catacomb from which bits must be removed with a nut pick. Anyone trying to remove a nut from its husk will soon have brown fingers: No surprise that dyes and inks are made from the husks.

Bitternut Hickory

Carya cordiformis H to 115 ft (35 m)

The Bitternut Hickory is so named because the nuts are too bitter to eat—even for squirrels. But colonists extracted their oil to burn in lamps, and the husks yield a yellow-brown dye.

KEY FACTS

Leaf buds are yellow.

+ **leaves:** Alternate, compound, to 8 in (20 cm); 7–11 leaflets

+ **flowers/fruits:** Male flowers in catkins, female 1 or 2 on spikes, on same tree; fruits clustered, husk thin, opening along winged sutures; nut shell bony, tip sharp

+ **range:** Eastern U.S., southern Ontario and Quebec

The Bitternut Hickory is one of the largest hickories and the most abundant, found in a range of soil conditions and temperatures. Despite its bitter nuts, it is nonetheless important to our palates, though indirectly. Pecans are grafted onto the rootstock of this wide-ranging hickory so that pecans can be produced north of the Pecan tree's usual, rather southerly range. The Bitternut's dense, strong wood is used in furniture, tool handles, ladders, in woodworking, pulp, and as fuel. Native peoples made bows from the wood and used it in making birch-bark canoes. As with other hickory woods, it is used for smoking meat and to make charcoal.

Pignut Hickory

Carya glabra H 82–132 ft (25–40 m)

The Pignut Hickory, it is thought, got its common name from colonists who noticed that their free-range pigs, rooting in the woods for tree nuts, were especially fond of those of this species.

KEY FACTS

The bark is ridged but not shaggy.

+ **leaves:** Alternate, compound, to 10 in (25 cm), smooth; 5–7 leaflets

+ **flowers/fruits:** Male flowers in catkins, female few, in clusters, on same tree; fruits pear-shaped, husk thin, splitting partly; nut unwinged; unribbed

+ **range:** Eastern U.S., southwestern Ontario

A slow-growing, short-branched, medium-size tree, the Pignut Hickory is the most common hickory in the Appalachians and an important species in oak–hickory forests of the East. The nuts make up as much as 25 percent of the diet of squirrels and are important food items for chipmunks, raccoons, crows, and wood ducks. But for the record, the fruit of a hickory is not a nut but a *drupe:* Fleshy tissue surrounds a pit, which contains the seed. In the hickories, the fleshy tissue is the husk, the pit is the nut, and the seed is the nut meat. A peach is constructed in much the same way.

Pecan

Carya illinoinensis H 70–100 ft (21–30 m)

The word "pecan" is descended from an Algonquian word *pakan,* which meant a nut so hard that it had to be cracked with a rock. The word is similar in other native languages.

KEY FACTS

Huge limbs support an oval crown.

+ **leaves:** Alternate, compound, 12–20 in (30–51 cm); 9–15 leaflets

+ **flowers/fruits:** Male flowers in catkins, female 1 to few on spike, on same tree; clustered; fruits thin-husked; nut to 2 in (5 cm); shell thin, brown, often with black

+ **range:** Centered in lower Mississippi Valley.

It is hard to imagine anyone in the world who is not acquainted with the pecan, the soul of pralines, used in ice creams, salads, and scones, and more recently encrusting everything from salmon to seitan. Pecan trees, the largest hickory, are grown far and wide. Spanish and Portuguese explorers took the first trees to Europe in the 16th century. Nevertheless, the United States still supplies about 80 percent of the world's demand. Total U.S. production reached 294 million pounds (133 million kg) (unshelled) in 2010. The long-lived Pecan is also appreciated for its shade, and many cultivars are grown extensively in North America and abroad.

Big Shellbark Hickory/King Nut Hickory
Carya laciniosa H 60–80 ft (18–24 m)

The Big Shellbark Hickory is also called the King Nut Hickory, and deservedly so, because its fruit can be as big as 2.5 inches (6.5 cm) across—just slightly smaller than a tennis ball.

KEY FACTS

The twigs are orange.

+ **leaves:** Alternate, compound, to 24 in (61 cm); 7 leaflets

+ **flowers/fruits:** Male flowers in catkins, female in clusters, on same tree; fruit to 2.5 in (6.5 cm); husk splitting to base; nut large, flattened; shell hard, thick

+ **range:** Midwest, Ohio and upper Mississippi Valleys

The Big Shellbark Hickory is very slow growing, and its fruits are so large that dispersal is hampered; thus, it has never been common. The medium to large tree grows mostly in mixed stands, especially on floodplains and in river swamps. Its nuts are prized for their flavor. The species name, *laciniosa*, from the Latin for "flap," means "shaggy," and the Big Shellbark is similar to the Shagbark Hickory. Key differences distinguish them. The Big Shellbark's bark curls up in strips, though less than the Shagbark's; its leaves have 7 (vs. 5) leaflets, the leaflets have no hairs at the tip of the teeth, the husk of the fruit is thicker, and it prefers wetter habitats.

Shagbark Hickory
Carya ovata H to 150 ft (46 m)

Hickory bark syrup is made from shed Shagbark bark by oven-toasting the bark, then boiling, reducing the liquid's volume, and adding sugar. It is used on meats, in drinks, and on pancakes.

KEY FACTS

Bark on older trees is shaggy.

+ **leaves:** Alternate, compound, to 14 in (36 cm); 5 leaflets

+ **flowers/fruits:** Male flowers in catkins, female on short spike, on same tree; fruit to 1.6 in (4 cm); nut usually ridged, shell thick, meat sweet

+ **range:** Most of eastern U.S., southern Ontario and Quebec

The Shagbark Hickory is similar to the big shellbark but usually has 5 (vs. 7) leaflets, and the leaves have hairs at the tooth tips. The bark is in long, loose plates that curve out, shaggier than in the Big Shellbark. Shagbark nuts are the best of the hickory nuts and were a staple in the diet of native peoples in the tree's range. A food prepared from Shagbark nuts was called *pawcohiccora* in an Algonquian language and altered by settlers to "hickory." Pawcohiccora was a sort of nut milk made from nut meal added to boiling water, and the resulting rich, oily gruel was collected for use in breads and stews.

Mockernut Hickory

Carya tomentosa H to 100 ft (30 m)

The name "Mockernut Hickory" suggests that the nut directs ridicule at the would-be nut eater who, having finally cracked its extremely thick shell, is rewarded by just a small kernel of meat.

The slow-growing, tall, straight Mockernut is the most common hickory in the South, and it is a key species in oak–hickory forests of the East. Its wood is among the best of the hickories, used to make tool handles, wood splints, furniture, and charcoal. The species label *tomentosa* refers to the hairiness of the underside of the spicy-smelling leaves. The nut meat, though there is not much of it, is sweet, and competition for the nuts from birds and wild mammals is strong. Many cavity dwellers inhabit the tree, including woodpeckers, Chickadees, and Black Rat Snakes.

Sassafras

Sassafras albidum H 32–60 ft (10–18 m)

Safrole, which makes Sassafras aromatic, is used to manufacture the drug ecstasy. Sassafras is not a good source, but an Asian relative is, and safrole is illegal in Canada and the U.S.

Sassafras is planted for its fragrance and for its foliage, which turns yellow to red in the fall. All parts are aromatic. The bark of the root was used to make a tea and to perfume soaps; the shoots were used to make root beer. But the aromatic element, safrole, turned out to be carcinogenic, and these products are now artificially flavored. The Choctaw, native from Florida to Louisiana, have long dried the mucilaginous leaves and ground them to a powder for flavoring and thickening dishes. Named *filé* by French-speaking settlers, the powder is used in Creole and Cajun kitchens to thicken gumbo, soups, and gravy. The leaves are safrole-free.

Northern Spicebush

Lindera benzoin H 6–12 ft (1.8–3.7 m)

The species label *benzoin* refers to the spicy, fragrant leaves, recalling benzoin, a resin obtained from unrelated plants and used in incense (devoid of the compound also called benzoin).

KEY FACTS

The shrub often has several stems.

+ **leaves:** Alternate, simple, entire, 3–6 in (7.6–15 cm), oval, aromatic

+ **flowers/fruits:** Male, female flowers on different trees, yellow, showy in clusters on bare tree; fruit a red, spicy berrylike drupe

+ **range:** Texas, Oklahoma, eastern U.S., south Ontario

The flowers of the Spicebush are among the first to appear in the spring before the trees leaf out. Though small, the flowers are spectacular if there are enough of this small, graceful tree in the understory. The common name derives from its fruit, which, dried and ground, is used like allspice. The fruit, leaves, and twigs are made into an herbal tea. The Spicebush is the main host (the related Sassafras is another) of the Spicebush Swallowtail, a butterfly that lays its eggs on the plant, in whose foliage the eventual caterpillars will undergo metamorphosis. Spicebush is also a host of the Eastern Tiger Swallowtail.

California Laurel/Oregon Myrtle

Umbellularia californica H 23–100 ft (7–30 m)

When the bank in North Bend, Oregon, closed in 1933, coins worth up to $10 were made of myrtlewood, letting commerce continue. The coins are still accepted but most are in collections.

KEY FACTS

Plant is shrubby in exposed situations.

+ **leaves:** Evergreen, alternate, to 4 in (10 cm), lance shaped

+ **flowers/fruits:** Flowers perfect, to 0.6 in (1.4 cm), yellow, in umbrella-like clusters (thus the genus name); fruit a berry, to 1 in (2.5 cm), ripening to purplish

+ **range:** Southern California to Oregon

The California Laurel is an attractive, medium-size tree with a short trunk that soon divides. This is the only species in the genus, but it has many common names, including California Bay. Its peppery leaves (it is also called Pepperwood) are used as the familiar bay leaf, but they have a stronger flavor. For native tribes, this plant treated many maladies, including headaches, though it can also cause headaches and so is called Headache Tree. The olive-like fruit, called the California Bay nut, is like a tiny avocado (the plants are close relatives). It has a relatively large pit surrounded by oily flesh, both of which are eaten.

Tuliptree/Tulip Poplar/Yellow Poplar

Liriodendron tulipifera H 98–165 ft (30–50 m)

Liriodendron derives from the Greek words for "lily" and "tree"; *tulipifera* means "tulip bearing." The Tuliptree is in the Magnolia family, as study of the flower suggests. It is not a poplar.

KEY FACTS

The twigs turn upward.

+ leaves: Alternate, uniquely 4-lobed; yellow in fall

+ flowers/fruits: Flowers perfect, to 3 in (7.6 cm), at branch tips, tulip-like; 3 green sepals; 6 upright petals, yellow-green, orange at base; fruit a cone-like cluster of winged seeds on a stalk

+ range: Eastern U.S.

The majestic Tuliptree is one of our biggest hardwoods, with a single, straight trunk, which is often half bare. Native peoples used the trunks to make dugout canoes. It is an important tree in eastern hardwood forests, especially at lower elevations. The tree is valued for its light, straight-grained wood, used in interior detail, furniture, general construction, plywood, and pulp. Many animals depend on the tree for seeds, browse, sap, or cover, but bees especially avail themselves of this important honey plant: The flowers from one 20-year-old Tuliptree can provide them with nectar sufficient for 4 pounds (1.8 kg) of honey.

Cucumber-tree

Magnolia acuminata H 60–95 ft (20–30 m)

The name "Cucumber-tree" describes the unripe fruit, a composite of many follicles with their developing seeds. When green, this "cone" is shaped much like a cucumber.

KEY FACTS

The bark is furrowed.

+ leaves: Alternate, simple, entire, 6–10 in (15–25 cm), oval, pointed; gold to maroon in fall

+ flowers/fruits: Flowers perfect, to 3 in (7.6 cm) across; 9 yellow tepals, outer 3 bent back; fruit to 3 in (7.6 cm), cone-like; seeds red

+ range: Louisiana to New York, southern Ontario

The Cucumber-tree has a wide distribution, but it is never abundant, usually scattered in the eastern oak–hickory forests. It is most often found in the mountains, and the largest specimens grow in the southern Appalachians. The hardiest magnolia, it is the only one native to Canada, found, but rare, in Ontario. Its slightly fragrant flower is smaller and not as showy as that of other magnolias, and it is the only *Magnolia* species that has yellow sepals. A number of cultivars have been developed. The wood is similar to Tuliptree wood and is sold alongside it. It is used in crates, cabinets, and some paneling.

Southern Magnolia

Magnolia grandiflora H 65–95 ft (20–29 m)

The Southern Magnolia had traveled to England by 1726, via Mark Catesby, an English naturalist who published and illustrated the first book on the flora and fauna of North America.

KEY FACTS

The bark is smooth, gray-brown.

+ **leaves:** Evergreen, alternate, simple, entire, 3–8 in (7.6–20 cm), oval, shiny, stiff, rusty below

+ **flowers/fruits:** Flowers perfect, to 8 in (20 cm) across, showy, fragrant; 6–12 broad, creamy white tepals; fruit compound, dry; seeds red

+ **range:** Coastal Plain, North Carolina to Texas

This small to medium-size tree of the lowlands is botanically and culturally identified with the South, growing largest in moist and well-drained soils and popular as an ornamental far from its native range. More than 100 cultivars have been developed. Other than the seed, the Southern Magnolia is not used much by wildlife, and the wood is of limited importance, but all parts of the tree have yielded compounds of potential pharmaceutical value. The genus name (and that of the family, Magnoliaceae) was assigned by botanist Carl Linnaeus in honor of French botanist Pierre Magnol of Montpellier, who conceived of the family as an organizing unit in plant taxonomy.

Sweetbay

Magnolia virginiana H to 90 ft (27 m)

Sweetbay is often called "tardily deciduous," but it varies from deciduous in its more northern extent to evergreen in its southern. The tree also grows larger in the South.

KEY FACTS

The bark is smooth.

+ **leaves:** Semi-evergreen, alternate, simple, entire, to 6 in (15 cm), blunt-tipped, silvery below

+ **flowers/fruits:** Flowers perfect, showy, fragrant, to 5.5 in (14 cm) across; 9–12 creamy-white tepals; fruit and seeds red

+ **range:** Coastal Plain and Piedmont, eastern U.S.

A graceful, small to medium-size tree, the Sweetbay prefers low elevations and wet, sandy, acid soils. Its foliage and twigs are aromatic, and its flower has a lemony scent. When a breeze catches the foliage, briefly flashing the leaves' silvery white undersides, the tree ripples with light. Primarily southeastern, it ranges north to Massachusetts, where the discovery in 1806 of a Sweetbay swamp in Gloucester caused a stir among botanists. Jacob Bigelow made the find public in 1814 in his *Plants of Boston*, calling it "our only species of this superb genus." The stand is now the focus of an annual "Save the Sweetbay" event in Gloucester.

Chinaberry Tree
Melia azedarach H to 49 ft (15 m)

The fruit of the Chinaberry contains very hard, reddish, five-grooved seeds that were widely used to make rosaries and other beaded objects before the advent of plastics.

KEY FACTS

The clustered fruits are persistent and toxic.

+ **leaves:** Alternate, once or twice compound, to 2.2 in (6 cm); toothed or lobed, smooth leaflets

+ **flowers/fruits:** Perfect, small, fragrant, in loose clusters; 5 petals, lavender, with purple tubular corona; fruit a yellow-brown drupe

+ **range:** Nonnative; widespread

Native to the Himalaya and eastern Asia, the Chinaberry (in the Mahogany family) was brought to South Carolina in the late 1700s as an ornamental and a shade tree. It escaped and is now invasive, especially common in thickets in old fields and on disturbed sites. The Chinaberry is ideally suited for invasion. It is fast growing and short lived, and produces huge numbers of seeds, which are dispersed by birds. Its seedpods persist on the tree into winter; it forms colonies by sprouting; and it is drought tolerant and a soil generalist with essentially no pests. The weak, soft wood has been used in furniture and cabinetry, as well as for firewood.

Osage Orange
Maclura pomifera H 40–60 ft (12–18 m)

A female tree will bear normal-looking fruit even if there is no male nearby to provide pollen; the fruit will not, however, contain seeds. The tree is not an orange but is in the Mulberry family.

KEY FACTS

The bark has an orange cast.

+ **leaves:** Alternate, simple, to 5 in (13 cm), oval, pointed; leaf axils with 1-in (2.5-cm) spine

+ **flowers/fruits:** Male, female flowers on separate trees; fruits orange-like, bumpy, to 5 in (10–13 cm) across; skin with milky sap

+ **range:** South-central U.S.

The Osage Orange is a medium-size tree of rich bottomlands but tolerates a range of conditions. Though of a rather narrow natural range, it has been planted and has naturalized through most of the eastern United States and in Ontario. It is planted in hedges and, because of its spines, was used in cattle "fences" before the advent of barbed wire. The bright orange wood is not commercially important, though it is good firewood. Squirrels may excavate the fruit to eat its seeds, but the tree is not important to wildlife. A large, extinct animal, such as the Mammoth or a horselike mammal, may have been the natural disperser of Osage Orange seeds.

White Mulberry
Morus alba H 50 ft (15 m)

The fastest known movement in plants is that of a mechanism in the White Mulberry's male flowers that ejects pollen into the air. The structure triggers at more than half the speed of sound.

KEY FACTS

The leaves are usually smooth above.

+ **leaves:** Alternate, oval, to 4 in (10 cm), serrate, often deeply lobed

+ **flowers/fruits:** Male and female catkins on same or different trees; fruit aggregates of drupes, ¾ in (2 cm), ovoid, red turning white, pink, or purple-black

+ **range:** Eastern U.S. and Ontario

The White Mulberry has been cultivated for several thousand years in China as a host for the silkworm. This spreading tree was introduced to North America for the same purpose, but the attempt failed—by no fault of the trees. They are naturalized across the eastern United States and into Ontario, often seen on roadsides and in abandoned fields. The invasive nonnative also crosses with the native Red Mulberry. Distinguishing the White Mulberry from the Red can be tricky, as the species overlap in leaf shape and hairiness. The White has leaves that are most often glossy above, often curling up.

Red Mulberry
Morus rubra H to 66 ft (20 m)

Choctaw women wore cloaks of cloth woven of fiber cord processed from the inner bark of the young shoots of Red Mulberry. Other garments were likewise made of the cloth or of the cord.

KEY FACTS

The leaves are often rough above.

+ **leaves:** Alternate, to 5.5 in (14 cm), entire or lobed, rounded and long-pointed

+ **flowers/fruits:** Male and female catkins on separate trees; fruits aggregates of drupes, 1.2 in (3 cm), cylindrical, red to purple to almost black

+ **range:** Eastern U.S., southern Ontario

The Red Mulberry is our most common native mulberry. It is also often planted in yards, but in summer the ground under a female tree will be covered with overripe berries, so some people buy only male trees. Those people will not enjoy the berries, which for centuries have been consumed raw and in beverages, preserves, cakes, and dumplings. The fruits are usually used immediately, and are seldom seen for sale because they do not keep for long. Genetic pollution of *Morus rubra* by the nonnative *M. alba*, the White Mulberry, is recognized as a problem. In Canada, the problem is considered serious enough that the Red Mulberry is listed as endangered.

California Wax Myrtle/Pacific Bayberry

Morella californica H to 26 ft (8 m)

Pacific Bayberry fruits are a source of wax and fragrance that have been used to make candles and soap. Dyes were also made from the berries.

KEY FACTS

The leaves are sticky.

+ **leaves:** Evergreen, alternate, to 4 in (10 cm), narrow, broader, sparsely toothed toward tip

+ **flowers/fruits:** Male, female, perfect flowers in catkins on same plant, in various combinations, inconspicuous; fruit a purple drupe with white wax coating

+ **range:** Lower British Columbia south

The aromatic Pacific Bayberry is a shrub or small tree of the coast. It is often planted as an ornamental, for its evergreen foliage, to attract wildlife, to serve as a hedge or screen, or to retard erosion. The roots contain bacteria that fix nitrogen from the air, making it usable as a plant nutrient. In 2007, a leaf blight, an infection by a fungus-like organism that is known also to damage apples and pears, was noted on the Pacific Bayberry in Oregon. The lower plant is affected first, but much of the plant is eventually defoliated. The leaves may regrow, but repeated defoliations could kill affected branches.

Wax Myrtle/Southern Bayberry

Morella cerifera H to 40 ft (12 m)

One of the birds that take cover in and eat the berries of the Wax Myrtle is *Setophaga coronata coronata*—the Myrtle Warbler, named for its association with the shrub.

KEY FACTS

The gray bark can be almost white.

+ **leaves:** Evergreen, alternate, to 4 in (10 cm), narrow, slightly toothed, broader near tip

+ **flowers/fruits:** Male and female catkins on separate plants; drupe green with light-blue wax coating, borne on female plants

+ **range:** Eastern and Gulf Coastal Plains

A large, evergreen shrub or small tree, the Wax Myrtle is a coastal plant of sandy to moist soils, often forming thickets. It is widely planted, and cultivars have been developed. *Cerifera* means "wax bearing," and all bayberries have been used to make candles. The berries are boiled and the melted wax skimmed from the surface of the water and molded or dipped to make fragrant candles. *Bayberry* means *"berryberry"*: In this usage, the word "bay" comes from the French *baie*, or "berry." But the word also hints at the leaf's aroma, invoking the bay leaf, a culinary herb for which aromatic Wax Myrtle leaves have long been substituted by native peoples and Europeans to flavor seafood and other dishes.

Tasmanian Bluegum
Eucalyptus globulus H 263 ft (80 m)

An ingredient of cough drops and chest ointments, aromatic eucalyptus oil from Tasmanian Bluegum also has been used to treat arthritis and in mouthwashes and insect repellents.

KEY FACTS

The bark shreds in long strips.

+ **leaves:** Evergreen, alternate, to 7.1 in (18 cm), narrow, tapering to a long point, hanging

+ **flowers/fruits:** Flowers single, in upper leaf axils; 4 petals, fused into a cap that falls from the bud; fruit a woody capsule

+ **range:** Nonnative in far West

Thousands of acres of Bluegum were planted for timber in California in the late 1800s. A flop wood-wise but still grown as a windbreak or ornamental, it escaped and now thrives north into British Columbia. But Bluegum is most invasive in and poses the greatest threat to California. Its flammable oils and old bark that piles up beneath it make it incendiary. Of the energy behind the 1991 Oakland Hills firestorm, which killed 25 and burned 3,000 homes, 70 percent came from eucalyptus. (More than 100 other "eucs," including the River Redgum, *E. camaldulensis,* and the Forest Redgum, *E. tereticornis,* also grow in California.)

Water Tupelo
Nyssa aquatica H 115 ft (35 m)

As its common and scientific names suggest, the Water Tupelo often grows in wet places, including Baldcypress and other swamps and periodically flooded areas.

KEY FACTS

Its feet are wet to flooded.

+ **leaves:** Alternate, simple, to 8.5 in (22 cm), pointed, with few teeth; long stalk

+ **flowers/fruits:** Male flowers in heads, female solitary, some perfect, on same tree; fruit a purple drupe, 1.5 in (4 cm)

+ **range:** Virginia to Texas (except peninsular Florida), Mississippi Valley

The Water Tupelo is a long-lived, medium-size to large tree of the Coastal Plain. If it grows in very wet places, the trunk is swollen and buttressed, sometimes hugely so. Above that (or if not buttressed), the trunk is long, straight, and clear, supporting a narrow, open crown. Many fruits are produced, and the seeds are dispersed mainly by water. The fruits are also eaten by wildlife, including songbirds, turkeys, and groundhogs. The wood is weak and soft, used in paneling, boxes, and crates. In addition to its use in pulp, the wood of the swollen base is a preferred medium of wood carvers. Honey from this and other tupelos is prized.

Blackgum/Black Tupelo

Nyssa sylvatica H 131 ft (40 m)

The trunk of the Black Tupelo is straight, extending to the top of the tree. The leaves are brilliant yellow, red, orange, and purple in the fall. It is cultivated and popular as an ornamental tree.

KEY FACTS

Its feet are dry to moist.

+ **leaves:** Alternate, simple, to 6 in (15 cm), rounded or broadly pointed; stalk short

+ **flowers/fruits:** Male flowers in dense heads, female 2 to several in cluster, some perfect, on same tree; fruit drupe 0.5 in (1.2 cm), usually 3–5, blue

+ **range:** Eastern United States, south Ontario

The Black Tupelo is a widely distributed, medium-size to large tree of the open woods in uplands and well-drained alluvial bottoms, especially in acidic soils. The tree has dense foliage, and its crown is conic to flat. In old trees, the bark is blocky, resembling the skin of an alligator. Its fruits are eaten by many birds and mammals, and bees collect nectar from its flowers to make the popular tupelo honey. In addition to the usual uses as lumber, veneer, and pulp, Black Tupelo is used to make rollers, bowls, blocks, mallets, gun stocks, wheel hubs, and pistol grips.

White Ash

Fraxinus americana H to 82 ft (25 m)

This tree is called the White Ash because the underside of the leaflets, in contrast to the deep green above, are so light green as to appear white.

KEY FACTS

Its twigs are smooth.

+ **leaves:** Opposite, compound, to 12 in (30 cm); 5–9 leaflets, to 6 in (15 cm), white-green below; yellow to red to purple in fall

+ **flowers/fruits:** Unisexual clusters on separate trees; fruit a samara, broadest near tip

+ **range:** Eastern U.S.; southern Ontario to Cape Breton Island

Our most common native ash, the White Ash is a tree of well-drained, rich soils, usually growing in mixed hardwood stands. It is the most valuable ash for timber, going into furniture and paneling. Also, most baseball bats are made from its wood, as are snowshoes, electric guitar bodies, and lobster pots. All *Fraxinus* species are menaced by the Emerald Ash Borer, a metallic green beetle native to Asia that was first seen in North America in 2002, and has been reported in 18 states as of 2013. The borer destroys a tree's transport system, killing half the branches in a year and most of the crown in two. It has killed 60 million ashes so far.

Green Ash

Fraxinus pennsylvanica H to 66 ft (20 m)

Its broad tolerance for different types and acid levels of soils makes the Green Ash a popular and widely planted ornamental and shade tree, as well as a successful street tree.

KEY FACTS

The leaves are golden in fall.

+ leaves: Opposite, compound, to 12 in (30 cm); 7–9 leaflets, to 6 in (15 cm), on narrowly winged stalks

+ flowers/fruits: Unisexual clusters on separate trees; fruit a samara, broadest at or above middle

+ range: Alberta to Cape Breton, to Texas and northern Florida

The Green Ash is the most widely distributed ash in North America, inhabiting wet uplands, floodplains, and stream banks, growing in mixed-hardwood forests or pure stands. It is most common in the Mississippi Valley. Small to medium-size, it has a tall, slender trunk, and its crown is irregular to rounded. It produces a large seed crop that is important to turkeys, quail, squirrels, and other small mammals. Moose and deer browse on the shoots and leaves. Its pale brown hardwood is used for similar purposes as that of the White Ash, and it, too, is affected by the Emerald Ash Borer.

Empress Tree/Princess Tree/Royal Paulownia

Paulownia tomentosa H to 60 ft (18 m)

Paulownia was named for Anna Pavlovna, queen of the Netherlands, daughter of Emperor Paul I of Russia, and granddaughter of Catherine the Great.

KEY FACTS

The dry pods are persistent.

+ leaves: Opposite, to 12 in (30 cm), sometimes 3-lobed, hairy above, densely so below; long-pointed; base heart-shaped

+ flowers/fruits: Perfect, to 2.5 in (6 cm), tubular, purple, fragrant, in pyramidal clusters to 14 in (35 cm); fruit capsules large

+ range: Widely invasive

The Empress Tree, a native of China, has been cultivated in Japan and Europe for several hundred years. It was introduced to North America in 1834 for its appealing clusters of purple flowers. Now naturalized and invasive, and cold hardy as far north as Massachusetts, it is seen planted in gardens and parks, but also in vacant lots where it has seeded itself. Each capsule contains 2,000 seeds; one tree can make 20 million in a crop. For a different vision of how prolific it is, consider that in the 1800s, when porcelain was shipped out of China, instead of excelsior, they packed it in the tiny, fluffy-winged seeds of the Empress Tree.

American Sycamore
Platanus occidentalis H 98–130 ft (30–40 m)

English settlers named the tree because its leaves recalled the Sycamore Maple (*Acer pseudoplatanus*), from England. Leaves are similar, but sycamores and maples are not related.

KEY FACTS

The bark falls in plaques.

+ **leaves:** Alternate, simple, 4–8 in (10–20 cm); 3–5 very shallow lobes

+ **flowers/fruits:** Male and female heads on same tree; fruit a narrow achene, with tuft aiding dispersal, in dense sphere from long stem, to 1.4 in (3.5 cm), eventually shattering

+ **range:** Eastern U.S.

The American Sycamore, the most massive tree of eastern North America, can be identified from far away by its thick single trunk, with its distinctive, irregular patches of green, tan, and white resulting from shedding, older bark. Found among hardwoods of the bottomlands in nature, the tree is grown for shade and is planted on city streets. It is sometimes infected with plane anthracnose, which causes leaf dieback and is host to the Eastern Mistletoe. It was under one of these trees on Wall Street that the New York Stock Exchange began in 1792 with 24 parties signing what became known as the Buttonwood Agreement. The Sycamore is also called Buttonwood.

London Plane Tree
Platanus × acerifolia H 66–115 ft (20–35 m)

Half the planted trees in London are the London Plane Tree. First used in England in the mid-1600s, it became *the* tree in the capital because it could withstand the challenges of urban life.

KEY FACTS

Fruiting balls appear threaded.

+ **leaves:** Alternate, simple, 4–8 in (10–20 cm); 3–7 shallow lobes

+ **flowers/fruits:** Male and female heads dense, on same tree; fruit a narrow achene with tuft aiding dispersal, in dense spheres (mostly 2) on stem, eventually shattering

+ **range:** Hybrid; grown widely in North America

The London Plane Tree is easily confused with the American Sycamore, especially because of their similar patchy, multishaded trunks. The leaves and the fruits distinguish the two. This tree is a hybrid between the American Sycamore and the Oriental Plane Tree (*P. orientalis*). The latter developed resistance to plane anthracnose, a disease from Asia, with which it evolved. Resistance inherited from the Oriental Plane Tree has made the London hybrid popular in eastern North America, where anthracnose has marred many American Sycamores, which have no resistance. But its resistance varies.

California Sycamore
Platanus racemosa H to 115 ft (35 m)

The largest California Sycamores are found in the canyons, though the species also grows on stream banks and in other moist areas in the central and southern areas of the state.

KEY FACTS

The tree often leans.

+ **leaves:** Alternate, simple, to 10 in (25 cm); 3–5 lobes, half as long as leaf

+ **flowers/fruits:** Flowers tiny, male and female heads on same tree; fruit a narrow achene, with tuft aiding dispersal, in dense maroon spheres, 2–7 appear to be strung zigzag on a stem

+ **range:** California

The California Sycamore, native in the foothills and Coast Ranges, is similar to the American Sycamore, but with more strongly lobed leaves. Its bark, too, is shed in plaques, and the trunk is an irregular patchwork of brown, gray, and white, but on older trees, it is furrowed with broad ridges. The species also often has multiple trunks. Native peoples used its larger leaves to wrap baking bread, the inner bark was used medicinally and for food, and the branches were used to build homes. Plane anthracnose can cause leaf dieback, and the tree is a host to the Bigleaf Mistletoe, also known as Sycamore Mistletoe.

California Buckthorn/California Coffeeberry
Rhamnus californica H 20 ft (6 m)

The California Buckthorn is often called Coffeeberry for its fruit, which progresses from green, to red, to black, much as the coffee plant's fruit. It does not, however, provide a coffee substitute.

KEY FACTS

This species often has multiple trunks.

+ **leaves:** Evergreen or nearly so, alternate, simple, to 3 in (7.6 cm), lustrous

+ **flowers/fruits:** Flowers perfect, small, inconspicuous, star-shaped, greenish, in small clusters in leaf axils; fruit is a berry, 0.5 in (1.3 cm)

+ **range:** California

The California Buckthorn grows as a shrub or small tree on coastal chaparral, on hillsides, and in ravines. It also occurs in parts of Oregon, Arizona, and New Mexico. The plant flowers and sets seed profusely and so is valuable in erosion control. Horticulturally, it is planted for its colorful berries and their contrast with the foliage. The berries are eaten by goats, deer, black bear, and livestock, as well as birds. They are sweet but laxative; native peoples in the Buckthorn's range used it medicinally for that reason, as well as to soothe toothaches by holding a warmed root against the gum and to treat various skin ailments.

Red Mangrove
Rhizophora mangle H to 82 ft (25 m)

Red Mangrove seeds germinate when still attached to the tree. Seedlings grow to 12 in (30 cm), resembling hanging pods. They fall and float until reaching a suitable substrate, then take root.

KEY FACTS

Aerial roots arise from the stems.

+ **leaves:** Evergreen, opposite, simple, to 6 in (15 cm), leathery, shiny

+ **flowers/fruits:** Perfect, to 1 in (2.5 cm) across, pale yellow, 2 or 3 at leaf axils; fruit egg-shaped, seeds germinating and seedlings elongating before falling

+ **range:** Southernmost Florida

Some 70 trees and shrubs in more than 20 families are called mangrove, uniquely adapted to the tropical intertidal zone. Only one, the Rhizophoraceae, is the Mangrove family, and only one of its species is native to North America, the Red Mangrove. Its mangrove look is due to its odd seedlings and to its arching aerial roots, which admit oxygen, often lacking in the stagnant sediments. The White Mangrove, *Laguncularia racemosa* (family Combretaceae), inhabiting higher ground, is also native to Florida. It, too, has aerial roots and precocious seedlings. The United Nations estimates that 20 percent of mangrove ecosystems were lost worldwide from 1980 to 2010.

Saskatoon/Pacific Serviceberry
Amelanchier alnifolia H to 40 ft (12 m)

The city of Saskatoon, Saskatchewan, was named after the berry (and not vice versa), in 1882. The word "Saskatoon" comes from a Cree phrase meaning "early berry."

KEY FACTS

Fall leaves are orange to red.

+ **leaves:** Alternate, to 2.4 in (6 cm), oval to round, toothed at tip

+ **flowers/fruits:** Flowers perfect, to 1.2 in (3 cm), fragrant, 3–20 at branch ends; 5 white petals; fruit a blue, sweet berry-like pome, to 0.4 in (1.1 cm).

+ **range:** Alaska to north California, to Wisconsin, Ontario

The word "Saskatoon" refers to both the berry and the plant. This shrub or small tree often forms thickets by sprouting from its extensive root crown and rhizome network, which makes it useful in erosion control. Twigs, foliage, and bark make the plant important to large game species, such as deer and moose, especially in winter, and the fruits provide food for many birds and mammals, including the black bear. The fruit looks like a blueberry. Native peoples have long eaten them, including in fermented form, and added to pemmican for its flavor and as a preservative. It is raised commercially, and growers market it as a superfruit, like blueberries and acai berries.

Hawthorn

Crataegus species H 16–49 ft (5–15 m)

People born on Manitoulin Island, Ontario, in Lake Huron, call themselves haweaters, because the island is home to many Hawthorns.

KEY FACTS

Fruits recall tiny apples.

+ **leaves:** Alternate, to 4 in (10 cm), entire, toothed, or lobed, even in same species

+ **flowers/fruits:** Flowers perfect, to 1 in (2.5 cm) across, single or clustered; 5 petals, white to red; fruit a pome, to 1 in (2.5 cm), yellow, red, or nearly black

+ **range:** All but northernmost Canadian provinces

While a tree may rather easily be identified as a Hawthorn, determining its species is often difficult. Variation within a species can sometimes be greater than that between two species, and they have a complicated breeding biology. As a result, more than 1,700 names have been published worldwide for what are now believed to represent around 200 species. Here, we are considering the genus only. The thorns are modified branches that can be 3 inches (7.6 cm) long. Densely branched, Hawthorns provide excellent cover and nesting sites for songbirds, and many, especially waxwings and thrushes, eat the fruits, called "haws."

American Plum

Prunus americana H to 36 ft (11 m)

The American Plum was likely cultivated in the Great Plains before European contact. Its branches figure in the Cheyenne Sun Dance, and the Navajo made a reddish purple dye with the roots.

KEY FACTS

The tree is spiny.

+ **leaves:** Alternate, to 4 in (10 cm), pointed, tapered at base, finely toothed

+ **flowers/fruits:** Flowers perfect, to 1 in (2.5 cm) across, clustered 2–5, with 5 white petals; fruit drupe to 1 in (2.5 cm), orange to red

+ **range:** Eastern and midwestern North America

The American Plum is the most widely ranging wild plum in North America, a tree of mixed hardwood forests and edge habitats. It is probably not native to Canada, but its range has been expanded around the world by cultivation. Like many woody species of the rose family, this one readily sends up suckers, which makes it valuable in erosion control. Though its small size makes its wood unimportant, its showy, fragrant flowers have made it a popular choice as an ornamental tree in parks and yards, and it is raised in orchards for its fruits, which are made into jellies and pies.

Mexican Plum
Prunus mexicana H to 35 ft (10.6 m)

Although found in northern Mexico, the Mexican Plum is a tree of the south-central states. It was named in 1882 from a specimen collected in the northeastern Mexican state of Coahuila.

KEY FACTS

The bark is furrowed.

+ **leaves:** Alternate, simple, to 4 in (10 cm), twice fine-toothed, hairy below; hairy stalk

+ **flowers/fruits:** Flowers perfect, to 1 in (2.5 cm) across, clustered 2–4; 5 white petals; fruit drupe, to 1.2 in (3 cm), red to purple

+ **range:** Midwest, south-central states

The Mexican Plum is sometimes confused with the American Plum. The former's hairy leaf-stalks and larger fruits can help distinguish it. It is common, though usually scattered, in a variety of habitats, including woodland edges and open fields, as well as richer woodlands. It seldom sends up suckers and is mostly single trunked, which makes it useful for grafting cultivated plum varieties. Its branching pattern and dark, peeling bark make it a favorite in bonsai. Though the juicy flesh is tart, it is improved with sugar and is often made into jelly and jam. It is often planted as an ornamental.

Black Cherry
Prunus serotina H to 110 ft (34 m)

The Black Cherry was likely being grown as an ornamental in Paris as early as 1630. Introductions in Europe continued, and it is now called a forest pest in Denmark, Germany, and the Netherlands.

KEY FACTS

The bark is plated in age.

+ **leaves:** Alternate, to 6 in (15 cm), finely blunt-toothed

+ **flowers/fruits:** Flowers perfect, showy in pendulous spike to 6 in (15 cm); 5 white petals; fruit drupe, to 0.4 in (1 cm), on flower spike, almost black

+ **range:** East, to Nova Scotia; New Mexico, into Arizona

The Black Cherry is our largest and most widely distributed cherry. A fast-growing pioneer in abandoned fields, it is also found in mixed stands of hardwoods and conifers. Its red-brown wood is hard, polishes well, and rivals walnut for use in cabinetry. Many birds and small animals eat its fruits, as do humans. The small cherries are bitter and so are made into sweet jams, but they are also used in soft drinks and ice cream because their flavor is more intense than that of sweet cherries. The tree is often damaged by tent caterpillars and Japanese beetles. The species label, *serotina*, means "late"; this species blooms after other cherries.

Klamath Plum

Prunus subcordata H to 26 ft (8 m)

The first European settlers in Pacific states soon began to cultivate the Klamath Plum, which they considered the region's most useful fruit. It took botanists until the mid-1800s to discover it.

KEY FACTS

The tree is spiny.

+ **leaves:** Alternate, to 3 in (7.6 cm), almost round; margin once or twice toothed

+ **flowers/fruits:** Flowers perfect, small, clustered 2–7; petals white or pinkish; fruit drupe, to 1.2 in (3 cm), dark red or yellow, flesh juicy

+ **range:** Pacific states

The Klamath Plum is the only plum tree native to the Pacific coast, a large, thicket-forming shrub or a small tree growing in pure stands or with other hardwoods or evergreens. It is found on the eastern slopes of the Coastal Ranges and Sierra Nevada of California and in the dry valleys east of the Cascade Mountains of Oregon and Washington. Although the plum is tart, it is eaten both fresh and dried, and a number of cultivars have been developed. The species name, *subcordata*, means that the base of the leaf is almost cordate, or heart shaped. The tree is too small for its wood to have commercial value.

Chokecherry

Prunus virginiana H to 26 ft (8 m)

Chokecherry fruit is unappealing (thus the common name) but is used in jellies, pies, and wine. Some cultivars have nonbitter fruit, and researchers hope to improve the species as a food plant.

KEY FACTS

The fruit is astringent.

+ **leaves:** Alternate, to 4 in (10 cm); finely, sharply toothed

+ **flowers/fruits:** Flowers in loose clusters to 6 in (15 cm), fragrant, 5 white petals; fruit drupe, to 0.4 in (1 cm), red or purple, thick-skinned, juicy

+ **range:** British Columbia to Newfoundland, northern U.S.

Native peoples used the Chokecherry to treat many ailments; the U.S. Pharmacopeia listed the bark as recently as 1970. The very nutritious fruits have always been eaten, appearing in stews, in pemmican, or alongside salmon. Some people believe that the Chokecherry holds more promise even than the Saskatoon for large-scale culture because of the cherry's hardiness and fruit yield. At present, Chokecherries come from wild trees, and not even 5 percent are harvested. The Chokecherry became North Dakota's state fruit, because archaeologists have found its pits at many sites there. Birds are so fond of the fruit that the tree is sometimes called the Virginia Bird Cherry.

Black Cottonwood

Populus trichocarpa H to 164 ft (50 m)

The Black Cottonwood is the first woody species to have had its genome mapped. It was important to map all the genes of a tree because of wood's economic importance.

KEY FACTS

The trunk is clean.

+ **leaves:** Alternate, to 4 in (10 cm), pointed, finely toothed; leafstalk not flat

+ **flowers/fruits:** Male catkins dense, to 2 in (5 cm), female loose, to 3 in (7.6 cm), on separate trees; fruit capsule 3-parted, to 0.5 in (1.3 cm)

+ **range:** Alaska through California, Idaho

The Black Cottonwood is the largest hardwood in the West and our biggest poplar. This majestic tree is especially successful on deep, rich soils built up by river silting, often developing pure stands. It also grows alongside other hardwoods and conifers. The capsule fruits of cottonwoods develop in the pendent female catkin, appearing as though attached to a thread. The seeds are tiny and bear cottony tufts that aid in their dispersal and that give the trees their name. Native tribes used the inner bark medicinally. It contains salicin, a compound related to aspirin that can reduce inflammation and break a fever.

Eastern Cottonwood

Populus deltoides ssp. *deltoides* H 72–100 ft (22–30 m)

The common name refers to the fluffy-tufted seeds. Another name is Necklace Poplar, a reference to the female catkin's bearing mature fruits and resembling a strand of beads.

KEY FACTS

The trunk branches are massive.

+ **leaves:** Alternate, simple, to 7 in (17.8 cm), triangular, tapering to tip; margin toothed, wavy; leaf-stalk slender, flat

+ **flowers/fruits:** Male and female catkins to 3 in (7.6 cm), on different trees; fruit capsule, to 0.5 in (1.3 cm), in 3–4 parts

+ **range:** Eastern U.S.

The Eastern Cottonwood is a tall tree of moist, rich lowland forests and is therefore largely absent from the higher Appalachians. The trunk soon branches, creating a wide, spreading crown, and the upright limbs arch at their ends for a vase-like form. The Eastern Cottonwood is often planted for quick shade, but any planting should be well thought out because the extensive roots can invade pipes and buckle sidewalks. Many people plant the male tree, which does not produce cotton. The species label, *deltoides*, refers to the triangular shape of the leaf. The flattened and extra-flexible leaf stem makes the leaf shake in even a calm breeze.

Plains Cottonwood
Populus deltoides ssp. *monilifera* H to 88 ft (27 m)

Isolated populations west of the Rocky Mountains in the Pacific Northwest were recently determined to be the Plains Cottonwood, previously believed to grow only east of the Rockies.

KEY FACTS

This is the largest tree of the Plains.

+ **leaves:** Alternate, to 3.5 in (8.9 cm); tapering to pointed tip; margin coarsely toothed, wavy; leaf-stalk slender, flat

+ **flowers/fruits:** Male and female catkins on different trees; fruit capsule, to 0.4 in (1.1 cm), conical

+ **range:** Manitoba to Texas Panhandle

The Plains Cottonwood has been considered a separate species, *P. sargentii,* but it is now usually thought of as a subspecies of *P. deltoides;* the subspecies has smaller leaves that have fewer teeth. It is the largest tree and the most abundant cottonwood of the Great Plains, found in pure stands in rich riverside habitats but otherwise mixed with other hardwoods. As in most catkin-bearing trees, cottonwood catkins form before the trees leaf out, the hallmark of pollination by wind. The Dakota ate the inner bark of the spring sprouts and also fed them to their horses. The subspecific label, *monilifera,* means "resembling a string of beads."

Quaking Aspen
Populus tremuloides H 39–60 ft (12–18 m)

The reason the trees quake has long been pondered. In a recent study, leaf stems were splinted to prevent quaking. Insects damaged nonquaking leaves 27 percent more than unsplinted leaves.

KEY FACTS

The white bark has black scars.

+ **leaves:** Alternate, simple, to 3.2 in (8 cm); round, finely toothed, whitish beneath

+ **flowers/fruits:** Male and female catkins on separate trees; fruit capsule, 70–100 in catkin 4 in (10 cm) long

+ **range:** Alaska to Labrador, south to Mexico, Nebraska, Missouri, Virginia

In the Quaking Aspen, suckering from roots and rhizomes produces more than just a thicket; it establishes dense clonal stands. In fact, the world's heaviest (and oldest) "organism" is a clone of aspens in Utah that is 80,000 years old, weighs 13.2 million pounds (6 million kg), and covers more than 106 acres (43 ha). Named Pando (Latin for "I spread"), the stand of 43,000 male trees is in decline, possibly stressed from disease, insect damage, and drought. The aspen is remarkable visually. Its leaves have a flattened stem, which is flexible, allowing the leaves to quake in the slightest breeze. They turn brilliant yellow in fall.

Pussy Willow

Salix discolor H to 30 ft (9.1 m)

Many of us first saw the Pussy Willow not outdoors but in an arrangement of its bare spring branches with their pearl-gray male catkins, which are said to look like cats' feet.

KEY FACTS

The catkins are unstalked.

+ **leaves:** Alternate, to 4.8 in (12 cm), oval, irregularly toothed, whitish green below

+ **flowers/fruits:** Male and female catkins on different trees, dense, fuzzy; fruit capsule beaked; seeds with cottony down

+ **range:** British Columbia to Newfoundland, south into Smoky Mountains

The Pussy Willow is a small tree or a shrub with many tall stems found mostly in pure stands, and especially in wet areas. Because it tends to be shrubby, it is often planted as a hedge. There are male and female trees, and the trademark catkins are male, emerging very early in the spring; a male tree must be planted if these catkins are to be expected. They are also eaten by birds. The specific label, *discolor*, means "two-colored" (not discolored), referring to the difference in color between the green upper leaf surface and the much lighter, whitish underside.

Bebb Willow

Salix bebbiana H 15–25 ft (4.5–7.5 m)

Bebb Willows are the premier diamond willows that, probably because of a fungus, develop large, diamond-shaped depressions on the trunk—a striking motif when the wood is carved.

KEY FACTS

The bark is gray-maroon.

+ **leaves:** Alternate, to 3.5 in (9 cm), oval, white below; veins netlike; leafstalks to 0.3 in (8 mm)

+ **flowers/fruits:** Catkins unisexual, on different trees, to 3 in (7.5 cm); fruit capsule beaked, seeds with threads

+ **range:** Coast to coast in Canada and northern U.S.

The Bebb Willow is the most common willow tree in Alaska and Canada. As is common among the willows, its buds, shoots, bark, and wood provide food for many mammals. A large shrub or a small, bushy tree, it has a short trunk and a rounded crown and is found most often in moist, rich soils. It is fast growing and sprouts from the roots, creating clonal thickets, and thus is an excellent pioneer species in areas that provide the plants with sufficient moisture. It is often found with other willows. The wood is used to make baseball bats and wicker furniture.

Black Willow

Salix nigra H 30–60 ft (9–18 m)

Wood of the Black Willow is light, does not splinter readily, keeps its shape, glues well, and is easy to work. It was at one time used in the manufacture of artificial limbs, or wooden legs.

KEY FACTS

The trunk is massive.

+ leaves: Alternate, to 6 in (15 cm), narrow, finely toothed, green above and below

+ flowers/fruits: Catkins unisexual, on separate trees, to 3 in (7.6 cm), upright at branchlet ends; fruit capsule ovoid, seeds tiny and silky

+ range: Maine to Minnesota, south to Texas

The Black Willow is our largest native willow and our only commercially important one. A species of wet sites, it is largest in the lower Mississippi River region. It is fast growing and forms dense root networks and so is a good pioneer species, used to reduce erosion on stream banks. Medicinal use of willow bark dates to the time of Hippocrates (400 B.C.), when the bark was chewed to relieve inflammation. An active ingredient is salicin, a chemical compound related to aspirin (acetylsalicylic acid). Both words echo the scientific name of the family, Salicaceae, and genus, *Salix*. Other substances give willow bark antioxidant, fever-reducing, immune-boosting properties.

Weeping Willow

Salix babylonica H to 60 ft (18 m)

The father of taxonomy, Linnaeus, erred in naming the Weeping Willow *babylonica* for the willow of Babylon, mentioned in Psalm 137. That tree was the Euphrates poplar, *Populus euphratica.*

KEY FACTS

The twigs are flexible and greenish gold.

+ leaves: Alternate, to 7.1 in (18 cm), narrow, finely toothed

+ flowers/fruits: Catkins unisexual on separate trees, to 1 in (2.5 cm), upright; fruit capsule on catkin, to 1 in (2.5 cm)

+ range: Nonnative; planted widely

The graceful Weeping Willow is a native to China and one of the world's most loved trees. It was first brought to North America from Europe in 1730 for use as an ornamental. The slender, flexible branches hang vertically to the ground, in curtain-like masses. Like other willows, it grows best in moist sites and is often almost stereotypically seen growing beside a lake. Its hanging twigs are greenish gold, and its leaves turn bright yellow in the fall. It has a short trunk, with pale to dark gray, roughly ridged bark often marred with burls from which grow pale yellow shoots.

Bigleaf Maple
Acer macrophyllum H to 98 ft (30 m)

In the Bigleaf Maple, researchers hunt unique traits to amplify them in cultivars. For example, trees that bear red leaves in northern California, and some in Washington, produce triple samaras.

KEY FACTS

The leaf is the largest of any maple.

+ **leaves:** Opposite, to 12 in (30 cm); 5 lobes, deep, pointed; teeth few, irregular

+ **flowers/fruits:** Flowers unisexual, green-yellow, fragrant, on same hanging clusters; fruits paired samaras, in long clusters

+ **range:** Coastal, British Columbia to southern California

In the Pacific Northwest, the only maple tree that is medium size to large is the Bigleaf Maple. It grows mostly in mixed stands on moist sites but is planted for shade in cities. Its trunk is straight and its branches stout, and it is a source of commercial lumber, the wood used in musical instruments, furniture, and cabinets. Older trees produce burls that lend special beauty to veneers. Because the bark of the Bigleaf Maple holds moisture, in places like the Quinault Rain Forest in Washington's Olympic Mountains, the tree is often densely clothed in epiphytic mosses, ferns, and liverworts.

Box Elder/Manitoba Maple
Acer negundo H to 65.6 ft (20 m)

"Box Elder" refers both to the poor-quality wood, suitable for nothing more elegant than boxes and crates, and to the supposed similarity of the leaves to those of elders *(Sambucus)*.

KEY FACTS

Fruiting is often prolific.

+ **leaves:** Opposite, compound, smooth; 3–9 leaflets, to 4.8 in (12 cm); teeth pointed, irregular

+ **flowers/fruits:** Clusters unisexual on separate trees; paired samara fruits, often in hanging clusters

+ **range:** Eastern United States, northwest to Alberta, to Ontario

The Box Elder, or Manitoba Maple, is our only compound-leaved maple. It is often grown ornamentally, and the cultivars include a seedless variety. It is not universally popular, however, naturalizing easily and sometimes invasive, even in North America. When it has three leaflets, it can recall Poison Ivy, but the latter has alternate leaves (the Box Elder's are opposite) and always has three leaflets (the Box Elder can have as many as nine), and its twigs are not green (the Box Elder's are). The tree is the main host of the Boxelder Bug, a black and red insect, but the bugs are more annoying than harmful.

Red Maple

Acer rubrum H to 92 ft (28 m)

The often showy Red Maple earned its species label, *rubrum,* meaning "red." The flowers, fruit, stems, buds, and fall leaves are all red or reddish.

KEY FACTS

The samaras are red, pink, or yellow.

+ **leaves:** Opposite, pale below; 3–5 lobes, pointed, once or twice toothed

+ **flowers/fruits:** Perfect or unisexual (sometimes in unisexual clusters, on same or different trees), on slender stalk; fruits double samaras on slender stalk

+ **range:** Eastern U.S., Quebec

The Red Maple is one of the most abundant trees in eastern North America, growing on many types of soil, in many tree assemblages, and from sea level well into the Appalachian Mountains. It is often planted for shade in yards and on streets. Cultivars have been developed that are more adapted to the urban environment or that enhance red and orange foliage in the fall. The fruits are eaten by squirrels and birds, and deer and rabbits browse the shoots. The wood is used in flooring, cabinets and furniture. Syrup is made from the sap, but it is inferior to that of the Sugar Maple.

Silver Maple

Acer saccharinum H to 98 ft (30 m)

The Silver Maple is tapped and its sap used to make syrup and sugar. For commercial success, the sugar level is too low, despite its species name, *saccharinum* from the Latin for "sugar."

KEY FACTS

The trunk often branches low.

+ **leaves:** Opposite, to 7.9 in (20 cm), silvery below; 5 lobes; sinuses narrow, margin irregularly toothed

+ **flowers/fruits:** Unisexual clusters on same or different trees; fruits double samaras on slender stalk, prominently veined

+ **range:** Eastern U.S., New Brunswick

This abundant tree of wet lowlands is related to the Red Maple but lacks the red. Its leaves are intricately divided, and they seem to flicker in a breeze, their silvery undersides (thus the common name) coming in and out of view. The fruits are the largest produced by a native maple and provide food for squirrels and birds. The tree was once planted widely as an ornamental and street tree, but now people think better of it; it splits, breaks, and sheds twigs, it rots and suckers, the seeds are a nuisance, the roots can ruin pipes, and its pests excrete a sticky sap on the ground, or car, beneath.

Sugar Maple
Acer saccharum H to 98 ft (30 m)

Many maple species grow in Canada, but many Canadians consider the Sugar Maple the national tree. An iconic, five-lobed leaf of a maple is the red focus of the nation's flag.

KEY FACTS

The fall foliage is brilliant.

+ leaves: Opposite, to 8 in (20 cm), paler below; 3–5 lobes, pointed; teeth few, coarse

+ flowers/fruits: Flowers perfect or unisexual, on slim stalk to 3 in (7.5 cm), in clusters on same tree; fruits double samaras

+ range: Nova Scotia to southern Manitoba, south to Tennessee

The magnificent Sugar Maple is one of eastern North America's most famous trees. European colonists learned from the native peoples that the tree's sap could be collected and concentrated to make maple syrup, and concentrated some more to make sugar. In addition to its syrup, throughout its range every fall, the tree attracts thousands of people who want to see the expanse of bright color, blinding bright yellow, orange, and red—sometimes all on the same tree. A key species of the eastern hardwood forests, it is also planted in yards, on streets, and in parks. Many cultivars have been developed.

Norway Maple
Acer platanoides H to 98 ft (30 m)

The Norway Maple is one of North America's most common street trees. The popular European import has been bred for improved color, leaf shape, growth form, and urban performance.

KEY FACTS

The leaf stems exude milky sap when broken.

+ leaves: Opposite, to 6.3 in (16 cm), pointed teeth on margin; 5 lobes, mostly pointed

+ flowers/fruits: Flowers perfect or unisexual, sexes on different clusters; fruits double samaras, wings almost in a straight line

+ range: Nonnative; planted widely

Introduced to the northeast by Philadelphia botanist John Bartram in 1756, the Norway Maple is one of the most popular trees in North America, widely planted, especially in cities and towns, from Ontario to Newfoundland, from Maine to Minnesota, and south to North Carolina and Tennessee, as well as from British Columbia to Oregon, east to Idaho and Montana. Sadly, the popular nonnative has jumped its bounds to invade forests in the northeast and the Pacific Northwest, outcompeting native species and overshading understory species. The Norway is nevertheless still widely sold in nurseries.

Ohio Buckeye/Fetid Buckeye

Aesculus glabra H 30–49 ft (9–15 m)

The seed of *Aesculus glabra* is called a buckeye, because it recalls the eye of a male deer. Carried in the pocket as a good luck charm, it was also charged with preventing rheumatism.

KEY FACTS

All parts are rank when crushed.

+ **leaves:** Opposite, compound; 5–7 leaflets, oval, pointed, to 6 in (15 cm)

+ **flowers/fruits:** Flowers perfect or unisexual, yellow, in terminal clusters to 7 in (17.8 cm); fruit a spiny capsule, to 2 in (5 cm); seeds brown, shiny, to 1.5 in (3.8 cm)

+ **range:** East-central United States

A medium-size tree, the Ohio Buckeye fares best and takes on a more pleasing form when it grows in river bottoms or on stream banks, although it grows with oaks and hickories in drier locations. Buckeye trees are not heavily used by wildlife, although squirrels will sometimes eat young buckeyes. All parts of the plant contain a toxic alkaloid that may deter would-be browsers. The Ohio Buckeye is planted often as an ornamental for its fruit and because of its bright orange fall foliage. Its light wood is used in woodcarving and was once used to manufacture artificial limbs.

Tree of Heaven/Ailanthus

Ailanthus altissima H 82 ft (25 m)

Ailanthus occurs in Ontario, Quebec, and 43 states, classed as a noxious or invasive plant in many. It should not be planted. Eliminating trees, especially females, reduces seed production.

KEY FACTS

It damages the native environment severely.

+ **leaves:** Alternate, compound, to 35.5 in (90 cm); 11–41 leaflets, paired along stem

+ **flowers/fruits:** Flowers mostly unisexual, sexes on separate trees, tiny, yellow-green, in large, conical clusters at tip of twigs; fruits double, twisted samaras in clusters

+ **range:** Nonnative; widely invasive

Called "tree from hell" and the "kudzu of trees," this native of China seems biologically engineered to invade. Ailanthus spreads by its profusely produced seeds and by sprouting from its roots. It produces a chemical that causes allelopathy, preventing most other plants from growing beneath it. It may be the continent's fastest growing tree, adding 3 to 6 feet (1–2 m) a year when young. What's more, its value to wildlife is virtually nil, and its wood is useless. First brought to North America in 1784, it was planted extensively as a street tree in the 1800s, faring well under the stresses of city life. But the opportunist soon turns a clearing into a dense, impenetrable mass of stems.

American Basswood

Tilia americana H 65–82 ft (20–25 m)

The American Basswood has been widely used medicinally—for example, to treat headache or digestive ailments. It is now said that overconsumption can cause heart problems.

KEY FACTS

The flower stalk has a long, leafy bract.

+ **leaves:** Alternate, to 8 in (20 cm), toothed; base unequally heart-shaped

+ **flowers/fruits:** Flowers perfect, to 0.6 in (1.5 cm), creamy, fragrant, 5 petals, few on stalk; fruit a nutlet, to 0.4 in (1.2 cm) across

+ **range:** Central, eastern U.S.; southern Ontario, Quebec

This stately tree of the lowland woods is the best known of our native basswoods, although the European species (lindens) are better suited for life in cities. The American Basswood has a straight trunk, unbranched for half its length, and a broad crown. Found mostly in mixed stands in rich soils, it is important to wildlife: It naturally develops cavities that attract cavity nesters like woodpeckers and wood ducks, and squirrels and quail eat its seeds. The timber is important in the Great Lakes region, used in cabinetry, musical instruments, boxes, excelsior, and pulp. Native peoples made thread and cord from its tough inner bark.

Common Hackberry

Celtis occidentalis H 32–60 ft (10–18 m)

Detracting from the Common Hackberry's beauty are "witches' brooms," abnormal but harmless tufts of short twigs often formed in its branches, probably caused by a fungus and a mite.

KEY FACTS

The bark has corky ridges.

+ **leaves:** Alternate, to 3.5 in (8.9 cm), toothed near pointed tip, base unevenly rounded

+ **flowers/fruits:** Flowers perfect or unisexual, on same plant; fruit a drupe, to 0.8 in (2 cm), orange turning purple

+ **range:** Southern Ontario and Quebec; midwestern, northeastern U.S.

The small to medium-size Common Hackberry excels in moist, fertile places but is drought tolerant and thrives in sandy soils, too. It has a straight trunk with thick branches. Its common name refers to the fruit and derives from "hagberry," a British word for a tree similar to the Chokecherry. The fruit is edible and sweet, though usually out of reach. The fruits are somewhat important to wildlife, including raccoons, squirrels, and game birds. The weak, soft wood is nonetheless used in some furniture and in fences, posts, and boxes. The Hackberry is planted in gardens and as a street tree.

American Elm

Ulmus americana H 132 ft (40 m)

American Elms lined many streets, especially in northeastern states, but Dutch Elm disease, a fungus carried by Elm Bark Beetles, has ravaged the trees. All elm species are susceptible.

KEY FACTS

The tree is vase-shaped.

+ **leaves:** Alternate, to 6 in (15 cm), rough above, toothed, pointed, uneven at base

+ **flowers/fruits:** Flowers perfect, small, 3–5 on slim, drooping stalk; fruits flat samaras, to 0.5 in (1.3 cm), deeply notched at tip; seeds with papery wing

+ **range:** East; southern Canada south

Many native peoples selected a towering American Elm as the site for councils and other important events, and the stately, graceful tree held Europeans similarly in awe. In many other ways, this is among the most important trees of eastern North America. In the wild, it tolerates a range of soils, usually growing alongside other hardwoods. Its buds, flowers, and fruits feed mammals and birds, and its twigs provide forage for larger mammals. Some cultivars have been developed that show some resistance to Dutch Elm disease. It is hoped that the same can be done for another deadly threat facing elms—elm yellows (see below).

Slippery Elm

Ulmus rubra H 49–82 ft (15–25 m)

This tree is named for its mucilaginous inner bark. The bark, a cough remedy before European contact, stayed in the U.S. Pharmacopeia until 1960 and is still used in cough preparations.

KEY FACTS

The trunk branches high in the tree.

+ **leaves:** Alternate, to 6.3 in (16 cm), dark green, hairy below

+ **flowers/fruits:** Flowers perfect, small, few, in crowded clusters on short stalk; fruit a samara, to 0.5 in (1.3 cm) across, slightly notched at tip, hairy

+ **range:** Southern Ontario through most of eastern U.S.

This lowland tree usually grows with other hardwoods. Its seeds are not of great importance to wildlife, but it provides some browse for deer and rabbits. The scientific name means "red elm," another of its common names, reflecting the red-brown wood. It is used for boxes, crates, and baskets. Native peoples used its bark for canoe shells when birch was not available. The Slippery Elm (and the American) is threatened by elm yellows, a disease spread by leafhoppers that attacks the trees' transport tissues. The first symptoms are yellowing, drooping, or early loss of leaves, but by the time they appear, it is too late.

Rock Elm

Ulmus thomasii H 82–115 ft (25–35 m)

Even though the Rock Elm's claim to fame is its hard wood, it is often called Cork Elm because its older branches can bear three or four thick, irregular corky wings.

KEY FACTS

The fruits are a key trait.

+ **leaves:** Alternate, to 4 in (10 cm), coarsely toothed, on hairy stalk

+ **flowers/fruits:** Flowers perfect, small, reddish, 2–4 in dangling clusters; fruit samara, flat, broad winged, slightly notched at tip

+ **range:** Northern U.S. Midwest, east to northeastern U.S., southern Canada

The Rock Elm is not a tree of pure stands but is usually found growing alongside such species as American Elm, Basswood, Sugar Maple, and White Ash. While it prefers a moist, well-drained loam, it also grows on drier, upland sites, including those on limestone. Some argue that growth on a rocky substrate earned it its common name, but most say it refers to the wood. The heaviest, hardest elm wood, it has few knots, bends well, and can take shock. That makes it superior when strength is mandatory, so it has seen use in ships' timbers, curved sections of furniture, early refrigerators and car bodies, hockey sticks, and ax handles.

Creosote Bush

Larrea tridentata H 3.3–10 ft (1–3 m)

For native peoples, the resinous, fragrant Creosote Bush served many medical purposes, but the resin also served as a glue for mending broken pottery or cementing a projectile point to its shaft.

KEY FACTS

Nodes make the stem look jointed.

+ **leaves:** Evergreen, opposite, compound; 2 leaflets, to 0.7 in (1.8 cm), resinous, joined basally

+ **flowers/fruits:** Flowers to 1 in (2.5 cm) across, velvety, axillary; 5 yellow petals; capsule 5-parted, fuzzy

+ **range:** Southern California, east to Utah and Texas

The Creosote Bush is a trademark of the Mojave Desert. In full bloom, the flowers lend the plant a yellowish cast, and after a rain, the resin exudes a tarry aroma. Its main value to wildlife is as shelter. Kangaroo Rats and Desert Tortoises are two animals that bed down under the plant. The Creosote Bush Grasshopper eats this plant exclusively, having evolved digestive processes allowing it to process the mostly inedible resin. A Creosote Bush can live 100 years, but when it dies, stems sprout at its edge, creating a Creosote Bush ring. One ring, called King Clone, has been aged with radiocarbon dating at 11,700 years. Its diameter averages 45 feet (13.7 m).

Joshua Tree
Yucca brevifolia H to 50 ft (15 m)

Other members of the genus *Yucca* are probably more familiar to the reader, and it helps to recall their leaves (living and dead) and flowers when contemplating the Joshua Tree.

KEY FACTS

Plants can sprout from rhizomes.

+ **leaves:** Evergreen, to 9 in (23 cm), spear-like, sharply toothed, in rosettes at branch ends

+ **flowers/fruits:** Flowers perfect, white, clustered at branch tips; 6 tepals; fruits to 5 in (13 cm), 3-angled, light red to yellow-brown

+ **range:** Endemic to Mojave Desert

The striking Joshua Tree seems to branch wildly, but closer examination reveals that each branching results in two equivalent, stout branches more or less at a right angle to each other. The branches are armed with horrific spines and are tipped with clusters of bladelike leaves that recall a Common Yucca plant. The tree remains unbranched, though, until it blooms the first time (or until its tip is somehow damaged). Old leaves persist, folded back and shaggy along the branches, which bear a thick, furrowed bark. In a sort of give-and-take, Yucca Moths pollinate the flowers, which open at night, and lay their eggs inside them, where eventually their larvae will undergo metamorphosis.

Cuban Royal Palm/Florida Royal Palm
Roystonea regia H to 131 ft (40 m)

Palms are indispensable to people who need to use them for food, fiber, sugar, wax, and wood. Palms are commercially important, too, as the source of rattan, oils, coconuts, and acai.

KEY FACTS

The crownshaft is conspicuous.

+ **leaves:** To 13 ft (4 m), compound, featherlike; many strap-shaped leaflets

+ **flowers/fruits:** Flowers unisexual, on same tree, to 0.3 in (8 mm) across, white, fragrant, in hanging clusters to 2 ft (0.6 m); drupes, to 0.6 in (1.5 cm), black

+ **range:** Southern Florida

In addition to its long, feathery leaves, the Cuban Royal Palm is distinguished from others by the occasional bulge along its trunk and, just below its canopy, by the bright glossy green "crownshaft." The crownshaft is made up of the bases of the leaves, the older ones closer to the outside. When a leaf has died, its base will eventually separate from the tree, leaving a new ringed scar just below the crownshaft on the light gray trunk. The tree is a tall, unbranched column with a spreading crown of leaves. The flowers are pollinated by bees and bats, and bats, as well as birds, eat the fruit and disperse the seeds.

Cabbage Palmetto

Sabal palmetto H to 82 ft (25 m)

This palmetto is named for its terminal bud, or heart of palm, where new fronds originate. "Swamp Cabbage" may be eaten raw or cooked. Once the bud is taken, the tree stops growing.

KEY FACTS

The trunk is straight and unbranched.

+ **leaves:** Leafstalk to 7.5 ft (2.3 m), serving as midrib; blade fan-like, to 6.6 ft (2 m)

+ **flowers/fruits:** Flowers perfect, small, white, in clusters to 6.6 ft (2 m); drupes, to 0.5 in (1.2 cm), black

+ **range:** Florida Keys, in Coastal Plain into North Carolina

The Cabbage Palmetto is a tree of the marshes, woodlands, and sandy soils of the Coastal Plain and is tolerant of salt spray. It is a straight, slender tree that is often grown in yards and along streets. This plant does not have a crownshaft; its leaves are produced by a terminal meristem. Native peoples harvested the fibers from the leafstalks and used them for scrubbing and to make straps and baskets. The fibers were not just used locally, though. They have been found among artifacts from the Winnebago in Wisconsin and the Iroquois of New York, as far as 700 miles (1,100 km) north of the Cabbage Palmetto's northern limit.

Saw Palmetto

Serenoa repens H to 23 ft (7 m)

Extracts of Saw Palmetto fruit are the top herbal treatment for benign prostatic hyperplasia, and research has shown that the extract can indeed improve urinary symptoms and flow volume.

KEY FACTS

The "saw" is the spiny leafstalk.

+ **leaves:** Leafstalk to 5 ft (1.5 m), ending in fanlike blade to 3.3 ft (1 m)

+ **flowers/fruits:** Flowers perfect, small, white, in clusters to 3.3 ft (1 m); drupes, to 0.8 in (2 cm), blue-black

+ **range:** South Carolina to southern Florida, west to Louisiana

The Saw Palmetto is not an erect palm. Instead, it sprawls along the ground, with fronds growing at the ends of long, branching, horizontal stems running as far as 15 feet (4.6 m) at or just below ground level. The result is a dense understory in pine flatwoods and scrub habitats. Some species associated with Saw Palmetto are the Crested Caracara, Sand Skink, Florida Mouse, and endangered Florida Scrub Jay. When the fruits are ripe, black bears will wander from their home range to eat them. Honey made from Saw Palmetto flowers is of high quality and commercial value.

Key Thatch Palm/Brittle Thatch Palm

Leucothrinax morrisii H to 40 ft (12 m)

Plant names are not static. The Key Thatch Palm, long called *Thrinax morrisii,* was given its own genus, *Leucothrinax,* after taxonomists found that it differed genetically from other *Thrinax* species.

KEY FACTS

The fronds are green and silver.

+ **leaves:** Leafstalk to 3.3 ft (1 m), smooth; blade fan-shaped, to 32 in (80 cm)

+ **flowers/fruits:** Flowers perfect, ivory-white, fragrant, tubular, in elongate, branching cluster to 6.6 ft (2 m); drupes, to 0.3 in (7 mm), white

+ **range:** Southern Florida

The slow-growing Key Thatch Palm has a slender, columnar trunk, with smooth, gray to brownish bark. Capped with a rounded, open crown of magnificent green and silver fronds, it is often planted as a specimen tree in the landscape or in small groups. As in other species, when grown in full sun, it will develop a denser, rounder crown, and in shadier conditions, the canopy is much more open. In nature, this tree is found in a range of habitats including the seashore and pineland sand. The Key Thatch Palm's leaves have long been used to make brooms and mats, though today these are more for decoration than for utilitarian purposes.

California Washingtonia/California Fan Palm

Washingtonia filifera H to 50 ft (15 m)

The iconic palms of Los Angeles were planted for the 1932 Summer Olympics. They will be replaced someday with native broadleafs. The palms are nonnative Mexican Washingtonias.

KEY FACTS

The leaflet edges bear threads.

+ **leaves:** Leafstalk to 5 ft (1.5 m), with hooked spines, bearing fan to 5 ft (1.5 m)

+ **flowers/fruits:** Flowers perfect, small, white, fragrant, in clusters to 16.5 ft (5 m); drupes, to 0.4 in (9 mm), almost black

+ **range:** Southwestern Arizona and southeastern California

The California Washingtonia is a columnar palm that is well known but rare, restricted in nature to streams and canyons that offer more water than the adjacent desert. It is planted in tropical areas around the world, especially along city boulevards like those in south Florida. As beautiful as the California species is, the Mexican Washingtonia (*W. robusta*) is prettier and faster growing. Other traits that make it stand out are its slimmer trunk, its height (to 75 feet/23 m vs. 50 feet/15 m in the California), its bright green leaves (vs. gray-green), and its less thready leaflet edges (*filifera* is Latin for "making threads" or "thread-bearing").

An Arizona slot canyon features Navajo sandstone, formed
from windblown sands of an ancient desert.

3 | Rocks & Minerals
Earth's Beauty and Power

With all the bustle and show of living things, it can be easy to overlook the ground beneath our feet. But the Earth's rocks and minerals are among the most fascinating aspects of the natural world. They are the record of billions of years of Earth's activity. Within their structures, from the smallest crystal to the largest mountain face, the beauty and dynamism of our planet are revealed.

■ What are Minerals?

Minerals are the solid substances that make up rocks. There are thousands of different minerals on Earth, some rare and valuable like diamond and gold, others, such as quartz, as common as beach sand. Minerals are defined by a few characteristics: They are solids. They are naturally occurring. They have an ordered atomic arrangement. They have a well-defined chemical composition.

Most minerals exist as small, irregular grains within rocks. Occasionally, isolated minerals have beautiful geometric shapes and smooth planar surfaces. These shapes and surfaces are crystals, and they are the outward expressions of the mineral's internal atomic arrangement. Crystals form when a mineral has sufficient time and space in which to grow.

Gems, or gemstones, are materials cut and polished, used for jewelry or other decorative items. Most gems, unless they are synthetically produced, are minerals with gem-specific names. For example, amethyst is a purple variety of quartz, and emerald is a green variety of beryl.

Identifying minerals can be both fun and challenging. It takes close observation, a few basic tests, and the process of elimination—and it definitely gets easier as you gain experience. You don't need many tools: A hand lens and a small hammer will get you started. Mineral identification kits containing streak plates, hardness tools, and a small bottle of diluted acid are helpful.

■ Three Types of Rocks

Rocks are the solid materials that make up the Earth. There are three basic types: igneous, sedimentary, and metamorphic.

Igneous rocks form when molten material, called magma, cools and solidifies. Igneous rocks can be either intrusive, meaning the magma solidified underground, or extrusive, meaning the magma erupted onto the Earth's surface before solidification. (Erupting magma is called lava.) Intrusive igneous rocks cool slowly, allowing minerals to grow into visible grains that form an interlocking, crystalline texture. Extrusive rocks cool quickly, resulting in a fine-grained volcanic texture. Some igneous rocks have two grain sizes with large, well-formed crystals surrounded by a groundmass of finer grains. This is called a porphyritic texture, and the large grains are known as phenocrysts.

In addition to texture, igneous rocks are categorized by chemistry. Mafic rocks are rich in iron and magnesium (minerals such as biotite and hornblende) and poor in silica (such as quartz and feldspar), and they tend to be dark in color. Felsic rocks are light in color, containing more silica-rich minerals and fewer that are rich in iron and magnesium.

Sedimentary rocks form from an accumulation of sediments. The sediments can be preexisting minerals or rock fragments—collectively called clasts—that have been transported by wind, water, or ice; or they can be chemically precipitated or biologically generated in place, as with many types of limestones.

During accumulation, sediments become compacted by the weight of material above. Finally, groundwater or other fluids passing through the formation precipitates a cement between the sediment grains—commonly silica or calcium carbonate—hardening the sediment pile into stone. Clastic sedimentary rocks such as conglomerates, sandstones, and mudstones are classified by grain size.

Metamorphic rocks form when preexisting rocks are transformed by heat and pressure, a process called metamorphism. This process involves physical and chemical changes that occur in the solid state—changes such as the growth of new minerals, and the physical rotation or recrystallization of existing minerals. The parent rock that becomes a metamorphic rock is called a protolith. Limestone, for example, is the protolith for the

KEY MINERAL PROPERTIES

+ **Hardness:** See page 184.

+ **Streak:** Streak is the color of the mineral when it is ground into a fine powder. Many mineral identification kits contain small white porcelain plates called streak plates. Scratching an edge of a specimen against this plate will produce a streak. If the mineral is clear, or if it is harder than the porcelain plate, the streak will be white.

+ **Crystal habit or aggregation:** The form of a mineral specimen can be helpful for identification. If the specimen has well-developed crystal faces, the shape of the crystal, known as its habit, can be diagnostic. Many specimens, however, do not have well-developed crystal faces and are instead aggregates of small mineral grains. A variety of terms describe crystal habit and state of aggregation:

    Prismatic means that the crystal is long in one direction with well-developed faces. Some prismatic crystals have blunt ends; they can also be topped with pyramid shapes.

    Columnar means that the crystal is long in one direction and resembles the shape of a rounded column.

    Tabular and **platy** habits refer to crystals that have two dimensions relatively equal in length and one that is short.

    Bladed habit refers to crystals that are both elongate and flat, like a blade.

    Fibrous means that the crystal grows in the shape of threads or filaments.

    Massive means that the mineral specimen does not have crystal faces.

    Granular describes aggregates of mineral grains that are all approximately the same size and dimension.

    Compact specimens are very fine grained so individual grains cannot be distinguished.

    Mammillary, from the Latin *mamma*, meaning "breast," describes specimens that are smooth and rounded.

    Botryoidal is the term for specimens that are smooth and globular, with smaller spherical shapes than mammillary specimens.

+ **Luster:** The way light is reflected from the surface of a mineral is called luster. Terms describing luster include metallic, glassy or vitreous, pearly, waxy, earthy, and dull. One mineral can display different types of luster depending on its crystal faces or habit.

+ **Color:** Color is an important property used in mineral identification, but it can be misleading because most minerals come in a variety of hues.

+ **Cleavage & fracture:** Some crystals tend to break along one or more smooth, flat surfaces known as cleavage planes. The orientation of these planes is helpful for identifying minerals. Breaks along irregular surfaces instead of cleavage planes are called fractures. Conchoidal fracture is a type of fracture with a smooth, scooped, or curved pattern, common in quartz and obsidian.

In a microscopic view of granite, polarized light interacts with mineral grains, revealing brilliant colors.

metamorphic rock marble, and granite is a protolith for some gneisses. Metamorphism occurs when rocks are subjected to high pressures and/or temperatures—conditions that happen during movements of the Earth's plates. When plates collide and push up large mountain ranges, for example, rocks are squeezed and heated in the thickening crust. Metamorphism also occurs in shallow regions of the crust by the heat given off of large magmatic intrusions.

■ Landforms & Structures

The study of the Earth is not just about identifying rocks and minerals. In fact, what most people notice first are not individual rocks, but rather the shapes and textures of the land—the natural features that make up the landscape. Why does a valley have a certain profile, for example, or why is one coastline rimmed with cliffs and another

|||||||||||||||||||||||||||||||||||||||||||||||||||||||||||

To him was given
Full many a glimpse
(but sparingly bestowed
On timid man) of Nature's processes
Upon the exalted hills.
—WILLIAM WORDSWORTH

|||||||||||||||||||||||||||||||||||||||||||||||||||||||||||

gentle beaches? These are landforms, and they can tell us a great deal about how our planet works. Most landforms are the result of a combination of features and processes—the structures of the rocks combined with the effects of weathering and erosion. Weathering is the breakdown of rocks by either physical or chemical means. Erosion is the removal and transport of rock materials by the energy of wind, water, or ice.

■ Fossils

Fossils are the preserved remains or impressions of ancient living things. Fossils form when organisms are buried in sediment and preserved via processes like compaction, chemical alteration, or replacement with minerals. Fossils are extremely important because they are our primary clues about past lifeforms and previous environments that existed on our planet—the key to our understanding of evolution. The best places to look for fossils are areas where sedimentary rocks are exposed along road cuts, hillsides, or along streams and rivers. Fossils are most common in fine-grained sedimentary rocks such as mudstones and shales, and marine sedimentary rocks such as limestones.

■ Use Caution

Collecting rocks, minerals, and fossils is an exciting hobby, but one that carries significant risks and responsibilities. Be sure to check the status of a prospective site beforehand. Gain permission if it is private land, or conform to the rules and regulations regarding prospecting and collecting if the land is public. Information can be obtained from state geological surveys, state parks, the U.S. Forest Service, and the Bureau of Land Management. Always put your safety first. Wear protective gear, especially when using hammers, chisels, or other digging equipment, and be aware of risks like rockfall and unstable slopes. Never enter closed or abandoned mines.

MINERAL HARDNESS SCALE

+ **Hardness:** The resistance of a mineral to scratching or abrasion is called its hardness. It is measured on a relative scale from 1 to 10 in increasing hardness. A mineral with a lower number can be scratched by a mineral with a higher number. Minerals with similar numbers can scratch each other. Each level of hardness is designated by a mineral standard.

HARDNESS	DESCRIPTION	MINERAL STANDARD
1	Very soft, falls apart in fingers	Talc
2	Easily scratched by fingernails	Gypsum
3	Can be scratched by a copper penny	Calcite
4	Can be scratched by a knife	Fluorite
5	Can be scratched by a knife with difficulty; can be scratched by glass	Apatite
6	Cannot be scratched by a knife; will scratch glass	Orthoclase
7	Scratches glass easily	Quartz
8	Scratches glass very easily	Topaz
9	Can cut glass; can be scratched by diamond	Corundum
10	Will cut glass; no other minerals will scratch it	Diamond

ROCKS & MINERALS

||||||||||||||||||||||||||||||||||||||||||||||||||||||||||||||||||||||||||||||||||||||||||||||||||||||||||||||||||||||||

Quartz

Class: Silicate Chemical formula: SiO_2

Quartz is one of the most common minerals in the Earth's crust, found in many sedimentary, igneous, and metamorphic rocks. Collectible varieties include amethyst and smoky quartz.

KEY FACTS

Well-formed crystals are hexagonal prisms with pyramid-shaped tops; quartz comes in various colors, usually clear, gray, or milky white; transparent to translucent.

+ hardness: 7

+ streak: White

+ locations: This is a common rock-forming mineral; gem-quality crystals are found in veins and granitic pegmatites.

Quartz, an important building block of the Earth's crust, consists of silica and oxygen atoms linked in a strong, three-dimensional framework. Quartz is distinguished by its hardness and, in well-formed crystals, its hexagonal-shaped prisms with pyramidal ends. Traces of other elements substituting for silicon create different varieties. Iron in the structure forms yellow quartz crystals known as citrine, or purple crystals known as amethyst. Smoky quartz contains trace amounts of aluminum. Quartz has a vitreous or greasy luster and no cleavage. Microcrystalline forms and aggregates have conchoidal fracture.

Chalcedony

Class: Silicate Chemical formula: SiO_2

Chalcedony is a variety of quartz made up of crystals so small they cannot be seen with the naked eye. It comes in beautiful colors and patterns, some valued as semiprecious stones.

KEY FACTS

This attractive mineral is hard and compact, translucent to opaque, with a waxy or vitreous luster and a conchoidal or uneven fracture.

+ hardness: 6–7

+ streak: White

+ locations: Cavities in volcanic rocks and as nodules in sedimentary rocks; clasts of chalcedony can be found in some ocean beach and river gravels.

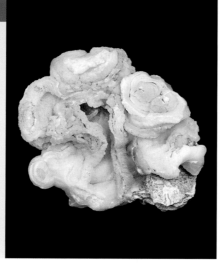

Chalcedony is microcrystalline or cryptocrystalline quartz. Microcrystalline crystals are so small that they can be identified only under a microscope; cryptocrystalline crystals, from the Greek *krypto*, meaning hidden, are so small that an optical microscope cannot resolve them. Chalcedony can form in rock cavities, most often in fine-grained volcanic rocks. It is commonly found in globular forms—botryoidal or mammillary shapes—and as the outer shell of geodes and the material replacing plant parts in petrified wood. Collectible varieties include agate, bloodstone, carnelian, onyx, and petrified wood.

Agate

Class: Silicate Chemical formula: SiO_2

A concentrically banded form of chalcedony, agate is commonly found filling rounded cavities in volcanic rocks. Cut and polished agates are sold as ornaments and used in jewelry.

KEY FACTS

This microcrystalline to cryptocrystalline quartz is banded typically parallel to its cavity; wide range of natural colors; many bright colors of commercial specimens are artificial.

+ hardness: 6–7

+ streak: White

+ locations: Rounded nodules in various rock types; cavities in volcanic rocks; eroded agates in river deposits

Agate is known for its beautiful colors and concentric bands, which are often enhanced by dyes and polish in commercial samples. Most agates form from silica-rich fluids that have filled open pockets or seams in volcanic rocks. The silica is deposited in bands of tiny, parallel fibrous crystals around the cavity's inside wall. During crystallization, impurities collect along the bands, creating alternating colors and rings. Look for well-developed crystals (often amethyst or smoky quartz) at centers of some agate nodules. Collectors use the term "agate" for nonbanded chalcedony such as moss agate and flame agate.

Opal

Class: Silicate Chemical formula: $SiO_2 \cdot nH_2O$

Opal is a form of cryptocrystalline quartz with water present. It is made of tiny spheres packed together that interact with light to create a distinctive "opalescent" rainbow shimmer.

KEY FACTS

Variable in color, opal exists as compact masses, crusts, and veins with vitreous or pearly luster.

+ hardness: 5.5–6

+ streak: White

+ locations: Near geysers and hot springs; also as pseudomorphs of marine shells or wood, meaning it has replaced the original material while retaining the item's shape and form

Opal is hardened silica gel in the form of tiny, tightly packed spheres with a significant amount of water (5 to 10 percent) in its pores. It is fragile and can easily crack and lose its water content over time, which diminishes its opalescence. Opal is usually white, but can also be green, brown, black, yellow, or gray due to various impurities such as iron and manganese. Opal is deposited by silica-rich waters at relatively low temperatures. It forms in small cavities in basalt, in cracks and small veins, and as pseudomorphs of wood and shells. The word "opal" comes from Greek *opallios*, meaning "precious stone."

Jasper

Class: Silicate Chemical formula: SiO_2

Jasper is a granular form of microcrystalline quartz, distinguished by its opacity, rich colors, and abundant impurities. It is typically deep red, yellow, or brown due to traces of iron oxide.

KEY FACTS

With hues of rich red, yellow, or brown, jasper is opaque and can have an angular, broken appearance or fine color banding.

+ hardness: 6–7

+ streak: White

+ locations: In association with various sedimentary rocks, banded iron formations, and fault zones; jasper pebbles and cobbles also occur in stream deposits.

Jasper is evenly colored or has banding or brecciation, an amalgamation of angular, broken shards. Polished jasper is traded as an ornamental semiprecious stone. Banded jasper differs from agate by its opacity (agate is typically translucent or semitranslucent). Classification of chalcedony, agate, jasper, chert, and flint is inconsistent. Agate and jasper are commonly considered to be varieties of chalcedony, which is cryptocrystalline or microcrystalline quartz. Chert and flint are often classified as sedimentary rocks composed of microcrystalline quartz. Jasper can also be classified as a colorful form of chert.

Chert/Flint

Class: Silicate Chemical formula: SiO_2

Chert, also known as flint, is made of cryptocrystalline or microcrystalline quartz. Chert occurs as layered formations or as nodules in marine sedimentary rocks.

KEY FACTS

Nodules and beds are opaque, hard, and dense masses with conchoidal or irregular fracture; white and gray varieties are often called flint.

+ hardness: 7

+ streak: White

+ locations: Found as nodules in chalk and other marine sedimentary rocks; also in association with Precambrian banded iron formations

Chert, also known as flint, is a hard, compact form of cryptocrystalline or microcrystalline quartz primarily marine in origin. Chert is commonly gray, white, yellow, black, or brown. It forms irregular sedimentary bands or layers, or nodular masses, in fine-grained limestones and chalk. It forms by accumulation of silica, sometimes from organic sources such as needlelike structures in marine sponges. Some cherts contain small fossils. Flint is the name commonly used to describe nodular chert. Chert is usually classified as a rock, but it is included here with other forms of microcrystalline quartz.

Feldspars

Class: Silicate Chemical formula: $XAl_{(1-2)}Si_{(3-2)}O_8$

Feldspars are a group of minerals made of silicon, oxygen, and aluminum, plus various amounts of sodium, calcium, and/or potassium. They are common constituents of the Earth's crust.

KEY FACTS

These common rock-forming minerals are light in color and are often blocky. They display two cleavage planes that intersect at an angle of approximately 90 degrees.

+ **hardness: 6–7**

+ **streak:** Most feldspars have a white streak.

+ **locations:** Widespread across all rock types; common in intrusive igneous rocks

Feldspars are building blocks of many rocks, especially granites and other intrusive igneous rocks. They are also found in clastic sedimentary rocks and as grains in metamorphic rocks. A key component of the Earth's crust, feldspars are also found in meteorites and rocks from the moon. Feldspars come in various colors, commonly white, pink, gray, or other light shades. They are distinguished from other minerals by their hardness and cleavage planes. Feldspar is mined for use in glassmaking and ceramics, and is used in products such as fiberglass, floor tiles, tableware, sinks, toilets, and paints.

Potassium Feldspars

Class: Silicate Chemical formula: $KAlSi_3O_8$

The potassium feldspars are a subgroup of feldspar minerals that contain potassium in their structures. They are important components of igneous rocks.

KEY FACTS

These feldspars with salmon pink grains are commonly found in granite; however, they can be white, gray, or yellow, and are often difficult to distinguish from plagioclase.

+ **hardness: 6–7**

+ **streak:** White

+ **locations:** Granite and other intrusive igneous rocks; well-formed crystals in granitic pegmatites

Potassium feldspars (also known as K-feldspar or K-spar) include microcline, orthoclase, and sanidine. They form short tabular or prismatic crystals. Microcline is milky white, pink, or blue-green and found in metamorphic and igneous rocks. Well-formed microcline crystals occur in granitic pegmatites. Orthoclase is common in granite and other intrusive igneous rocks; sanidine is a high-temperature form of potassium feldspar common in rhyolite and other extrusive igneous rocks. Orthoclase with fine intergrowths of sodium feldspar (albite) is known as moonstone and displays a unique iridescent luster.

Garnet

Class: Silicate Chemical formula: $X_3Y_2(SiO_4)_3$

Garnets are colorful minerals common in igneous and metamorphic rocks. Individual types of garnet are defined by their chemistry—how different elements fit into their crystal structure.

KEY FACTS

Although usually deep red, garnets can be other colors. The mineral can be distinguished by its hardness and its well-formed 12-sided crystals.

+ hardness: 6.5–7.5

+ streak: White or colorless

+ locations: Most often occurs as small crystals in schists; also found in granitic rocks, pegmatites, and marbles

Garnet types include almandine, pyrope, spessartine, and grossular. Specimens commonly have well-developed crystal faces. In schists and other rocks, the symmetrical, colorful crystals are distinctive, especially when surrounded by flattened grains of other minerals. The most recognizable garnets are deep red with a vitreous luster and high hardness, but they can be orange, yellow, green, blue, purple, pink, brown, and black. Garnets are used as gems and industrial abrasives. Their presence can give geologists a clue to where the rock came from; for example, some garnets indicate origins in the Earth's mantle.

Andalusite

Class: Silicate Chemical formula: Al_2SiO_5

Andalusite is one of three aluminosilicate minerals with the same chemistry but different structures: andalusite, kyanite, and sillimanite. It is found in metamorphic rocks.

KEY FACTS

Prismatic pink, violet, green, or gray crystals have nearly square cross sections; the crystals can be twinned. Andalusite has one perfect cleavage plane and an uneven or conchoidal fracture.

+ hardness: 6.5–7.5

+ streak: White to colorless

+ locations: Can be found in schists and other aluminous metamorphic rocks

Three aluminosilicate minerals have the same chemistry but different crystal structures, known as polymorphs: andalusite, kyanite, and sillimanite. Each polymorph forms under different pressure and temperature conditions, so finding one of these minerals is an important clue about the history of the rock. Andalusite forms under high temperatures at relatively low pressures. Nearly square cross sections of prismatic andalusite crystals are distinctive, but andalusite can also have a massive or compact form. A variety of andalusite called chiastolite contains dark inclusions that create the shape of a cross.

Kyanite

Class: Silicate Chemical formula: Al_2SiO_5

Kyanite is an aluminosilicate mineral commonly forming long tabular or bladed crystals. It is an indicator of medium- to high-pressure metamorphism.

KEY FACTS

Although it is best known for its cyan-blue color, other colors exist in kyanite, including clear, white, gray, and yellow.

+ hardness: 6.5–7 perpendicular to long axis; 4–5 parallel to long axis

+ streak: White

+ locations: Found in mica schists and gneisses; also occurs in some hydrothermal quartz veins

Kyanite is the medium- to high-pressure polymorph of the three aluminosilicates: andalusite, kyanite, and sillimanite. Kyanite specimens commonly display elongate, bladed, or tabular crystals. Kyanite also displays two different levels of hardness corresponding to different regions of the crystal. They are hard (hardness of 6.5 to 7) in the direction perpendicular to the long axis of the crystal and soft (hardness of 4 to 5) parallel to that axis. Kyanite is often found in association with staurolite and garnet in metamorphic rocks. Kyanite is used in a variety of refractory and abrasive products.

Sillimanite

Class: Silicate Chemical formula: Al_2SiO_5

Sillimanite is one of the three aluminosilicates, along with andalusite and kyanite, that have the same chemistry but different structures. Sillimanite grains are often fibrous or needlelike.

KEY FACTS

Characteristic forms of sillimanite include colorless, white, or gray needles as well as fibrous masses. Variations are brownish or blue-green.

+ hardness: 7

+ streak: White

+ locations: Just as its close chemical relatives andalusite and kyanite, this aluminosilicate polymorph occurs in mica schists and gneisses.

Sillimanite is the high-temperature polymorph of the three aluminosilicates: andalusite, kyanite, and sillimanite. It is commonly colorless or white, but can also be brown, yellowish brown, or blue-green. Sillimanite crystals are usually prismatic and fibrous with vitreous luster. Fibrous varieties are also known as fibrolite. It is found in aluminous metamorphic rocks, commonly associated with corundum, cordierite, andalusite, and kyanite. It is the state mineral of Delaware, where it is common in the schists of the Delaware Piedmont, in places as large masses and rounded boulders.

Staurolite

Class: Silicate Chemical formula: $Fe_2Al_9Si_4O_{22}(OH)_2$

Staurolite is a hard, medium- to dark-colored mineral found in some metamorphic rocks. Specimens often have well-developed crystal faces creating prismatic shapes.

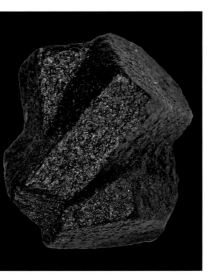

KEY FACTS

Two crystals of staurolite are often found grown together, some at right angles. Right-angled "fairy crosses" or "fairy stones" are thought by some people to bring good luck.

+ **hardness: 7–7.5**

+ **streak:** White to gray

+ **locations:** Like andalusite and kyanite, it can be found in mica schists and gneisses.

Staurolite is distinguished by its hardness and its propensity for growing two "twin" crystals at the same time. The twins are commonly oriented perpendicularly to each other, forming a cross. Some collectors call these "fairy crosses," and they are considered good-luck charms. Other twins intersect at about 60-degree angles. Staurolite crystals are medium to dark in color, usually reddish brown, tan, brown, or black. Prismatic crystals have a vitreous luster and are hexagonal or diamond shaped in cross section. Crystals with well-developed faces have a glassy luster and are hexagonal or diamond shaped in cross section.

Epidote

Class: Silicate Chemical formula: $Ca_2(Al,Fe)_3(SiO_4)_3(OH)$

Epidote is a common mineral with a characteristic pistachio green color. It is usually a secondary mineral, grown during metamorphism or alteration of a rock.

KEY FACTS

The green colors of epidote are distinctive. This mineral has one direction of perfect cleavage—a characteristic distinguishing it from green amphiboles, which have two cleavage planes.

+ **hardness: 6–7**

+ **streak:** Colorless to gray

+ **locations:** Found in metamorphic rocks, pegmatites, and small cavities in basalt

Epidote is a common mineral found in rocks that have undergone metamorphism or alteration, often by the heat of a nearby magma chamber. Epidote forms well-developed columnar prisms or thick, tabular crystals with grooved faces. It also forms thin crusts and seams, filling vesicles (small cavities) and fractures in basalts. Epidote has a characteristic green color, often pistachio green, yellow-green, or greenish black. It has a vitreous luster, and it is transparent to translucent. Epidote crystals are pleochroic, meaning that different colors appear through the crystal prism as the prism is rotated.

Beryl

Class: Silicate Chemical formula: $Be_3Al_2Si_6O_8$

Beryl is best known for its gemstone varieties emerald and aquamarine. It is an aluminosilicate mineral with beryllium in its crystal structure. It is most often found in granitic pegmatites.

KEY FACTS

The beautiful colors of highly prized gem-quality beryl include red, pink, green, and blue. The crystals form hexagonal prisms, which add to the mineral's popular appeal.

+ hardness: 7.5–8

+ streak: White

+ locations: Although usually found in granitic pegmatites, beryl also occurs in some metamorphic rocks.

Beryl is often recognized by its hexagonal crystals. Beryls form beautiful prisms in granitic pegmatites. The beryl crystal structure contains relatively large spaces that can accommodate additional atoms of different elements, especially manganese, iron, and chromium, resulting in a wide variety of colors. Beryl with iron in the beryllium site forms the gem aquamarine. Manganese in the aluminum site forms the pink gem morganite. Chromium beryls are emeralds. Beryl crystals have a vitreous to greasy luster and are harder than quartz. In addition to granitic pegmatites, beryls can also be found in some metamorphic rocks.

Tourmaline

Class: Silicate Chemical formula: $Na(Mg,Fe)_3Al_6(BO_3)_3(Si_6O_{18})(OH,F)_4$

Tourmaline forms beautiful columnar crystals with rounded triangular cross sections. Its colors include black, brown, blue, pink, and green, sometimes with several colors in one crystal.

KEY FACTS

Distinguishing characteristics of tourmaline include well-formed columnar crystals with pyramidal faces on one end. Some are multicolored, including a type popularly called "watermelon tourmaline."

+ hardness: 7–7.5

+ streak: White

+ locations: Found in granites, pegmatites, quartz veins, and some metamorphic rocks

Tourmaline crystals have rough, vertical striations and rounded to triangular cross sections. Crystals often have pyramidal faces on one end. Tourmaline is most commonly the black, vitreous variety known as schorl. A popular gem called "watermelon tourmaline" displays concentrically zoned colors with red or pink inside and green outside, formed by chemical changes during crystal growth. Red or pink tourmalines contain manganese; green tourmalines result from iron. The largest crystals are found in pegmatites, though tourmaline is also found in schists and gneisses. Gem-quality tourmalines are collected in Maine, Connecticut, and California.

Pyroxene

Class: Silicate Chemical formula: $XY(\text{Si,Al})_2O_6$

Pyroxenes are a group of important rock-forming minerals found in igneous and metamorphic rocks. The most common pyroxene mineral is augite, a component of some volcanic rocks.

KEY FACTS

These silicates are similar to amphiboles but have cleavage planes intersecting at right angles.

+ **hardness:** 5–6.5

+ **streak:** Variable; augite light green to colorless.

+ **locations:** Mafic and ultramafic igneous and metamorphic rocks; forms large crystals in porphyritic volcanic rocks; also occurs as granular masses

The pyroxene group contains various minerals defined by crystal structures and chemical compositions, including aegirine, augite, diopside, jadeite, and enstatite. These are high-temperature minerals similar to amphiboles but without water content. Pyroxenes form prismatic crystals with vitreous luster, in elongate and short varieties. Most pyroxenes require a hand lens to identify. They differ from amphiboles by their cleavage angles: pyroxene intersecting at right angles; amphibole intersecting in wedge or diamond shapes. Earth's upper mantle is made of olivine and pyroxene. Augite is found in volcanic dikes.

Amphibole

Class: Silicate Chemical formula: $XY_2Z_5(\text{Si,Al,Ti})_8O_{22}(\text{OH,F})_2$

Amphiboles are a group of important rock-forming minerals. They are typically dark in color and are distinguished by their diamond- or wedge-shaped cleavage intersections.

KEY FACTS

Elongate crystals are generally flatter than pyroxenes, with oblique, perfect cleavage planes that intersect at approximately 120 degrees.

+ **hardness:** 5–6

+ **streak:** Variable; usually white or colorless

+ **locations:** Found in many igneous and metamorphic rocks including granite, basalt, gneisses, schists, and amphibolite

Like pyroxenes, amphiboles are minerals rich in iron and magnesium. Unlike pyroxenes, amphiboles are hydrous, meaning they have water in their structure. Geologists sometimes call amphiboles "garbage-can minerals" because many elements fit into the crystal structure. The most common amphibole is hornblende, a constituent of granitic rocks. Others include tremolite and actinolite in schists and metamorphosed impure limestones. Amphiboles form elongate, prismatic crystals, sometimes in radiating aggregates, typically gray (tremolite) to dark green (actinolite). Darker colors indicate increasing iron content.

Hornblende

Class: Silicate Chemical formula: $(Ca,Na,K)_{2-3}(Mg,Fe,Al)_5(Si,Al)_8O_{22}(OH)_2$

Hornblende is the name for various iron- and magnesium-rich amphiboles found in many igneous and metamorphic rocks. The dark-colored mineral in granitic rocks is usually hornblende or biotite.

KEY FACTS

A distinguishing characteristic of hornblende is its wedge-shaped amphibole cleavage. Its most common colors are black and dark brown.

+ hardness: 5–6

+ streak: Colorless to white, gray, or brown

+ locations: Found in many igneous and metamorphic rocks; large, well-formed crystals occur in granitic pegmatites

Hornblende is usually black or dark brown, but can be dark green. It is distinguished from other dark minerals such as biotite, tourmaline, and pyroxenes by its wedge-shaped cleavage angles. Cleavage angles are usually identified under a hand lens. Well-formed crystals are commonly short columnar prisms with six-sided cross sections and a vitreous to dull luster. Hornblende crystals can have a bladed or even fibrous habit and form massive aggregates. Hornblende is a common constituent of many rocks including granite, syenite, diorite, gabbro, basalt, schist, gneiss, and amphibolite.

Talc

Class: Silicate Chemical formula: $Mg_3Si_4O_{10}(OH)_2$

Talc is a soft, sheetlike mineral. It is the main component of soapstone and is also used in lubricants, cosmetics, and in industrial applications as a heat-resistant material.

KEY FACTS

Its greasy touch and low hardness easily identify talc. It is soft enough to be scratched by a fingernail, although talc schist is used as a building material.

+ hardness: 1

+ streak: White

+ locations: Metamorphic rocks including schists, marbles, and metaperidotites; associated with serpentine, pyroxene, and olivine

Talc is an important commercial and industrial material, used for its heat-resistant and lubricant properties. It is found in metamorphic terrains associated with ocean floor and ultramafic rocks such as marbles and peridotites. The common building material soapstone is a talc schist. Talc forms white to light green aggregate masses with a distinctive greasy touch and low hardness. It rarely forms individual crystals, but can replace crystals of other minerals during alteration, and can take on their shape, known as pseudomorphs. Talc is found, among other places, in the Appalachians, California, and Texas.

Muscovite

Class: Silicate Chemical formula: $KAl_3Si_3O_{10}(OH,F)_2$

Muscovite is one of the most common micas, a group of sheetlike minerals called phyllosilicates. It is the clear, light brown, or gray mica found in schists, gneisses, and sometimes in granites.

KEY FACTS

This widespread form of mica has thin, flexible, transparent cleavage sheets and is generally light gray or silvery. It has important uses in electronics.

+ hardness: 2–3

+ streak: Colorless or white

+ locations: Granitic rocks, pegmatites, schists, and gneisses; large crystals found in various areas of the U.S.

Muscovite is a common mica, easily identified by its flaky habit and perfect single cleavage that creates easy-to-peel, thin, flexible sheets. Muscovite is distinguished from biotite by its light colors, usually clear, white, gray, or silvery. It typically forms larger crystal masses than biotite and is more common. Large sheets of muscovite are mined from granitic pegmatites for use in the electronics industry. The name refers to the Muscovy region in Russia where large sheets of muscovite were once used as window glass. Large crystals are common in North Carolina, New England, Colorado, and South Dakota.

Biotite

Class: Silicate Chemical formula: $K(Mg,Fe)_3(Al,Fe)Si_3O_{18}(OH,F)_2$

Biotite, along with muscovite, is part of the mica group of phyllosilicates, or sheetlike minerals. It is a platy black mineral found in some granites and schists.

KEY FACTS

Black or deep brown plates (which are tabular crystals) have a distinctive flaky cleavage. Sometimes a gold tint to the plate gives this mineral a resemblance to pyrite.

+ hardness: 2–3

+ streak: Colorless or white

+ locations: Commonly a component of schists, but also can be found in biotite granites

Biotite is a black or deep brown mica. Gold-tinted biotite can be distinguished from pyrite by its platy cleavage and low hardness. Large biotite "books" can be found in granitic pegmatites. Well-formed crystals can have a hexagonal cross section, and they can easily peel into thin, flexible sheets. Most biotite, however, is found as small, embedded dark-colored grains. Biotite is an important mineral in geological research, used for analyzing the temperature histories of metamorphic rocks and, in some cases, for determining the approximate ages of igneous rocks.

Serpentine

Class: Silicate Chemical formula: $Mg_3Si_2O_5(OH)_4$

Serpentines are a group of minerals that form by the alteration of olivine and pyroxene. Serpentines are associated with rocks that originated in the oceanic crust and upper mantle.

KEY FACTS

This silicate is commonly dark green with a greasy to waxy luster and feel; compact and massive. The name suggests its resemblance to snake skin.

+ hardness: 2–5

+ streak: White

+ locations: Zones of altered rocks, associated with other magnesium-rich minerals. It is commonly found along California's coast ranges.

Serpentine minerals have three main varieties: antigorite, chrysotile, and lizardite. Antigorite is most common, occurring in compact masses that are typically dark green with a greasy luster. Lizardite is fine grained and scaly, and is found associated with altered marbles. Chrysotile is a type of asbestos with a fine fibrous form and very low hardness. Masses of serpentine make up rocks called serpentinites; these can have irregular green patterns and scaly surfaces resembling a snake skin, hence the name "serpent rock." Rocks of serpentine masses and serpentine breccia are used as decorative building stones.

Kaolinite

Class: Silicate Chemical formula: $Al_2Si_2O_5(OH)_4$

Kaolinite is the main component of clay. It forms compact masses that can be layered. It is either a primary mineral, formed in place, or it is the weathering product of minerals such as feldspar.

KEY FACTS

Compact, flourlike masses are usually white or earthy red, and they have an earthy odor. This clay has a great variety of uses, ranging from cosmetics to china to construction bricks.

+ hardness: 2

+ streak: White

+ locations: Found in clay beds and in place of altered feldspars in granites and pegmatites

Kaolinite is a type of clay found in areas of chemical weathering. It forms microscopic platy crystals that make up compact, earthy masses. It has a dull luster, is relatively soft, has a distinct earthy odor, powdery feel, and is usually white. Illite and montmorillonite are other common clay minerals in soils. It is difficult to distinguish these without laboratory equipment. Kaolinite is one of the most important nonmetallic minerals, used in many commercial and industrial applications including cosmetics, paints, adhesives, and glossy paper. Kaolinite clay is also used to make bricks, tiles, fine pottery, and china.

Calcite

Class: Carbonates Chemical formula: CaCO$_3$

Calcite is the primary mineral component of limestone and marble. It is the stable form of calcium carbonate at surface temperatures and pressures, and often has a biological origin.

KEY FACTS

Its "fizzing" in dilute acids (for example, hydrochloric acid) is a characteristic feature of calcite. Its hardness is also diagnostic.

+ hardness: 3

+ streak: White

+ location: Component of carbonate rocks including limestone, travertine, and marble; also forms thin veins and the linings of small cavities

Two common forms of calcium carbonate are calcite and aragonite. Calcite is the stable form at the Earth's surface. It is typically clear or white, but can be yellow or gray, with a pearly or vitreous luster. It is found in many different crystal habits and shapes: rhombohedrons, prisms, granular masses, and microcrystalline masses. Crystals display three perfect cleavage planes. Calcite differs from other white or clear minerals by its hardness—unlike quartz, it can be scratched by a knife—and its fizzing reaction in dilute acids. Dolomite, on the other hand, will fizz only under a concentrated hydrochloric acid solution.

Dolomite

Class: Carbonates Chemical formula: CaMg(Co$_3$)$_2$

Dolomite is an important rock-forming mineral in sedimentary and some metamorphic rocks. It is the primary component of the rock known as dolomite or dolostone.

KEY FACTS

Although it is similar to calcite, dolomite reacts only to concentrated hydrochloric acid.

+ hardness: 3.5–4

+ streak: White

+ location: Besides its occurrence in the rock dolomite, the mineral also occurs with calcite in limestone and some marbles. It can be found in hydrothermal veins and small cavities where it forms crystals.

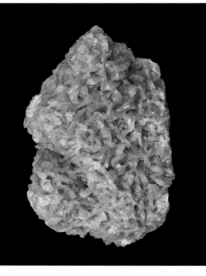

Dolomite is similar to calcite but with magnesium in its structure as well as calcium. Dolomite crystals are commonly rhombohedrons that are curved or slightly saddle shaped; like calcite, it forms granular and microgranular masses. Dolomite is typically white, gray, brown, or yellow. Its hardness, luster, and cleavage are similar to calcite. Calcite, however, will fizz under a diluted hydrochloric acid solution, whereas dolomite requires concentrated acid or a scratched or powdered surface to cause a fizzing reaction.

Malachite

Class: Carbonates Chemical formula: $Cu_2CO_3(OH)_2$

Malachite is a beautiful green, copper carbonate mineral. It is often associated with copper ore deposits and coexists with azurite, a less common copper carbonate that is brilliant blue.

KEY FACTS

This attractive carbonate is an opaque, green, banded mineral. It can be cut and polished to a high luster and is often used in jewelry.

+ hardness: 3.5–4

+ streak: Green

+ location: Occurs in association with copper ores, especially in the copper mining region of the American Southwest. It is also found in limestones.

Malachite is an opaque, striking green mineral, often displaying color bands of different shades of green. It is typically found in massive aggregates with a globular or bubbly surface (geologists call this botryoidal or mammillary habit), but can also form thin coatings or stalactites in caverns, or veins/veinlets along fractures in limestone. Well-developed crystals are rarely true malachite crystals; instead, they are pseudomorphs that form by replacing crystals of azurite. Malachite masses take a beautiful high polish. Pure specimens are cut and polished into various forms and collected as semi-precious or decorative stones.

Gypsum & Anhydrite

Class: Sulfates Chemical formula: $CaSO_4 \cdot 2H_2O$

Gypsum and anhydrite are calcium sulfate minerals, gypsum as the hydrous form and anhydrite the form without water ($CaSO_4$). They are typically found in evaporite deposits.

KEY FACTS

Gypsum forms flattened, glassy crystals that are sometimes clear; anhydrite is usually found as veins or larger masses.

+ hardness: 1.5–2 (gypsum); 3–3.5 (anhydrite)

+ streak: White

+ location: Found in evaporite deposits as massive sedimentary beds; also in association with clay beds. A vast example is New Mexico's White Sands.

Gypsum and anhydrite occur in deposits formed when lakes and shallow seas evaporate, and from waters circulating through sandstones and clays. Crystals are colorless to white; can be tabular or diamond shaped/rhombic. Radiating blades of gypsum coated in sand are called desert roses. Glassy gypsum crystals are called selenite. Massive gypsum beds are known as alabaster. Gypsum is mined to produce cement, Sheetrock, and fertilizer. It makes up the dramatic dunes of White Sands National Monument in New Mexico. Anhydrite is harder and has three good cleavage planes that create cubic-like fragments.

Galena

Class: Sulfides Chemical formula: PbS

Galena is a gray metallic mineral typically found in hydrothermal ore deposits. Galena is made of atoms of lead and sulfur packed together in a cubic structure and is an important lead ore.

KEY FACTS

This important ore is a distinctive lead-gray color with a metallic luster, and a cubic or octahedral shape. A special characteristic is how heavy it feels in the hand.

+ hardness: 2.5

+ streak: Gray

+ location: Found in hydrothermal deposits, often in association with sphalerite, pyrite, and chalcopyrite

Galena is a major lead ore and a source of silver, bismuth, and thallium. It commonly forms perfect cubes or octahedrons. Galena is opaque gray with a shiny metallic luster when fresh. Specimens easily tarnish with exposure to air, giving some samples a dull luster. Hardness is a key identifying characteristic—it can be scratched with a fingernail—yet it has a high specific gravity, so it feels heavy in the hand. The distinctive gray streak is also diagnostic. Galena is common in silver and lead mining areas of the U.S. and Canada. It is the state mineral of Missouri and Wisconsin.

Chalcopyrite

Class: Sulfides Chemical formula: $CuFeS_2$

Chalcopyrite, a copper iron sulfide, is an important copper ore. It is a brassy and golden mineral found in hydrothermal ore deposits and is a common specimen in rock and mineral collections.

KEY FACTS

Brassy color differs from the bright golden hue of gold and pyrite. Specimens are easily tarnished. Chalcopyrite is softer than pyrite and harder than gold.

+ hardness: 3.5

+ streak: Green-black

+ location: Ore deposits, primarily in hydrothermal veins. An important location is in the greenstone belt of Ontario.

Chalcopyrite is distinguished by its greenish brassy color, distinct from the metallic gold of both pyrite and gold. Its hardness is also distinctive. It is softer than pyrite, whereas gold is softer than both chalcopyrite and pyrite, and is more malleable. Well-formed chalcopyrite crystals are rare and typically have a tetrahedral shape, which distinguishes them from the cubic form of pyrite. Chalcopyrite occurs in sulfide ore deposits formed by deposition of copper during the circulation of medium-temperature fluids in volcanic environments. It occurs as mineral veins, granular masses, and nodules.

Pyrite

Class: Sulfides Chemical formula: FeS_2

Pyrite, commonly known as fool's gold, is made of iron sulfide. It forms distinctive golden, metallic cubic crystals that are popular with collectors.

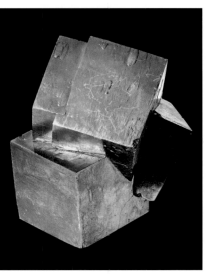

KEY FACTS

Pyrite's hardness is diagnostic—much harder (but lighter) than gold, and it can scratch glass. Its streak is also a helpful identifier.

+ hardness: 6–6.5

+ streak: Greenish black

+ location: Hydrothermal veins and other sites in sedimentary, metamorphic, and igneous environments. Small crystals can be found in shale.

Pyrite is known as fool's gold because of its gold color and high metallic luster. Unlike true gold, however, pyrite is relatively hard—similar in hardness to quartz—and is commonly found in the form of distinctive crystal cubes or crystals with 12 sides in the shape of pentagons. It is also significantly lighter than gold. Crystal faces of pyrite commonly display striations. Another diagnostic feature of pyrite is that it gives off a sulfur smell when broken or hammered. Pyrite is the most common sulfide mineral and is used to produce sulfuric acid. It is sometimes found as a replacement mineral in fossils.

Halite

Class: Halides Chemical formula: NaCl

Halite is the naturally occurring form of table salt and is also known as rock salt. It is found in evaporite deposits and can form thick beds in sedimentary sequences.

KEY FACTS

Distinguished by its salty taste and solubility in water, halite has long been familiar as table salt; however, current grocery store salt is primarily synthetic.

+ hardness: 2.5

+ streak: White

+ location: Evaporite deposits and in salt crusts along shorelines in arid environments

Halite is used as road salt and is important in the chemical industry—for example, to produce hydrochloric acid. Salt crystals are cubic, and are typically clear, white, or sometimes bluish or with blue spots. They have a vitreous or greasy luster. Halite precipitates out of seawater when the water evaporates to less than 10 percent of its original volume. Because of halite's low density, salt beds can rise under pressure through weaknesses in overlying rocks. These disruptions of sedimentary strata create interesting structures in the deserts of the U.S. Southwest.

Fluorite

Class: Halides Chemical formula: CaF_2

Fluorite is a colorful mineral that forms cubic or octahedral crystals. Fluorite fluoresces under ultraviolet light, and the phenomenon is named after this property of fluorite.

KEY FACTS

Transparent to translucent crystals of fluorite come in various colors, including pale purple, pink, green, and yellow. Its fluorescence under ultraviolet light is a famous characteristic.

+ hardness: 4

+ streak: White

+ location: Occurs in hydrothermal veins; also in cavities of granitic pegmatites

Fluorite is usually found as well-formed crystals with cubic and octahedral faces. It can be colorless, but is more often colorful, including dark blue, violet, pink, green, and yellow varieties. The colors are influenced by concentrations of rare earth elements in the crystal structure. Crystals have a vitreous luster. Hardness is a good diagnostic tool. It is softer than quartz but harder than calcite. Some specimens show cleavage planes intersecting at 60 degrees. Fluorite forms in medium- and low-temperature hydrothermal deposits, in crystal pockets of granites and granitic pegmatites, and as a component of some metamorphic rocks.

Silver

Class: Native elements Chemical formula: Ag

Silver is a rare native element that occurs in hydrothermal deposits and in association with other ores. Silver is distinctively malleable and is an important precious metal.

KEY FACTS

This prized metal has a silver-white color and a high metallic luster when it is fresh; however, as is well known, its surface oxidizes or tarnishes easily to dark gray and black.

+ hardness: 2.5

+ streak: Silver-white

+ location: Hydrothermal veins or by alteration of other minerals; also placer deposits

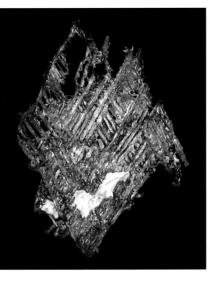

Native silver is rare, found in hydrothermal deposits and alteration zones (where high-temperature fluids have either deposited material or changed existing rocks). It occurs in thin plates, as well as in granular habits and skeletal and wire-like, branching forms. It is distinguished by its opaque, silver-white color and metallic luster that easily tarnishes. It is distinctly malleable and is relatively soft. Most silver is produced by refining other compounds instead of mined in its pure form. It is used as currency and in jewelry, and is important industrially for its conductivity and malleability.

Gold

Class: Native elements Chemical formula: Au

The precious metal gold is used for currency, jewelry, electronics, and more. It occurs in hydro-thermal deposits and in quartz veins, but is mostly found as grains in river and beach sands.

KEY FACTS

Known for its distinctive color and metallic luster, gold resists oxidation. Most gold is discovered in accumulations of particles in river and beach sands.

+ hardness: 2.5–3

+ streak: Golden

+ location: Hydrothermal deposits, quartz veins in large granitic bodies, and secondary weathering deposits

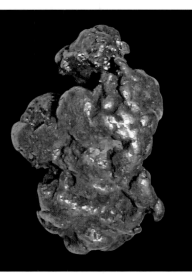

Gold rarely forms distinct crystals; instead, it is found as golden yellow platelets, branching, wiry forms, and as inclusions in quartz. Gold is soft, but has a very high density. It is malleable, and it maintains its metallic luster and color without oxidation. Gold differs from pyrite by its hardness (gold is softer) and by a lack of sulfur smell. Flecks of yellowish mica can resemble gold, but mica has a vitreous, not metallic, luster and is brittle. Gold is extremely unreactive and survives chemical weathering. Most gold is found as particles in gravels and sands that have weathered from gold-bearing veins.

Copper

Class: Native elements Chemical formula: Cu

Native copper occurs as irregular nodules, veins, and wirelike or platelike forms. It is an important industrial metal because of its high thermal and electrical conductivity.

KEY FACTS

The original copper-red color is often oxidized to black, blue, and green. Native copper is usually found in irregular masses, often with strange shapes.

+ hardness: 2.5–3

+ streak: Copper-red

+ location: Hydrothermal veins; also in rare lava flows including along the mid-continent rift in northern Michigan

Most industrial copper is extracted from other minerals such as chalcopyrite, but native copper can be found. Copper rarely exists as well-formed crystals and is more often found as irregular masses, sometimes with branching or wirelike shapes, and as the matrix surrounding volcanic clasts. A key to identifying copper is its copper-red streak and low hardness—it can be scratched by a knife. Copper specimens commonly display thin black, blue, or green oxidation coatings. One of the world's largest deposits of native copper is in Michigan's Upper Peninsula, where it is associated with a thick series of lava flows.

Diamond

Class: Native elements Chemical formula: C

Diamond, the hardest mineral, consists of pure carbon, and in nature, forms only at high pressures and temperatures deep in the Earth. Diamonds come to the Earth's surface via volcanic vents.

KEY FACTS

A diamond's famed hardness is diagnostic—nothing will scratch it. The various colors of diamonds are results of impurities such as the elements boron and nitrogen.

+ hardness: 10

+ streak: None

+ location: Rare; found in association with ultramafic (iron- and magnesium-rich) igneous rocks in continental shields

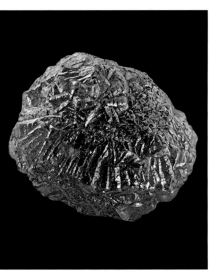

Diamonds are a high-pressure crystalline form of carbon with individual atoms packed together by strong chemical bonds. Graphite is also pure carbon, but is a low-pressure mineral in the form of carbon sheets with weak bonds. Diamonds are unstable at the Earth's surface temperatures and pressures, but they cannot convert to graphite on human timescales. Small fragments of quartz can look like diamonds but are distinguished by testing hardness. Diamond crystals are usually octahedrons or cubic forms. Impurities give diamonds different colors: Boron, for example, creates a blue tint; nitrogen casts a yellow hue.

Magnetite

Class: Oxides Chemical formula: Fe_3O_4

Magnetite is a form of iron oxide found in some metamorphic and igneous rocks as well as secondary deposits in black sands. True to its name, its magnetism is diagnostic.

KEY FACTS

This magnetic oxide is black, with a semi-metallic luster. Black sands often found along rivers and on ocean beaches are grains of magnetite that have eroded out of rocks.

+ hardness: 5.5–6

+ streak: Dark gray to black

+ location: Common accessory mineral in a variety of metamorphic and igneous rocks

Magnetite is a major iron ore, found as octahedral crystals, granular aggregates, and in tiny veins. It is dark gray to black with metallic luster. Its black streak and magnetism are distinguishing characteristics. Magnetite is important to geologists because it records information about the Earth's magnetic field when it crystallizes. Analyzing magnetite-bearing rocks of different ages allows geologists to understand movement of the Earth's tectonic plates. Large crystals of magnetite occur in pegmatites and in high-temperature veins.

Hematite

Class: Oxides Chemical formula: Fe_2O_3

Hematite is the most common form of iron oxide and an important iron ore. It occurs in various forms including black platy or tabular crystals and irregular masses with a brick red hue.

KEY FACTS

A red streak is diagnostic for this major source of iron, found as thick sedimentary beds in some formations. Crystals have a metallic luster, but masses are typically dull.

+ **hardness:** 5–6

+ **streak:** Bright to earthy red

+ **location:** A common component of many different rocks; large deposits are found in banded iron formations.

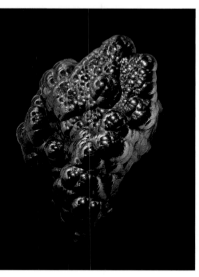

Hematite varies in form and color, appearing steely gray, black, brown, orange, or red. The red streak is a good method to differentiate it from similar minerals such as magnetite and ilmenite. Hematite can form circular aggregates of platy crystals known as iron roses. It also forms tabular and hexagonal crystals, and irregular, rounded masses. Clays with high hematite content are called red ochre, which is used as a pigment. Hematite forms as a deposit in quartz veins, as well as in metamorphic rocks. It is the primary component of banded iron formations.

Corundum

Class: Oxides Chemical formula: Al_2O_3

Corundum is an aluminum oxide known for its high hardness and beautiful gemstone varieties, including pink and red rubies, and blue, green, and yellow sapphires.

KEY FACTS

The hardness of corundum is diagnostic; only diamonds are harder. This mineral forms tabular and prismatic crystals. Impurities provide colored gems such as rubies and sapphires.

+ **hardness:** 9

+ **streak:** White

+ **location:** Pegmatites, gneisses, schists, and marbles; also in association with ultramafic igneous rocks

Corundum is an aluminum oxide mineral with a close-packed crystal structure and strong bonds between atoms, giving it high hardness and density. Corundum of pure aluminum oxide is clear, white, or gray. Impurities lead to various colors. Chromium creates pink or red rubies. Iron and titanium create blue sapphires; iron alone creates yellow sapphires. Corundum forms during alteration of volcanic and ultramafic igneous rocks and during metamorphism of aluminum-rich sedimentary and igneous rocks. Corundum is used in industry as an abrasive. Corundum and quartz never coexist in the same formation.

Granite
Type: Igneous

Granite is one of the Earth's most common rocks and an important building block of the continents. It is a light-colored intrusive igneous rock with a distinct granular texture.

KEY FACTS

A relatively homogeneous rock with interlocking minerals, granite can be grayish or light pink to brick red; exposed by erosion.

+ **grain size:** Medium to coarse

+ **texture:** Crystalline; uniform grain size to porphyritic

+ **composition:** Quartz + alkali feldspar + plagioclase feldspar ± hornblende ± biotite ± muscovite

Granite is an intrusive igneous rock containing at least 20 percent quartz and two types of feldspar—plagioclase and alkali feldspar in relatively equal amounts. It can be difficult to distinguish between the feldspars in the field, or to determine the amount of quartz; thus, many medium-grained, light-colored igneous rocks are called granite, even if they are technically something else. Granites form from felsic magma that slowly cools below the Earth's surface. Multiple intrusions of granitic magma build up large rock bodies called batholiths, which form in the cores of mountain belts and are exposed by erosion.

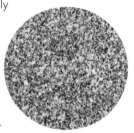

Porphyritic Granite
Type: Igneous

A porphyritic granite has mineral grains of two different sizes. Typically these are large feldspar grains, often with well-formed crystal faces, surrounded by smaller grains of other minerals.

KEY FACTS

This form of granite commonly has large pink or red alkali feldspar grains. It is often the material of monuments.

+ **grain size:** Medium to coarse

+ **texture:** Crystalline; magmatic alignments of large crystals are possible.

+ **composition:** Quartz + alkali feldspar + plagioclase feldspar ± hornblende ± biotite ± muscovite

The term "porphyritic" describes igneous rocks that have minerals with two grain sizes. The larger grains are called phenocrysts, from the Greek *phaino*, meaning "visible," and *cryst*, meaning "crystal." Porphyritic granites often have pink or red alkali feldspar phenocrysts surrounded by smaller grains of plagioclase, quartz, biotite, and/or hornblende. These phenocrysts form when the alkali feldspar is the first mineral to crystallize out of the magma. The feldspars grow before the other minerals crystallize and fill the remaining space. Porphyritic granites are used for countertops, wall claddings, and monuments.

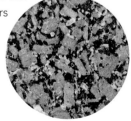

Rapakivi Granite
Type: Igneous

Rapakivi granites are granites with a striking texture of large, rounded feldspar grains with different colored rims. The name "rapakivi" is Finnish and means "crumbly stone."

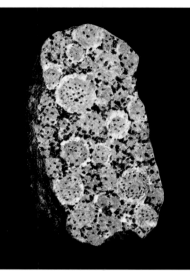

Rapakivi granites are common decorative stones with large pink or brown alkali feldspar grains surrounded by gray or white plagioclase rims. The large grains, called phenocrysts or megacrysts, typically have a rounded shape. Rapakivi granites are found on all continents, usually in the older continental cores. The building stone known as Baltic brown is a rapakivi granite from Finland that is about 1.6 billion years old. It is a popular stone used for kitchen countertops and flooring. The groundmass of this granitic rock is a dark color because it is rich in biotite and hornblende.

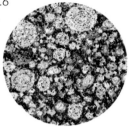

Aplite
Type: Igneous

Aplites are light-colored, fine-grained intrusive rocks that commonly form dikes. Most are found associated with crosscutting large granitic bodies and have a composition similar to granite.

Cutting across many exposures of granitic rocks are relatively narrow dikes that are conspicuously light in color and fine in grain size. These are called aplite dikes and they typically have bulk compositions that are similar to granite. Although the grains are small, quartz, feldspar, and sometimes muscovite are usually visible. Aplites are thought to be formed from the last remnants of the magma that created the surrounding granite. This magma must have cooled quickly to produce the fine texture. Aplites are often more resistant to weathering than the surrounding rock, and can project out as ridges.

Pegmatite

Type: Igneous

A pegmatite is an intrusive igneous body with very coarse-grained, interlocking crystals. Many pegmatites are granitic. Pegmatites are sites for finding rare mineral specimens.

KEY FACTS

Crosscutting dikes, veins, or irregular intrusions with very large grains are features of pegmatite.

+ **grain size:** Very coarse

+ **texture:** Crystalline; grain size and mineral distribution can be heterogeneous.

+ **composition:** Variable; most pegmatites are granitic.

Pegmatites are intrusive bodies of rock notable for very coarse grain size. The crystals vary from a few inches to several yards across. Pegmatite intrusions can be dikes, veins, and lenses that cut across preexistent rock. Geologists differ about how pegmatites form; most agree that they form from the last remaining melt in a magma chamber, which is saturated in water and other fluids. This melt must cool very slowly to grow such large crystals. Its concentration of rare elements leads to crystallization of exotic minerals, including many gems. Pegmatites are important sources for mining feldspar and mica.

Granodiorite

Type: Igneous

Granodiorite is an intrusive igneous rock that resembles granite but contains less alkali feldspar and more plagioclase than a true granite. It is a common building and decorative stone.

KEY FACTS

Usually light to medium gray overall, granodiorite has interlocking white, gray, and black mineral grains.

+ **grain size:** Medium to fine

+ **texture:** Crystalline

+ **composition:** Quartz + plagioclase + alkali feldspar ± hornblende ± biotite

Granodiorite has an intermediate composition between felsic granite and mafic gabbro. It has an overall light- to medium-gray appearance, dominated by white or gray grains of plagioclase and quartz, plus black grains of biotite and hornblende. Granodiorite is often lumped with granite and described as a granitic rock or granitoid. It is a common building and decorative stone, used for countertops, building facades, and structural blocks. The rocks of the Sierra Nevada are primarily granites and granodiorites, with granodiorite making up many iconic formations in Yosemite National Park.

Syenite
Type: Igneous

Syenite is an uncommon intrusive igneous rock similar to granite but with little or no quartz and abundant alkali feldspar. Nepheline syenites are syenites that contain nepheline and no quartz.

KEY FACTS

Syenite is dominated by alkali feldspar grains.

+ **grain size:** Medium to coarse

+ **texture:** Crystalline, sometimes porphyritic

+ **composition:** Alkali feldspar ± quartz or feldspathoid (nepheline) ± plagioclase ± amphibole or pyroxene ± biotite

Syenite is an intrusive igneous rock that is predominantly alkali feldspar. Quartz is either absent or present in small amounts. If the rock contains nepheline, which is never present with quartz, it is known as a nepheline syenite. Syenites are formed in the thick continental crust from magma derived from partially melting granitic rocks, and are associated with failed continental rift valleys. Syenites are mined for use in the glass and ceramics industries to lower the melting temperature of a glass or ceramic mixture. They are also used as decorative stones, especially a blue variety that gains its color from the blue mineral sodalite.

Diorite
Type: Igneous

Diorite is a medium- to dark-gray intrusive igneous rock made of plagioclase feldspar and a dark mineral, usually hornblende. It is commonly found in association with granite and granodiorite.

KEY FACTS

Diorite has a salt-and-pepper appearance with mixed light and dark grains.

+ **grain size:** Medium to coarse

+ **texture:** Crystalline; can be porphyritic

+ **composition:** Plagioclase + hornblende ± biotite

Diorite is an intrusive igneous rock made up of predominantly plagioclase feldspar plus an iron and magnesium mineral component, typically hornblende, sometimes with biotite. It contains less quartz than granodiorite and less quartz and alkali feldspar than granite. Because of its mineral assemblage, diorite is medium to dark gray in overall color, with a salt-and-pepper appearance created by the light plagioclase grains and black hornblende grains. Inclusions of diorite are common in some granitic bodies. Diorite is used as a structural stone, a countertop material, and a decorative stone.

Gabbro
Type: Igneous

Gabbro is a dark-gray, coarse-grained intrusive igneous rock made predominantly of plagioclase and pyroxene. It is an important component of oceanic crust.

KEY FACTS

Gabbro is a plutonic rock that is dark gray to black.

+ grain size: Coarse

+ texture: Crystalline

+ composition: Plagioclase + pyroxene ± olivine

Gabbro is a mafic rock, meaning it is silica-poor and iron- and magnesium-rich. It contains plagioclase feldspar plus pyroxene, usually augite, sometimes with olivine and/or hornblende. Gabbros with predominant orthopyroxene are called norites. Plagioclase in gabbro tends to be darker gray than in granitic rocks. Gabbro is coarse grained because it cools slowly at depth instead of erupting near or on the surface like basalt. It is an important component of oceanic crust. Exposures of gabbro can indicate rock formations that originated as ocean floor but were thrust onto continents by movement of the Earth's tectonic plates.

Anorthosite
Type: Igneous

Anorthosite is more than 90 percent plagioclase feldspar. It is uncommon, found in isolated exposures in some mountain belts and in ancient cores of the continents.

KEY FACTS

Anorthosite has interlocking grains of plagioclase feldspar, some with iridescent labradorite.

+ grain size: Coarse

+ texture: Crystalline; some with plagioclase grains in a dark-colored matrix

+ composition: 90 percent plagioclase, 10 percent ferromagnesian minerals

Anorthosite is an uncommon plutonic rock made almost entirely of plagioclase feldspar. Depending on the feldspar's color, anorthosites can be light gray or bluish. Anorthosite crystals are typically quite large. Some rock samples brought back from the moon are anorthosite. The origins of anorthosite are still debated among geologists. One hypothesis is that it formed from accumulation of plagioclase crystals that floated to the top of a magma chamber and then rose to shallower levels in the crust as a buoyant crystal mush. Anorthosite over 1.1 billion years old is exposed in the Adirondack Mountains of New York.

Peridotite

Type: Igneous

Peridotite is the dominant rock of the Earth's upper mantle. It is a dense, iron- and magnesium-rich intrusive igneous rock made of olivine and pyroxene.

KEY FACTS

Most peridotites have a reddish weathering surface.

+ grain size: Medium to coarse

+ texture: Crystalline; some layered; olivine often altered to serpentine

+ composition: Olivine + pyroxene ± hornblende + accessory minerals, including chromite

Peridotite is an ultramafic rock, meaning it is poor in silica and rich in iron and magnesium, resulting in a dense, often dark-colored rock. Many peridotite outcrops are slabs or pieces of mantle rock thrust onto the continents by large-scale movement of the Earth's plates during the construction of mountains or brought to the surface by volcanic eruptions. Peridotite comes in several subvarieties and many different textures, depending on the relative amounts of olivine and pyroxene, their chemical makeup, and the rock's magmatic history. Most peridotite outcrops contain secondary minerals including serpentine and talc.

Dunite

Type: Igneous

Dunite is a type of peridotite made of more than 90 percent olivine. Fresh exposures are a beautiful green, but most outcrops have a tan or brown weathering surface.

KEY FACTS

Dunite is an intrusive igneous rock of equigranular olivine with black grains of chromite as a common accessory.

+ grain size: Medium to coarse

+ texture: Crystalline, typically equigranular

+ composition: More than 90 percent olivine

Dunite is an uncommon rock that is made almost entirely of olivine. Fresh, unaltered dunites have an equigranular texture of interlocking, glassy green olivine grains. These rocks may include black grains of chromite either dispersed with the olivine or aggregated in chromite layers. A significant supply of commercial chromium comes from chromite concentrations in dunite. Dunite is named after Dun Mountain in New Zealand, where it is part of a group of rocks that were formerly pieces of the ocean floor. Most dunite outcrops have a tan or brown weathering surface. In others, the olivine has been partially or completely altered to serpentine.

Rhyolite
Type: Igneous

Rhyolite is the extrusive equivalent of granite. It is a light-colored, fine-grained volcanic rock that forms from explosive eruptions of a highly viscous lava.

KEY FACTS

Rhyolite is a light to medium gray or pink volcanic rock with occasional flow banding; blocks are brittle with a flinty appearance.

+ grain size: Fine

+ texture: Can be glassy or porphyritic

+ composition: Same as granite

Rhyolite is an extrusive igneous rock that forms from the volcanic eruption of a felsic (silica-rich) magma. Rhyolitic lava is extremely viscous and sticky, so it does not flow very far or accumulate into typical volcanic cones; instead, it builds rounded lava domes. Rhyolitic eruptions are explosive and produce a lot of ash. Rhyolites' appearances include very fine-grained, glassy varieties, and porphyritic varieties with visible quartz. Rhyolite has few dark, iron- and magnesium-rich mineral phases, so it is typically light in color and weight. The countertop and tile stone known as *porfido trentino* is a rhyolite from Italy.

Pumice
Type: Igneous

Pumice is a volcanic rock characterized by many cavities of various sizes called vesicles. Pumice forms when the lava in a volcanic eruption turns frothy from the expulsion of water and gases.

KEY FACTS

A volcanic rock with many holes (vesicles), pumice floats in water. It is typically light in color.

+ grain size: None to very fine

+ texture: Rough surface, vesicular

+ composition: Varies; commonly rhyolite or andesite

Pumice is made of volcanic glass that is extremely porous due to its many gas bubbles, called vesicles. Pumice forms from magma that is saturated in gases and water. During volcanic eruptions, these are released, creating a frothy foam-like lava. This lava cools so quickly that no crystals are able to grow, and the resulting rock is called a glass. Pumice is usually light in color and typically has the same composition as a rhyolite or andesite, depending on the composition of the magma from which it formed. Pumice is used as an abrasive material, including a bath accessory for smoothing rough skin.

Obsidian
Type: Igneous

Obsidian is volcanic glass that forms from viscous lava, typically in rhyolite lava domes. The brittle nature of obsidian and its conchoidal fracture make it an ideal material for sharp tools.

KEY FACTS

This shiny volcanic glass has a conchoidal fracture; banding and inclusions are common; colors are variable.

+ grain size: None to very fine

+ texture: Glassy

+ composition: Can vary; commonly rhyolite

Obsidian is glassy and brittle with a conchoidal fracture, meaning it breaks irregularly along curved surfaces. Paleolithic hunter-gatherers exploited these properties, chipping conchoidal flakes from obsidian pieces to create sharp projectile points and other tools. Obsidian forms in lava domes from lava so sticky and viscous that it is cooled too rapidly for minerals to crystallize, resulting in a glass. Silica-rich lavas like rhyolite or dacite are the most viscous and, thus, most common form of obsidian. Obsidian is frequently opaque black, but it can have various colors and patterns, including intricately banded forms.

Tuff
Type: Igneous

Volcanic eruptions produce massive amounts of ash and other material known as tephra. If enough heat is retained after it falls to the ground, tephra will fuse together, forming tuff.

KEY FACTS

Tuff is a volcanic rock made of consolidated ash and other volcanic fragments; can be porous.

+ grain size: Variable

+ texture: Visible volcanic clasts; possible layering or bedding

+ composition: Can vary; commonly rhyolite or andesite

Material ejected into the atmosphere during a volcanic eruption is classified by size. Particles smaller than 0.08 in (2 mm) are known as ash, particles between 0.08 and 2.5 in (2 and 64 mm) are called lapilli, and anything larger is a volcanic bomb. Collectively, this material is called "tephra," the Greek word for ash. When tephra falls to the Earth and solidifies, it forms a rock called tuff. Tuffs can also form when tephra becomes slowly compacted and cemented into a rock. Tuffs vary in their composition and appearance. Most have visible volcanic fragments of different sizes and shapes, including many that are angular.

Dacite
Type: Igneous

Dacite is a volcanic rock of intermediate composition between rhyolite and andesite. It is the volcanic equivalent of granodiorite. Dacite is associated with highly explosive volcanic eruptions.

KEY FACTS

Dacite often has visible quartz and/or plagioclase; it can be light gray or dark gray to black.

+ **grain size:** Fine to medium

+ **texture:** Commonly porphyritic with phenocrysts of plagioclase, quartz, biotite, or hornblende

+ **composition:** Same as granodiorite

Dacite is a volcanic rock that is slightly less rich in silica than rhyolite, but like rhyolite, comes from a highly viscous lava and produces violently explosive volcanic eruptions. Dacite forms thick rounded lava domes on the surface of volcanoes, including the lava dome at Mount St. Helens. Dacite is named after a locality in the mountains of Romania, part of a region known during the Roman Empire as Dacia. Dacite is typically light gray, but some can be dark gray or even black. Dacite is common in volcanoes that form along continental subduction zones—margins of continents beneath which oceanic slabs sink.

Andesite
Type: Igneous

Andesite is a volcanic rock similar to basalt but with more silica. It is the volcanic equivalent of diorite. Andesitic volcanoes are steep-sided cones, common around the Pacific Rim.

KEY FACTS

A common igneous rock, andesite is typically medium to dark gray; can have greenish or reddish hues.

+ **grain size:** Fine to medium

+ **texture:** Commonly porphyritic with phenocrysts of plagioclase and/or pyroxene or hornblende

+ **composition:** Same as diorite

Andesite is an extrusive igneous rock that forms from the lava flows of stratovolcanoes. Stratovolcanoes are a type of composite volcano built from alternating layers of lava, ash, and cinders, resulting in a steep-sided, symmetrical cone. Some of the world's most beautiful mountains are andesitic stratovolcanoes, including Mount Fuji in Japan and Mount Shasta, Mount Hood, and Mount Adams in the Pacific Northwest. Andesite is named for the Andes Mountains of South America where the rock is also common. Andesite is typically porphyritic, with visible crystals of plagioclase or pyroxene set in a fine-grained groundmass.

Diabase
Type: Igneous

Diabase, also called dolerite, has the same composition as basalt and gabbro. These are distinguished by their grain size: basalt fine, diabase medium, and gabbro coarse.

KEY FACTS

Diabase forms dikes and sills; most outcrops have a light-brown weathered surface; fresh surfaces are dark gray or green overall.

+ **grain size:** Medium

+ **texture:** Crystalline; commonly porphyritic

+ **composition:** Plagioclase + pyroxene + olivine

Diabase, also known as dolerite, is a dark-colored rock that forms dikes, sills, and other relatively small igneous bodies. Diabase is compositionally identical to basalt and gabbro, and all three rock types come from the same type of magma. Basalt, however, forms when the magma erupts onto the Earth's surface, and gabbro forms when the magma cools slowly at great depths. Some outcrops called trap-rock are made of diabase. Diabase traprock near Washington, D.C., is quarried and crushed for use in concrete and as road base material. The Palisades of the Hudson River is an enormous sill made of diabase.

Basalt
Type: Igneous

Basalt forms extensive lava flows worldwide. The oceanic crust is composed primarily of basalt, and the lunar "maria"—the dark regions visible on the moon—are also basalt.

KEY FACTS

A common dark-colored volcanic rock, basalt is typically black, gray, or brown; surfaces vary from smooth to sharp and cindery.

+ **grain size:** Fine

+ **texture:** Can be porphyritic

+ **composition:** Plagioclase + pyroxene + olivine

Basalt is one of the most common rock types. It is a mafic igneous rock with an aphanitic texture, meaning its grains are so fine that they cannot be distinguished with the naked eye; thus, the rock looks relatively uniform. Porphyritic basalts are exceptions; they contain visible plagioclase or olivine crystals surrounded by a fine-grained groundmass. Basalt lava flows cover much of the Earth's surface, including the ocean floor, volcanoes in Hawaii and Iceland, and large regions of continents known as flood basalts. Columbia River Basalts are flood basalts that extend across Washington, Oregon, and Idaho.

Sandstone
Type: Sedimentary

Sandstone is made of compacted and cemented sand-size particles. Sandstones vary depending on composition of the grains, type and amount of cement, and depositional environment.

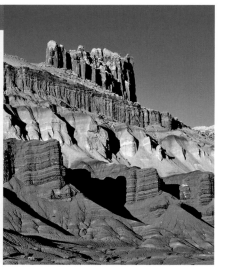

KEY FACTS

Sandstone is a sedimentary rock made of sand-sized grains.

+ grain size: Medium

+ texture: Clastic

+ composition: Variable; common components include quartz, feldspar, mica, and rock fragments.

Sandstone is one of the most common sedimentary rocks. It is made of particles or clasts that fall within the size range that classifies sand, 0.0008 to 0.08 in (0.02 to 2 mm). This is a broad definition—the only parameter being grain size—thus, the rocks that are identified as sandstones can vary widely. Depending on how rounded or angular their grains are and the amount of compaction of the sediment, sandstones can be quite porous and thus serve as important reservoirs worldwide for resources such as water and hydrocarbons. Sandstones are used for building facades and unpolished floor tiles. Decorative flagstone is typically sandstone.

Quartz Arenite
Type: Sedimentary

Quartz arenite is a sandstone composed almost entirely of quartz. Quartz arenites form from accumulations of quartz sand, commonly in windblown deserts, beaches, and high-energy rivers.

KEY FACTS

This sedimentary rock is made of sand-size grains of quartz; cement is often quartz but can also be calcite or hematite.

+ grain size: Medium

+ texture: Clastic

+ composition: More than 90 percent quartz

An arenite is a texturally mature sandstone, meaning it is of consistent grain size and shape, and contains few to no clay particles between the grains. Its grains are predominantly quartz. Sediments mature during weathering, transportation, and deposition, as less resistant minerals break down and more resistant minerals become rounded. Quartz arenites form from sediment deposited far from the source, typically in environments such as sand dunes, coastlines, and rivers. These sandstones can take on different colors, commonly cream, orange, or red, depending on small amounts of iron oxide coating the sand grains.

Graywacke
Type: Sedimentary

A wacke is a variety of sandstone with mud or clay in the matrix that surrounds the sand grains. A graywacke is a type in which many of the sand grains are rock fragments.

KEY FACTS

An immature, poorly sorted sandstone, graywacke is typically dark.

+ grain size: At least 50 percent sand size with fine-grained matrix

+ texture: Clastic, poorly sorted

+ composition: Rock fragments, feldspar, quartz, minerals rich in iron and magnesium, clay, and mud

Unlike arenites, wackes are sandstones that are not well sorted—they contain a mixture of grain sizes, from clay to sand, as well as grain shapes from angular to rounded. The term "graywacke" refers to wackes with a composition that includes numerous rock fragments as well as minerals such as micas. The minerals and fragments are fragile, so their preservation in a sandstone indicates that the sediment source cannot be far. Geologists interpret many graywackes as deposits of turbidity currents—strong currents in water moving down a slope. These rocks are sometimes called turbidites instead of graywackes.

Arkose
Type: Sedimentary

An arkose is a type of sandstone made up of at least 25 percent feldspar. Arkoses typically also contain quartz and mica grains and rock fragments.

KEY FACTS

A sedimentary rock, arkose is made of sand-size grains of feldspar and quartz; cement is commonly quartz or calcite.

+ grain size: Medium

+ texture: Clastic

+ composition: More than 25 percent feldspar

An arkose or arkosic sandstone contains a moderate to high percentage of feldspar grains. Feldspar is susceptible to breakdown via chemical weathering, so its presence in a sandstone indicates that the sediment source is not far from its depositional environment. Arkoses thus tend to form relatively close to mountains or other uplands from which feldspar-rich sediment, typically from granitic rocks, is shed. In fact, geologists sometimes use arkosic sandstones in the sedimentary record as an indication of previous mountain belts. Arkoses are often pink or red, but can also have gray or even greenish hues.

Limestone
Type: Sedimentary

Limestone is made up mainly of calcite (calcium carbonate). It is primarily a marine sedimentary rock, though some freshwater limestones are known to occur.

KEY FACTS

A light-colored sedimentary rock, limestone is often fine grained; some formations are rich in marine fossils. It fizzes in contact with hydrochloric acid.

+ **grain size:** Very fine to coarse

+ **texture:** Variable

+ **composition:** Primarily calcite

Calcium carbonate can chemically precipitate out of solution in sea or lake water, settling to the bottom, accumulating as a sediment, and forming limestone. Limestone may also form from accumulated animal shells made of calcium carbonate. Thus, fossils are common in limestones; some limestone formations are made entirely of fossil shells. Some limestones contain small, spherical beads of calcite called oolites that develop when calcium carbonate precipitates around a particle such as a sand grain. Limestone forms cliffs in arid environments, but forms caves and other karst landforms in humid environments.

Dolomite/Dolostone
Type: Sedimentary

Dolomite is similar to limestone but with a high percentage of the mineral dolomite—calcium magnesium carbonate—instead of calcite. Dolomite rock is sometimes called dolostone.

KEY FACTS

Although it is similar to limestone, dolomite does not fizz as actively in hydrochloric acid; dolomite's weathering surface is a buff color, whereas limestone tends to be grayer.

+ **grain size:** Fine to medium

+ **texture:** Compact, relatively homogeneous

+ **composition:** Primarily dolomite

Dolomite, also known as dolostone, is similar to limestone and is often found in association with limestone. Most dolomite originated as limestone or lime mud, and became dolomite when calcite (calcium carbonate) was replaced by dolomite (calcium magnesium carbonate), a process called dolomitization. Dolomite and limestone can be difficult to distinguish in the field. Dolomite tends to weather to yellowish beige or brown, whereas limestone tends to be gray. Dolomite is named after French geologist Déodat de Dolomieu, who first described it in the 1700s from cliffs in the Italian Alps now called the Dolomites.

Travertine
Type: Sedimentary

Travertine is a type of limestone formed via direct chemical precipitation of calcite out of solution. This commonly occurs in association with the waters around hot springs.

KEY FACTS

Deposits of travertine are associated with caves, hot springs, and geysers; typically layered with light colors; fossils are common.

+ grain size: Very fine

+ texture: Massive, nonclastic, porous

+ composition: Calcite (calcium carbonate)

Travertine is a beautiful ivory- to peach-colored stone that forms around hot springs when calcium carbonate falls out of solution. This occurs when carbon dioxide is released as heated waters rise from depth or are agitated as in waterfalls and cascades. The results are porous, finely layered deposits of calcium carbonate. Travertine is a popular building stone. Much of Rome, which is surrounded by volcanic thermal springs, was constructed with travertine, including the Colosseum. Travertine is also used to clad modern buildings including the Getty Center in Los Angeles and Lincoln Center in New York.

Chalk
Type: Sedimentary

Chalk is a variety of limestone made of tiny skeletons of minute marine organisms. The skeletons, composed of calcium carbonate, accumulate on the floors of shallow seas.

KEY FACTS

White, pure limestone chalk is relatively soft, made of microscopic shells with a carbonate mud matrix.

+ grain size: Fine

+ texture: Bioclastic

+ composition: Calcite (calcium carbonate)

Chalk is a relatively pure limestone made of accumulated shells of marine microorganisms cemented together with a carbonate mud. Many chalk deposits formed during the Cretaceous period, from about 142 to 65 million years ago, a time when global sea levels were high and parts of the continents were flooded with shallow seas. "Cretaceous" comes from the Latin word *creta*, meaning chalk. Many small, calcareous marine organisms thrived in these environments, and deposits of their skeletons built up on the sea floor to form layers of chalk. Chalk formations commonly contain nodules of chert known as flint.

Conglomerate
Type: Sedimentary

Conglomerates are sedimentary rocks composed of particles larger than sand grains, including pebbles, cobbles, and boulders. They form in various depositional environments.

KEY FACTS

Coarse-grained clasts of conglomerates are surrounded by a fine-grained matrix; relative proportions of clasts and matrix vary; size, shape, and composition of clasts also vary.

+ **grain size:** Larger than 0.08 in (2 mm)

+ **texture:** Clastic

+ **composition:** Variable

Sedimentary rocks made of particles larger than sand grains are known as conglomerates. The particles vary from rounded river pebbles to angular fragments of a debris flow. Conglomerates represent high-energy deposition including river- or beach-deposited gravel beds, alluvial fans, glacial till deposits, and mudslides. Conglomerates are classified by the relative proportion of clasts and matrix—the fine-grained material cementing the rock together. Conglomerates in which the clasts touch are known as clast supported. Those in which the clasts are surrounded by matrix and not touching are called matrix supported.

Breccia
Type: Sedimentary

Breccias contain coarse, angular particles surrounded by finer grains. The term "breccia" is used to describe sedimentary formations and rocks with fault-related or igneous origins.

KEY FACTS

Breccias show angular, broken fragments surrounded by a fine-grained matrix.

+ **grain size:** Larger than 0.08 in (2 mm)

+ **texture:** Clastic

+ **composition:** Variable

Breccias are similar to conglomerates but are distinguished by their angular, broken rock fragments instead of rounded clasts. The angular clasts can come from a single source or from multiple rock types. Like other sedimentary rocks, the cement is commonly calcareous or siliceous. Breccias form in various environments. Sedimentary breccias can form from angular fragments of an underwater debris flow, deposits of a landslide, or other mass-wasting events. Fault breccias form when rocks are fragmented during slip along a brittle fault zone. Volcanic breccias form from angular fragments ejected during an eruption.

Shale
Type: Sedimentary

Shale is a sedimentary rock made of silt- and clay-sized grains—particles smaller than sand. It is distinguished by its fissility, meaning its tendency to split into thin layers parallel to bedding.

KEY FACTS

A clastic sedimentary rock, shale splits easily into layers.

+ **grain size:** Very fine

+ **texture:** Fine grains, fissile

+ **composition:** Variable; common mineral grains include quartz, feldspars, mica, calcite, and clays.

Sedimentary rocks made of grains finer than sand—a mixture of silt and clay—are generally known as mudrocks. Shale is a mudrock characterized by its property of easily breaking in one direction parallel to bedding. This is called fissility, and is formed by alignment of platy minerals such as micas and clays. Shale forms from compaction of fine-grained sediment, a process that typically occurs in still or slow-moving water as in bogs, deltas, and deep regions of the continental shelf. Fossil preservation is enhanced in these sediments and quiet depositional environments; thus, shales are common sources of fossils.

Mudstone
Type: Sedimentary

Mudstone is a fine-grained sedimentary rock made up of a mixture of clay- and silt-size particles. Unlike shale, mudstone is massive and does not easily split into layers.

KEY FACTS

A fine-grained sedimentary rock, mudstone does not easily split into layers.

+ **grain size:** Very fine

+ **texture:** Massive, not laminated

+ **composition:** Variable; common mineral grains include quartz, feldspars, mica, calcite, and clays.

Mudstone is a common sedimentary rock. It is clastic, but its particles are generally too small to see with the naked eye. Both mudstone and shale are made of compacted mud; the difference lies in the way these rocks break. Mudstone has a tendency to break into irregular blocks, whereas shale consistently breaks along a plane that is parallel to bedding. Shales and mudstones come in a variety of colors, most commonly gray, but also red, green, and purple. The colors usually indicate the presence of iron—red when iron is oxidized, green when iron is partially reduced.

Gneiss
Type: Metamorphic

Gneiss (pronounced nice) is a metamorphic rock characterized by compositional bands of separated light and dark minerals. These bands define a coarse layering or foliation in the rock.

KEY FACTS

This foliated metamorphic rock is defined by compositional banding, with few platy or needlelike minerals.

+ **grain size:** Medium to coarse

+ **texture:** Crystalline; compositional banding

+ **composition:** Variable; quartz, feldspar, and hornblende are common components.

Gneisses form from metamorphism of preexisting rocks that are typically rich in quartz and/or feldspar, such as granitic rocks and sandy sedimentary formations. Gneisses with an igneous parent rock are classified as orthogneisses; those with a sedimentary parent are called paragneisses. Gneisses form when the parent rocks are exposed to high temperature and pressure in the Earth's crust, usually by burial during mountain building. Gneisses' compositional bands are layered segregations of light-colored minerals such as quartz and feldspar from dark-colored minerals such as hornblende.

Migmatite
Type: Metamorphic

Migmatites, or migmatitic gneisses, are rocks that have partially melted at high temperatures and pressures, resulting in a hybrid rock with both igneous and metamorphic properties.

KEY FACTS

These rocks are heterogeneous, with segregations of light- and dark-colored minerals usually in irregularly folded layers.

+ **grain size:** Medium to coarse

+ **texture:** Crystalline

+ **composition:** Variable; light layers commonly granitic; dark layers generally contain hornblende and biotite.

Migmatites are named from the Greek *migma*, meaning mixture. They are mixtures of igneous and metamorphic components, formed when metamorphism progresses past the point at which the rock begins to melt. The melting is incomplete, producing segregations of melt that become swirly, light-colored igneous layers surrounded by the dark-colored residual metamorphic rock. Migmatites form under very high temperatures and pressures deep in the Earth's crust, as in the cores of large mountain ranges. These rocks are used as decorative stones, and they are often incorrectly called granites.

Amphibolite
Type: Metamorphic

Amphibolite is a rock rich in amphibole—typically hornblende or actinolite—often with plagioclase. Some amphibolites are massive; others have fine foliation or bands.

KEY FACTS

Colored black or dark green, amphibolite is often associated with gneisses and/or schists.

+ **grain size:** Medium to coarse

+ **texture:** Variable; can be massive, foliated, or banded

+ **composition:** Hornblende or actinolite plus plagioclase, sometimes garnet, and other minerals

Amphibolites are widespread metamorphic rocks made predominantly of amphibole plus plagioclase. Some amphibolites display an easily recognizable salt-and-pepper appearance with grains of black hornblende and white plagioclase. Amphibole grains are typically aligned in a weak to moderate foliation. The protolith, or parent rock, of amphibolites is generally difficult to determine. Most amphibolites form by metamorphism of mafic igneous rocks such as basalt and gabbro, though some originate as sedimentary rocks. Amphibolite is used as a building and decorative stone, which is often called "black granite."

Mylonite
Type: Metamorphic

Mylonites are hard, compact, foliated rocks with very fine grains. They commonly display a strong linear fabric defined by elongate minerals.

KEY FACTS

Strong deformational fabrics, compact texture, and fine grain size distinguish mylonites.

+ **grain size:** Very fine; some coarse relict grains common

+ **texture:** Well-developed foliation and lineation

+ **composition:** Variable

Mylonites form when rocks are deformed by movements along a fault. They typically form in deep zones of faults where rocks deform in a ductile fashion, like warm plastic, rather than breaking brittlely. As a rock is sheared, its minerals are pulverized on a microscopic scale into smaller grains. Harder minerals like quartz and feldspars can resist complete pulverization, resulting in larger, rounded, or oval grains surrounded by the fine-grained material. If there is significant shearing, even the resistant grains will be reduced to the fine matrix. Most mylonites form along discrete shear zones.

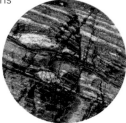

Schist

Type: Metamorphic

Schist is a common metamorphic rock characterized by its schistosity, which is a fine layering formed by the preferred orientation of platy and needlelike minerals, typically micas.

KEY FACTS

A finely layered (schistose) rock, schist has more than 50 percent sheetlike minerals, commonly muscovite, biotite, and other micas, and/or chlorite.

+ grain size: Medium to coarse

+ texture: Well-developed schistosity

+ composition: Mica- or chlorite-rich, otherwise variable

The term "schist" comes from the Greek *schistos*, meaning "divided." Schists typically split along parallel planes defined by the alignment of platy minerals such as micas. These planes are a type of foliation called schistosity. Schists form from metamorphism of fine-grained sedimentary rocks such as mudstones and shales. These rocks typically first form slate, then phyllite, an intermediate rock; eventually, as temperatures and pressures increase, they form schist. Schists can also develop from metamorphism of other fine-grained rocks such as volcanic tuffs. Some schists display tight chevron-type folds.

Garnet Schist

Type: Metamorphic

Schists can be classified according to their significant mineral components; for example, a biotite schist or a muscovite–chlorite schist. Garnet schist or garnet–mica schist features garnet.

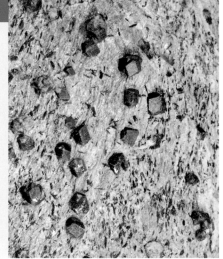

KEY FACTS

This form of schist has rounded garnet crystals, typically deep red, embedded in a scaly matrix of smaller, aligned micas.

+ grain size: Medium

+ texture: Well-developed schistosity; commonly larger than other minerals

+ composition Typically muscovite and garnet; can include biotite, staurolite, kyanite, or sillimanite

If a mudstone or shale is buried at depth by large-scale movements in the Earth's crust (for example, during the building of a mountain belt), the rock undergoes progressively higher pressures and temperatures. These conditions cause the rock to metamorphose into a schist. Types of minerals that grow in the schist depend on the peak pressure and temperature conditions reached by the rock. Thus, the mineral makeup of some schists is an important tool for scientists to understand the geologic history of a region. Garnet schists can indicate that a rock reached relatively high pressures and/or temperatures.

Greenschist

Type: Metamorphic

These fine-grained metamorphic rocks are rich in chlorite, epidote, or actinolite. The minerals give the rock an overall greenish hue.

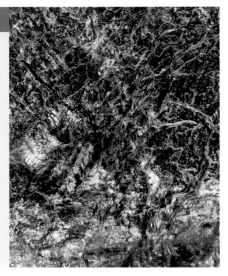

KEY FACTS

Greenschist is a chlorite schist or other foliated metamorphic rock with chlorite and/or actinolite.

+ grain size: Medium

+ texture: Foliated

+ composition: Chlorite, epidote, actinolite

The term "greenschist" usually describes a chlorite schist, though it can refer to other foliated metamorphic rocks that are rich in green-colored minerals such as chlorite, actinolite, and epidote. Greenschists have a foliation defined by the alignment of chlorite grains. Most greenschists form when a fine-grained mafic volcanic rock such as a basalt is metamorphosed at low temperatures and pressures. The term "greenschist" also describes metamorphic conditions. Greenschist metamorphism occurs during low to medium pressures and temperatures. For example, a slate is not a greenschist, but it develops under greenschist conditions.

Blueschist

Type: Metamorphic

Blueschist is a metamorphic rock that typically has a blue hue from the presence of the mineral glaucophane, a sodium-rich amphibole. The term also describes a set of metamorphic conditions.

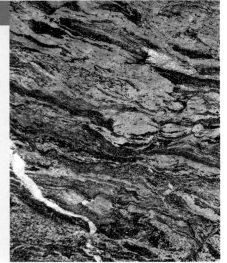

KEY FACTS

Blueschist is a foliated metamorphic rock with a bluish color.

+ grain size: Medium

+ texture: Foliated, not necessarily schistose

+ composition: Variable; blue color comes from glaucophane

A blueschist rock forms under blueschist metamorphic conditions, but not all rocks that experience blueschist metamorphism turn into blueschists. Confused yet? The complication comes from two different uses for the word "blueschist," a similar scenario to the use of the word "greenschist." The rock type called blueschist is a metamorphic rock that contains the bluish amphibole glaucophane. The term "blueschist" can also refer to a set of metamorphic conditions characterized by high pressures and low temperatures. These conditions occur along subduction zones, where one of the Earth's plates is pulled down beneath another.

Slate

Type: Metamorphic

Slate is a fine-grained metamorphic rock that splits easily along a well-developed foliation. Slate is commonly used for roofing tiles as well as for other decorative stones.

KEY FACTS

Typically gray or greenish, slate is a brittle rock with well-developed foliation.

+ **grain size:** Fine

+ **texture:** Compact, well foliated

+ **composition:** Rich in micas and quartz, though often too fine-grained to distinguish individual minerals

Slate is a common metamorphic rock characterized by its ability to split into very thin sheets or plates, some of which are quite large. Slate is actively quarried for use as floor tiles, hearths, and roofing tiles. The original blackboards in classrooms were made of single sheets of slate, though most blackboards today are composed of man-made materials. Slate forms via the metamorphism of mudstones, shales, or tuffs at low pressures and temperatures. If metamorphism progresses, mica crystals grow larger, turning a slate into a schist. Quartz veins and small faults are common in slate outcrops.

Marble

Type: Metamorphic

The metamorphism of limestone or dolomite results in formation of marble, a rock that is composed almost entirely of recrystallized, interlocking grains of calcite or dolomite.

KEY FACTS

A dense stone, marble is usually pure white, gray, or yellow; other varieties are pink, blue, or black.

+ **grain size:** Fine to coarse

+ **texture:** Crystalline; often massive; some show banding

+ **composition:** Calcite and/or dolomite; accessory minerals can include micas, graphite, and serpentine.

Some of the most famous sculptures and buildings are created out of marble, a type of metamorphic rock. Michelangelo's "David" is carved out of an Italian stone known as Carrara marble. The Taj Mahal in India is clad almost entirely in a marble known as White Makrana. In the United States, the Supreme Court Building and the Lincoln Memorial are built with slabs of Alabama marble. Marble is white when it is pure calcite. More commonly, marbles contain impurities and other minerals resulting in different colors and textures. Some coarse-grained marbles can look similar to quartzite, but marble is much softer.

Serpentinite
Type: Metamorphic

Serpentinite is composed of serpentine, often with chromite or magnetite and veins of calcite, dolomite, or talc. It forms by alteration and metamorphism of ultramafic rocks.

KEY FACTS

Light yellow-green, dark green, or black, serpentinite is typically brittle with a greasy or waxy feel.

+ grain size: Fine

+ texture: Variable; brecciated forms, veins common

+ composition: Primarily serpentine; other minerals may include chromite, magnetite, and talc.

Serpentinite is a striking rock, commonly streaked with various colors from yellow-green to black, with a glossy surface that is greasy or even soapy to the touch. Serpentinite forms when olivine-rich rocks like peridotite are altered in the presence of water, a process known as serpentinization. The formation of serpentinites occurs at low metamorphic grades. Serpentinite is a popular stone used for floor tiles and countertops in buildings, and as an ornamental stone. The popular decorative stone known as *verd antique* ("antique green") is a serpentinite with a brecciated or angular, broken texture.

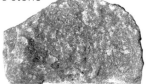

Quartzite
Type: Metamorphic

Quartzite is extremely hard and compact, formed by metamorphism of quartz-rich sandstones or chert at elevated temperatures and pressures.

KEY FACTS

Hard, compact, and quartz-rich, quartzite has a distinctive sugary appearance. It is commonly white or gray.

+ grain size: Medium

+ texture: Compact, interlocking grains

+ composition: Primarily quartz; other minerals may include micas and/or hematite.

Quartzite forms when quartz grains of the protolith—the parent rock before metamorphism—are heated and squeezed together until they form new grains. This process creates a homogeneous mosaic of interlocking grains, resulting in a dense, very hard stone. Textures from the protolith—rounded quartz grains surrounded by fine quartz cement, for example—are destroyed. Quartzites are distinguished by their hardness, conchoidal fracture, and sugary texture in outcrop. The decorative stone aventurine is quartzite. Green aventurine, used in jewelry, is colored by a chromium-rich mica known as fuchsite.

Plants

Plants are among the most diverse groups of organisms. Their fossils help scientists understand the evolution of life as well as changes in the Earth's climate through geologic time.

KEY FACTS

Preserved parts of plants include leaves, bark, seeds, spores, and pollen.

+ time period: About 450 million years ago to present

+ size: Variable

+ where to find: Fine-grained sedimentary rocks

Plant fossils are the geologic record of ancient plants. The first land plants appeared about 450 million years ago. Complete fossils are rare; most are fragments such as leaves, cones, and stems. Most plants decompose rapidly; to become fossilized, specific conditions have to occur. The plant must be buried by sediment quickly enough to shield it from decomposition, but gently enough to preserve its integrity. Thus, most plant fossils are found in fine-grained sedimentary rocks such as shales and mudstones. Fossil plants, especially leaves, spores, and pollen grains, are important to our understanding of past climates.

Petrified Wood

Petrified wood is a plant fossil preserved by the infiltration of minerals around the organic material, a process called permineralization. Most petrified wood is made of quartz or calcite.

KEY FACTS

Three-dimensional replicas of woody plants, including trunks, branches, and bark; harder and more brittle than unfossilized wood.

+ time period: About 400 million years ago to present

+ size: Variable

+ where to find: Eroded out of sedimentary rock formations

Ancient plants fossilized by permineralization are called petrified wood. The pore spaces around and within the plant's organic material are filled by minerals, usually microcrystalline quartz. The organisms retain their original structures, so permineralized fossils are three-dimensional forms, not casts or impressions. Petrified wood is found throughout the U.S. and Canada, especially in western states. In Arizona's Petrified Forest National Park, entire trees over 200 million years old were fossilized in buried logjams of ancient rivers. Petrified wood can be colorful due to the presence of iron, carbon, and manganese.

Graptolites

Graptolites are a group of small, free-floating marine organisms that flourished during the early and middle Paleozoic eras. Graptolite fossils are found worldwide, mostly in black marine shales.

KEY FACTS

These small, twiglike fossils have saw-blade shapes, some in spirals.

+ time period: 500–315 million years ago

+ size: Typically 1–4 in (2.5–10 cm); some as long as 8–10 in (20–25 cm)

+ where to find: Marine shales of Ordovician, Silurian, and Devonian periods

Graptolites are a group of relatively simple marine animals that lived during the early explosion of life in the Earth's oceans beginning about 500 million years ago. Graptolites are wormlike animals that lived in colonies floating through seawater like plankton. Graptolites constructed tubelike shells out of soft collagen, and the tubes were linked to form branching colonies. Many graptolite fossils resemble plant leaves or branches of plants. Others take the shape of tiny saws, tuning forks, and spirals. Graptolites are typically found in black marine shales associated with calm waters of the outer continental shelf.

Bryozoans

Bryozoans are colonial aquatic organisms that resemble small corals. They are made of many small individuals living in compartments of a calcareous skeleton

KEY FACTS

Lacy or fanlike fossils resemble moss or corals.

+ time period: 488 million years ago to present

+ size: Individuals are less than 1 mm; colonies range from millimeters to meters.

+ where to find: Shallow-marine limestone formations; also shales and mudstones

Bryozoans are colonial organisms that secrete calcareous shells with many compartments. Each compartment contains a body cavity with a retractable food-gathering arm called a lophophore. The lophophore is a group of tentacles armed with beating fibers called cilia that gather food particles from the seawater. Most bryozoans attach themselves to the sea bottom onto a rocky substrate or the discarded shell of another organism. They take many different forms, including sheetlike crusts, netted fans, and small branching trees. They build reefs. Bryozoan fossils are most common in shallow marine limestone formations.

Corals

Corals live in tropical marine settings and are the principal component of reefs. Corals are ancient animals, with fossil specimens dating back more than 500 million years.

KEY FACTS

Skeletons of coral have various shapes including horns, brains, spirals, branches, and disks.

+ time period: 540 million years ago to present

+ size: Variable; millimeters to meters

+ where to find: Limestones

Corals belong to the phylum Cnidaria (organisms with stinging cells), which includes sea anemones and jellyfish—simple, soft-bodied marine organisms. Corals construct a skeleton made of calcium carbonate and are well preserved as fossils. Living corals consist of a soft body cavity called a polyp with a single opening, surrounded by tentacles that gather food from the water. The polyps are attached to a hard skeleton that they construct throughout their life. Coral fossils display many shapes, including disks, cylinders, spirals, and brain-like domes. Because of their limited ecological tolerances, corals are indicators of ancient environmental conditions.

Trilobites

Trilobites are ancient animals, now extinct, that dominated the world's oceans during the Cambrian period. Related to crustaceans, they had a hard outer shell that is well preserved.

KEY FACTS

Body is segmented, and has an armored shell and crescent-shaped head.

+ time period: 540–20 million years ago

+ size: Typically 1–4 in (2.5–10 cm)

+ where to find: Fine-grained marine sedimentary rocks; famous locales are Burgess Shale in British Columbia, Beecher's Trilobite Bed in New York, and Wheeler Shale in Utah.

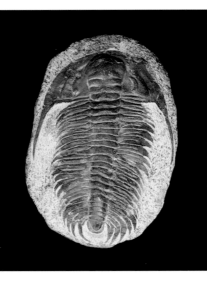

Trilobites are among the earliest arthropods—a classification of invertebrate animals with exoskeletons, including insects, spiders, and crustaceans. Most trilobites were 1 to 4 inches (2.5 to 10 cm) long, though one of the largest trilobites, found on the shore of Hudson Bay, measures 28 inches (71 cm). Trilobites were prolific and geographically dispersed, roaming worldwide Paleozoic oceans in deep and shallow water environments. These factors, plus the easy preservation of their shells, resulted in trilobite fossils being both common and widespread around the globe. Trilobite fossils are important time markers for sedimentary formations.

Brachiopods

Brachiopods, also known as lampshells, are marine animals with two shells connected at a hinge. At first glance, they appear similar to bivalves, though these animals are unrelated.

KEY FACTS

The two shells of brachiopods are symmetrical along the midline, perpendicular to the hinge, with the bottom shell typically larger than the top shell.

+ **time period:** 540 million years ago to present

+ **size:** Most are about 0.25–4 in (0.6–10 cm); largest up to 8–12 in (20–30 cm)

+ **where to find:** Limestones and marine shales

Brachiopods are marine organisms with two shells, similar to clams and mussels. Clams and mussels, however, are bivalves, unrelated to brachiopods and different in many ways. In brachiopod shells, the left half is a mirror image of the right; they are commonly asymmetrical top to bottom, with the bottom larger than the top. Though only a few hundred species live today, brachiopods dominated the world's oceans approximately 540 to 252 million years ago, when tens of thousands of species flourished. Many species died off during a mass extinction event approximately 250 million years ago at the end of the Permian period.

Bivalves

Bivalves are marine and freshwater organisms with two shells joined along a hinge. Fossil bivalves are found worldwide from all time periods since the Cambrian.

KEY FACTS

The two shells of bivalves are symmetrical parallel to their hinge; the valves are mirror images of each other.

+ **time period:** 540 million years ago to present

+ **size:** Most are 0.4–4 in (1–10 cm); some are as large as 3.3 ft (1 m) or more.

+ **where to find:** Limestones and shales

Bivalves are aquatic animals with two shells that meet along a hinge. These include living clams, oysters, mussels, and scallops. Bivalves have a muscular foot that emerges from the shell and allows some species to burrow into the sand or mud. Other bivalves anchor themselves to rocks or coral reefs. Others, such as scallops, swim by clapping their shells together, forcing out water to propel them. Fossil bivalves are common, and they come in a variety of shapes and sizes. The symmetry of their two shells is key to their identification. In contrast, brachiopod symmetry is perpendicular to the orientation of the hinge.

Gastropods

Gastropods are snails, most with coiled, hard shells made of calcium carbonate. They are found in a range of habitats, including marine and freshwater environments as well as terrestrial habitats.

KEY FACTS

Snails and sea slugs are gastropods. Snail shells come in various shapes with single chambers. Sea slugs have internal shells, and some lack a shell.

+ time period: About 540 million years ago to present

+ size: about 0.04–35 in (1 mm–90 cm)

+ where to find: Fine-grained clastic sedimentary rocks and limestones

Gastropods first appeared during the early Cambrian period about 540 million years ago. They are a large and diverse group, with species living in marine environments, freshwater, and terrestrial habitats. Most gastropods have a coiled shell with a single chamber. Unlike bivalves, gastropods have heads that are distinguished from their bodies. Like bivalves, they have a single, muscular foot that helps them move. Gastropod and bivalve fossils are abundant and widespread. Because of their excellent preservation and the diversity of their species, they are important to scientists for understanding evolutionary processes.

Ammonites

Ammonites are an extinct group of mollusks best known for their beautiful spiral shells. Ammonite fossils are prolific in the geologic record.

KEY FACTS

The shells of ammonites are divided into chambers.

+ time period: About 400–65 million years ago

+ size: Up to about 6.5 ft (2 m)

+ where to find: Marine sedimentary rock formations; key localities include Badlands National Park in South Dakota and Guadalupe Mountains National Park in Texas.

Ammonites are marine animals related to squid, cuttlefish, and octopuses, but went extinct about 65 million years ago, the same time as dinosaurs' demise. They are best known for their distinctive spiral shells, though a few species evolved with nonspiral forms. Ammonites were prolific and diverse, with many species. Their fossils are among the most abundant on Earth and serve as excellent time markers, allowing geologists to link the rock layers in which they are found to specific time periods. The name "ammonite" refers to the Egyptian god Ammon, who wore ram's horns resembling the spiral shape of the shells.

Echinoderms

Echinoderms are a diverse group of marine animals including sea stars, sea urchins, and sand dollars. Echinoderms' skeletal plates made of calcite are well preserved as fossils.

KEY FACTS

Fossils of echinoderms show a 5-point radial symmetry.

+ time period: 540 million years ago to present

+ size: About 0.08–8 in (0.2–20 cm); some crinoid stalks are much longer.

+ where to find: Limestones and marine shales

Echinoderms are a broad group of marine organisms that appeared in the early Cambrian period about 540 million years ago, which has many living members today. The most recognizable are sea stars, sea cucumbers, sea urchins, and sand dollars, which have a unique skeleton made of interlocking plates of calcite. These interior, or endoskeletons, called tests, enclose the organs and are covered with a spiny skin. The skeletons also display a radial symmetry, often with five points, such as the five limbs of a sea star. Echinoderm skeletons are widespread in the fossil record and are an important indicator of past marine conditions.

Crinoids

Crinoids are marine animals that look like flowers, with a flexible stalk topped with a head of waving arms. They are echinoderms, related to sea stars and sea urchins.

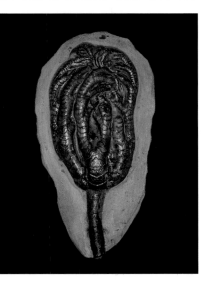

KEY FACTS

Fragmented stems look like small, circular plates; complete fossils are rare.

+ time period: 488 million years ago to present

+ size: Up to about 130 ft (40 m)

+ where to find: Limestones and marine shales; excellent localities are early Devonian sandstones in West Virginia.

Crinoids, like all echinoderms, have a skeleton interlocking calcite plates that house their vital organs. Crinoids are sessile—meaning that they live attached to some sort of substrate such as a rock, shell, or reef—or they are free-floating, planktonic organisms. Crinoids have three basic body parts: a stem, a cup or calyx, and arms. The cup and arms look like tassels or blossoms atop stems. Crinoid stems are flexible cylinders made of small, circular plates with a star-shaped interior canal. Fragments of stems are common fossils, especially in limestones in the western U.S. and Canada. Some limestone formations consist entirely of crinoid parts.

Fish

Animals classified as fish include modern bony fish, sharks, and rays, as well as many species that are extinct. Some mudstones and shales contain exquisitely well-preserved fish fossils.

KEY FACTS

Most fish have elongate bodies with fins, and are covered with scales or bony plates.

+ **time period:** 500 million years ago to present

+ **size:** Variable; some fossil fish reach over 65 ft (20 m) long.

+ **where to find:** Fine-grained sedimentary rocks; famous locales include Fossil Butte National Monument in Wyoming.

Fish are the first vertebrates in the fossil record, about 500 million years ago. Those fish were jawless, and bony plates covered all or most of their bodies, instead of a hard internal skeleton. Over time, different fish appeared, including sharks and rays about 400 million years ago, and eventually fish with internal bony skeletons about 200 million years ago. The Niobrara Formation in the central and western U.S. contains abundant fish fossils from an interior seaway that existed about 85 million years ago. Wyoming's Green River Formation, about 48 million years old, contains spectacular beds of fossil fish.

Shark Teeth

Shark teeth are relatively common fossils and a favorite of collectors. Sharks shed thousands of replaceable teeth during their lifetime—an abundant source for fossilization in ocean sediments.

KEY FACTS

Individual teeth with various shapes can be found.

+ **time period:** 450 million years ago to present

+ **size:** Up to about 7 in (18 cm)

+ **where to find:** Along beaches and eroding out of marine sedimentary rock formations, even far inland such as in many localities in Wyoming

Sharks are among a group of fish that includes skates and rays. They first appeared about 450 million years ago and are distinguished by an internal skeleton of cartilage, not bone. Without hard parts, sharks are rarely preserved as fossils, except their teeth. Most teeth become fossilized by minerals filling in the pores around the tooth material. This process takes thousands of years and is why many fossilized shark teeth are dark instead of the light whites and yellows of living sharks' teeth. A famous fossilized shark is the giant Megalodon, which lived approximately 20 million to 2 million years ago and whose teeth are up to about 7 inches (18 cm) long.

Insects

Insects are among the most diverse group of animals. Wingless insects first appeared about 400 million years ago, and flying insects appeared about 360 million years ago.

KEY FACTS

Wings with a network of veins are more often fossilized than other body parts; whole insects are found in amber.

+ time period:
About 400 million years ago to present

+ size: Up to about 2 ft (0.6 m)

+ where to find:
Fine-grained sedimentary rocks; famous locales include Florissant Fossil Beds in Colorado.

Insects, one of the most diverse groups of animals on the planet, were the first to develop flight, approximately 360 million years ago. Insects have segmented bodies: a head, a middle section called a thorax, and an abdomen. They have a hard outer skeleton, and their legs are jointed. Insects were particularly abundant about 305 to 270 million years ago, when gigantic dragonflies had wingspans over 2 feet (0.6 m) long. Insects are relatively poorly preserved as fossils because of their fragile bodies. One exception is amber, or fossilized tree resin, which preserves entire insects that became trapped in tree sap.

Amphibians & Reptiles

Amphibians and reptiles are among the earliest animals to inhabit the land. These groups evolved to include a wide diversity of species, both living and extinct.

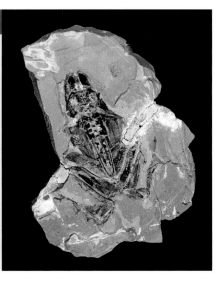

KEY FACTS

Tetrapods, named from the Greek words for "four-footed," are four-limbed vertebrates, and the earliest of these ancient animals able to walk on land were amphibians and reptiles.

+ time period:
368 million years ago to present

+ size: Up to about 130 ft (40 m) long

+ where to find:
Fine-grained sedimentary rocks

The first vertebrates to "migrate" from water to the land were amphibians known as the early tetrapods, or four-legged animals. A famous fossil is a 360-million-year-old skeleton of a large creature called Ichthyostega found in Greenland. It more resembled crocodiles than today's amphibians. Others include the bizarre Diplocaulus, a giant newt-like creature with a boomerang-shaped skull. Fossils resembling modern frogs, toads, newts, and salamanders are found in rocks younger than 200 million years old. Reptiles evolved later, and their abundant fossils include turtle shells, lizard skeletons, dinosaur bones, and reptile eggs.

Dinosaurs

Dinosaurs are part of the diapsid group of reptiles. They grew to enormous proportions and dominated most terrestrial habitats until their sudden demise about 65 million years ago.

KEY FACTS

Large bones are found in Triassic to Cretaceous period rocks; dinosaurs had two skull openings behind their eye sockets.

+ time period: 225–65 million years ago

+ size: Variable; up to about 190 ft (58 m) long

+ where to find: Sedimentary rocks; Dinosaur National Monument in Colorado and Utah is a famous site.

Dinosaurs are diverse reptiles that populated the Earth for 160 million years. There are two main groups: Ornithischia, "bird-hipped," with rear-facing pubic bones, and Saurischia, "reptile-hipped," with forward-facing pubic bones. Ornithischia include Triceratops and Stegosaurus. Saurischia include herbivorous sauropods such as Apatosaurus and carnivorous theropods such as Tyrannosaurus and Velociraptor. Paleontologists trace the ancestry of birds to theropods. Recent discoveries indicate that some dinosaurs had extensive feathered plumage. Dinosaur fossils range from footprints to fragments of bone to entire skeletons.

Birds

Bird fossils are of great interest to scientists tracing the evolutionary history from dinosaurs to modern birds. Bird fossils, however, are uncommon because of their fragile hollow bones.

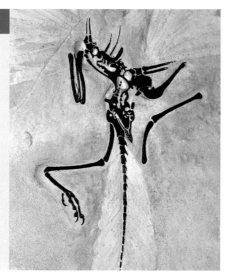

KEY FACTS

Bird fossils are uncommon; evidence for feathers is an important feature.

+ time period: About 150 million years ago to present

+ size: Variable

+ where to find: Sedimentary rock formations; uncommon

One of the oldest and most famous early bird fossils is Archaeopteryx, from a 150-million-year-old limestone quarry in Germany. Archaeopteryx has features of both dinosaurs and birds and has been interpreted as an intermediate evolutionary stage. Like modern birds, Archaeopteryx had wings and feathers, but like dinosaurs, it had teeth, claws, and a long tail. As more discoveries of feathered dinosaurs emerge, however, interpretations of Archaeopteryx and the transition from dinosaurs to modern birds are changing. Birds continued to diversify after non-avian dinosaurs' extinction, and by about 35 million years ago, most groups of modern birds had appeared.

Mammals

The 65-million-year period after dinosaurs' extinction is often considered the Age of Mammals. Their fossils help scientists understand climate changes and movements of the continents.

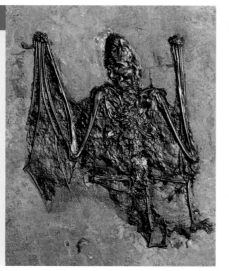

KEY FACTS

Mammal fossils range from bone fragments and teeth to entire skeletons and even hair.

+ **time period:**
About 150 million years ago to present

+ **size:** Variable

+ **where to find:**
Widespread; famous localities include La Brea Tar Pits in Los Angeles, Mammoth Site in South Dakota, and sites in western Nebraska.

Mammals are warm-blooded, have hair, and feed milk to their young. They appeared around the same time as dinosaurs, but mammals didn't flourish until after dinosaurs' demise about 65 million years ago. Early mammals were relatively small, looked similar to rodents, and fed on insects. As mammals diversified, they became larger and dominated nearly every terrestrial ecosystem. Mammal fossils, especially teeth, are widespread and well preserved. Mammals in famous fossil finds from the Pleistocene, the period of the last ice ages, include skeletons of mammoths, mastodons, saber-toothed cats, native horses, and camels.

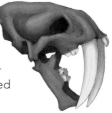

Trace Fossils

Trace fossils include footprints, tracks, burrows, bite marks—even fossilized dung. These are traces of living organisms that have been preserved in the geologic record.

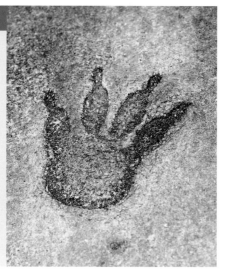

KEY FACTS

Burrow marks, footprints, feeding marks, and excrement are preserved primarily in sedimentary rocks.

+ **time period:**
About 542 million to 10,000 years ago

+ **size:** Variable

+ **where to find:**
Sedimentary rocks; locales include dinosaur tracks at Dinosaur State Park in Connecticut and Dinosaur Valley State Park in Texas.

Trace fossils preserve organisms' activities rather than their body parts. Invertebrates' trace fossils include impressions of feeding, burrowing, and boring, mostly preserved in fine-grained marine sedimentary rocks. Trace fossils from vertebrates, especially land animals, include bite marks and footprints, even pieces of fossilized dung known as coprolites. Trace fossils provide an understanding of how prehistoric organisms lived, moved, and ate. Sometimes it is difficult to distinguish a trace fossil from a pseudofossil, a geological formation that looks similar to something made by an organism but is inorganic.

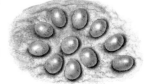

Lineation
Deformed rocks

Lineations are linear structures in deformed rocks. There are several types, including the hinges of folds and the long axes of aligned minerals and stretched clasts.

KEY FACTS

Lineations are structures that can be represented by a line: long in one dimension relative to others.

+ fact: Look for lineation along the surface of a foliation plane.

+ fact: Deformed rocks with strong linear fabrics and without planar fabrics (foliation) are L-tectonites.

+ fact: Common in mountain outcrops and ancient cores of continents

A lineation is a visible texture of linear elements that develops when a rock is deformed. For example, a conglomerate originally made of rounded cobbles develops a lineation if it is squeezed such that the cobbles are stretched out into elongate, cigar-like shapes (it develops a foliation if the clasts are flattened into pancakes). Other common lineations are defined by the alignment of elongate minerals, which can form as metamorphic minerals grow into a preferred orientation, or they can be original minerals that are rotated into a linear fabric during deformation.

Foliation
Deformed rocks

Deformed rocks that exhibit layering or any through-going textures that are planar are said to have foliation. These rocks may split or break along foliation planes, or may be more massive.

KEY FACTS

Foliation is an arrangement of planar or tabular features in a deformed rock.

+ fact: Rocks can have more than one type of foliation in more than one orientation.

+ fact: Layering in schists and gneisses is an easy way to identify foliation.

+ fact: Common in outcrops in mountain belts and the ancient cores of continents

A rock has a foliation if it has a visible texture or fabric created by planar or tabular elements. These form as a rock is squeezed and flattened by forces in the Earth's crust that are typical of mountain-building processes. Slates are metamorphic rocks that cleave or break along foliation planes. The fine layering in schists created by aligned plates of mica is another type. In higher-grade metamorphic rocks such as gneisses, light and dark layers of different minerals define a foliation. Sometimes rocks display more than one foliation.

Augen
Deformed rocks

Augen are large, lens-shaped minera grains found in deformed rocks. Augen are typically surrounded by fine grains of mica and other minerals aligned in wavy layering or foliation.

Augen, German for "eyes," are eye-shaped minerals or clusters of minerals in some deformed rocks. They can be relict crystals from the original rock that were rotated and rounded during high-temperature deformation, or they can be newly grown metamorphic crystals that developed during deformation. Augen are typically harder, compact minerals such as feldspar, quartz, and garnet. They are surrounded by micas and other platy crystals that are aligned in a wavy foliation. Augen form in areas of the Earth's crust called shear zones, where rocks move past each other at relatively high temperatures.

Boudinage
Deformed rocks

Boudinage forms when a relatively rigid rock layer is stretched and pulled apart into sausage shapes known as boudins. Surrounding layers appear to flow around the boudins.

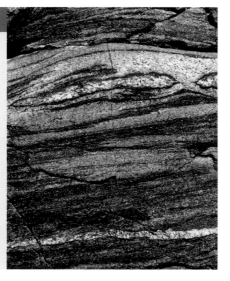

Boudinage is a structure in veins or rock layers that are more rigid or competent than the surrounding rock. When these are stretched, the competent vein or layer is pulled apart. The result is an effect that resembles a link of sausages. Boudinage is from the French *boudin,* a type of sausage. Boudinage structures range from thin, undulating ribbons in narrow veins, to larger, blocky, rotated tablets. Boudinage is an example of how different rocks react to the same forces: Rock around boudins behaves like heated plastic, flowing in a solid state to accommodate the strain, while the mechanically stronger boudin layer pulls apart like taffy.

Folds
Deformed rocks

Folds are bends in bodies of rock that form during shortening or contraction. Folds occur at all scales, from deformation of individual crystals to structures the size of mountain ranges.

KEY FACTS

This deformation refers to folded, bent, or crenulated features in a rock mass.

+ fact: Found in all types of rock

+ fact: Small folds in an outcrop can mirror larger folds on a landform scale.

+ fact: Large folds create parallel ridges and valleys in the Appalachian region of the U.S.

Folds are formed when rocks shorten or contract due to movements in the Earth's crust and upper mantle. Folds can be found at all scales, from bent individual mineral grains to enormous warps defining mountain ranges. Geologists have many terms to describe different kinds of folds. Folds in the shape of a lowercase *n* are called anticlines; folds in the shape of a lowercase *u* are called synclines. Folds can be upright, asymmetrical, or recumbent. Tight, Z-shaped folds are called crenulations or chevrons. Folded sedimentary rocks can form traps for accumulation of oil and natural gas.

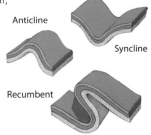

Anticline

Syncline

Recumbent

Faults
Deformed rocks

Faults are brittle fractures that accommodate movement in rocks. The body of rock on one side of a fault moves relative to the body on the other side by sliding along the fault plane.

KEY FACTS

In faults, movement is apparent by offsets or juxtaposition of different rock formations.

+ fact: Movement, or slip, releases energy as earthquakes.

+ fact: Faults are rarely single, clean fractures; they are typically networks of fractures and broken rock.

+ fact: Active faults are common along boundaries of the Earth's plates.

Three types of faults are thrust or reverse, normal, and strike-slip. Thrust or reverse faults are planes along which deeper rocks move upward relative to shallower rocks. These faults are caused by compressional forces in the crust. Normal faults are planes along which younger, shallower rocks drop down relative to older, deeper rocks. Normal faults accommodate stretching or thinning in the crust and are common in the U.S. Basin and Range province. Strike-slip faults are planes along which rocks slide laterally past each other without significant vertical movement. The San Andreas Fault is a famous strike-slip fault.

Joints
Deformed rocks

Joints are naturally occurring cracks found in almost all rock types. Joints are distinguished from faults by a lack of movement across the fracture plane.

KEY FACTS

To see a joint, look for brittle cracks that cut across rock faces.

+ fact: Joints form from stresses in the Earth's crust.

+ fact: Joints are important conduits for groundwater in aquifers.

+ fact: Joints can have various orientations, appearing in regular sets or as isolated fractures.

Many rock outcrops contain brittle cracks, called joints. There is no movement along joints, however. If there is evidence of sliding along a fracture, the crack is called a fault, not a joint. Joints form by forces that act upon a body of rock. As the Earth's plates move, forces are distributed through the crust, causing some rocks to crack. Rocks can also crack during changes in temperature, from uplift to shallower depths in the Earth's crust, or by cooling of magma. Joints occur as isolated cracks or in widespread fractures. They are affected by weathering and erosion, and are an important control on the evolution of a landscape.

Veins
Deformed rocks

Veins are narrow, sheetlike bodies made up of one or more minerals. Veins form when minerals crystallize out of water-rich fluids.

KEY FACTS

Veins appear as crosscutting stripes or stringers on the surface of an outcrop.

+ fact: The most common veins are white and are filled with quartz or calcite.

+ fact: Veins can also be host to rare minerals and ore deposits.

+ fact: Veins are found in all rock types.

Cutting across many outcrops are narrow, sheetlike bodies made up of one or more minerals. These are called veins, and they are filled with minerals that crystallized out of water-rich fluids. In some cases, veins form when fluids infiltrate existing cracks or joints in a rock formation. Veins following crack systems may form individual tabular bodies or spidery networks of irregular shapes. Some veins are tightly folded, indicating high-temperature deformation of the rock body after the vein crystallized. Veins are important sources of metals and other precious minerals, including gold and copper.

Tafoni
Weathering & erosion

Tafoni, or honeycombing, is a surface feature marked by rounded pits, hollows, and shallow caverns, often clustered in groups and separated by hardened ridges or visors.

KEY FACTS

These features are easiest to identify in cliff faces with a honeycomb appearance.

+ fact: Develop in many different rock types

+ fact: Appear as small pits clustered together or large, rounded hollows

+ fact: Found in sandstone cliff faces and some granitic bodies; common in coastal environments

Clusters of hollows, rounded pits, and shallow caverns on rock faces and outcrops are known as tafoni, honeycombing, or cavernous weathering. Geologists have posed many hypotheses for how cavernous weathering develops. Most agree that salt crystallization is key. Salts crystallize in tiny pits on the surface of a rock, transported by wind or water. The salt expands, breaking off adjacent grains of the host rock and expanding the pit. This process forms larger and larger rounded hollows. In some porous sandstones, cement hardening along internal joints or bedding planes may be important in channeling erosion.

Desert Varnish
Weathering & erosion

Desert varnish is a surface coating common on the sandstone rock walls of the desert Southwest, though it can form on many rock types and in different environments.

KEY FACTS

The varnish is a crust on the surface of a rock; dark colors vary from reddish brown to shiny black.

+ fact: Crust is usually less than 0.02 in (0.5 mm) thick.

+ fact: Common on exposed rock faces in arid environments

+ fact: Many petroglyphs are created by chipping through desert varnish to paler rock beneath.

Desert varnish is a dark, hard coating made of clay minerals, manganese oxide, and iron oxide. It is found on many different rock exposures. Geologists have debated the varnish's formation because manganese and iron oxides could indicate bacterial metabolism. Rocks on Mars appear to have desert varnish; however, most geologists agree that it can develop without the aid of life. Much of the varnish is composed of silica, which reaches the surface via clay minerals in windblown dust or by leaching from the rock's interior. Black, shiny desert varnish is rich in manganese oxide; dull reddish varnish is rich in iron oxide.

Spheroidal Weathering
Weathering & erosion

Spheroidal weathering, also known as onion-skin weathering, is a process that creates rounded outcrop formations and boulders in the shape of spheres.

KEY FACTS

Exposed rounded stones may be with or without surrounding disintegrating crusts.

+ **fact:** Can occur in many different rock types

+ **fact:** Rounded products of spheroidal weathering are called "core stones."

+ **fact:** In granite, the disintegrating rock around core stones is called "grus."

Spheroidal weathering is a type of weathering that results in rounded outcrop formations and spherical boulders. Spheroidal weathering occurs when a rock formation is broken into blocks by several sets of fractures called joints. Groundwater penetrates the rock formation along the joints, facilitating chemical weathering at a faster rate than where the rock is not fractured. Physical weathering is also accelerated along joints. Weathering progresses, rounding off the edges of two intersecting joints and the corners of three intersecting joints, producing curved rock formations and even perfect spheres.

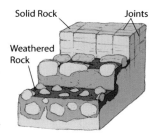

Sheeting or Exfoliation Joints
Weathering & erosion

Sheeting joints, also known as exfoliation joints, are curved fractures that parallel the surface of a body of rock. The joints separate sheets of rock, creating dome-like landforms.

KEY FACTS

These joints are most common in granite but can form in other rock types.

+ **fact:** Sheeting joints separate slabs of rock of various thickness.

+ **fact:** Weathering and erosion along sheeting joints create spectacular landforms such as Half Dome in Yosemite National Park.

+ **fact:** Rockfall can occur when slabs fail along the joints.

Sheeting joints, also known as exfoliation joints, are common features of granitic terrains. These are sets of curved fractures oriented roughly parallel to the surface of a rock. The joints separate concentric slabs or shells of rock that progressively weather and erode, creating domes and other rounded landforms. One explanation for sheeting joints is that they form during the release of pressure when rocks overlying a body of granite are removed by erosion—a removal called "unroofing." Another explanation is that sheeting joints form from mechanical stresses along the surface of a dome.

Dikes
Igneous terrain

Dikes are sheets of igneous rock that crosscut surrounding bedrock at a steep angle. Basalt dikes commonly occur in areas of extension in the Earth's crust.

KEY FACTS

This band or sheet of rock crosscuts surrounding rock at a steep angle.

+ fact: Dikes are made of igneous rock, but they can crosscut any rock type.

+ fact: Large numbers of dikes in a region are called dike swarms.

+ fact: Widespread; common in volcanic terrain, granitic regions, and mountain belts

Dikes are igneous intrusions that crosscut bedrock as steeply angled or vertical sheets. Different types of dikes form in different magmatic environments. Dikes form when magma rises in the Earth's crust until reaching neutral buoyancy—the same buoyancy as surrounding rock. Then, the magma spreads laterally, vertically, or horizontally depending on local stresses and weaknesses. Vertical orientations create dikes; lateral orientations create sills. Dikes with finer grain sizes along their margins than in their interior are called chilled margins. These form when magma cools more quickly in contact with surrounding rock.

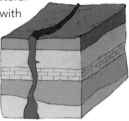

Sills
Igneous terrain

Sills are intrusive sheets of igneous rock oriented horizontally or parallel to the fabric of the surrounding bedrock. They are common in sedimentary basins intruded by magma.

KEY FACTS

This band or sheet of igneous rock parallels the structure of the surrounding bedrock.

+ fact: Sills are commonly thicker than dikes.

+ fact: Sills consist of igneous rock but can intrude any rock type.

+ fact: Widespread; common in volcanic terrain and regions of sedimentary bedrock

Sills are sheetlike intrusive bodies oriented parallel to the layering of surrounding rock, commonly horizontal in flat sedimentary basins. Sills are like dikes, forming from lenses of buoyant magma that spread laterally in the crust. In sills, stresses in the Earth's crust favor emplacement horizontally or parallel to existing structures. These generally form near the Earth's surface and can be thicker than dikes because pressures tend to be lower closer to the surface. The Palisades Sill in New York and New Jersey is a diabase sill that forms spectacular cliffs along the Hudson River.

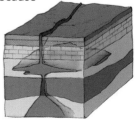

Inclusions
Igneous terrain

Some intrusive igneous rocks contain inclusions of other rocks. Inclusions can result from magma or from blocks of surrounding rock entrained in magma as it cooled.

KEY FACTS

Inclusions are angular blocks or rounded blobs of rock that are compositionally different from the surrounding rock.

+ fact: Inclusions are found in both intrusive and extrusive igneous rocks.

+ fact: Some inclusions are called xenoliths, from the Greek word *xeno*, meaning "foreign."

+ fact: Some xenoliths contain diamonds.

Large exposures of intrusive igneous rocks such as granites are heterogeneous. Granites often show variability in grain size, crystal orientation, and mineralogical makeup. Some granites (and other igneous bodies) contain inclusions of other rock types. Inclusions can be chunks of mantle rock brought to the surface by ascending magma. These inclusions, called xenoliths, can tell us about rocks deep in the Earth's interior. Inclusions of basalt and/or diorite are also common in granitic bodies. These are thought to be blobs of mafic magma that became entrained when the granite was not fully hardened.

Volcanic Bombs
Igneous terrain

Volcanic bombs are pieces of lava ejected during volcanic eruptions. These masses solidify in the air before landing and can have shapes like teardrops, ribbons, or balls.

KEY FACTS

Chunks of volcanic rock have distinctive shapes including teardrops, ribbons, and spindles.

+ fact: Volcanic bombs are 2.5 in (6.4 cm) to many feet in diameter.

+ fact: Can form out of different types of lava; basaltic bombs are common.

+ fact: Occur on flanks of volcanoes such as Mauna Kea in Hawaii

Volcanic bombs are ejected chunks of magma that form streamlined masses as they harden while flying through the air. Volcanic bombs are larger than about 2.5 inches (6.4 cm) in diameter. Smaller particles are called lapilli or ash. Bombs commonly form elongate spheroidal bodies, sometimes with twisted "tails" or spindle shapes. Some are not completely solidified when they hit the surface, and they flatten upon impact. Other shapes include teardrops and ribbons. Volcanic bombs are ejected along vents that liberate gases as large, bursting bubbles and indicate more explosive varieties of volcanoes.

A'a Lava
Igneous terrain

A'a (pronounced ah-ah) is a Hawaiian word for a type of lava flow characterized by sharp, angular rubble. It forms when the top parts of a flow harden and break into chunks.

KEY FACTS

These lava flows produce piles of sharp, rubbly volcanic rock.

+ fact: Develops where lava flows over steep slopes

+ fact: Flows typically advance faster than pahoehoe flows.

+ fact: A'a is common on the Hawaiian Islands.

Basaltic lava flows have three types: a'a, pahoehoe, and pillow lava. A'a creates piles of sharp, jagged rubble. A'a forms when the surface of a flow hardens into a solid and then breaks up into fragments because of the continued movement of lava beneath. The fragments tumble down the front of the flow and are entrained beneath it as molten lava in the interior continues to advance. This movement is similar to the tread of an advancing bulldozer. A'a flows create thick piles of sharp fragments and blocks that are difficult to walk on. Chunks of a'a are black or red volcanic rocks often used in landscaping.

Pahoehoe Lava
Igneous terrain

Pahoehoe (pronounced pa-hoy-hoy) is a Hawaiian word for a type of lava flow characterized by smooth surfaces with wrinkles and ropelike ridges.

KEY FACTS

Pahoehoe forms hardened lava flows with smooth surfaces and curved ridges.

+ fact: Pahoehoe can transition to a'a, but a'a does not turn into pahoehoe.

+ fact: Footpaths on volcanic rock often follow pahoehoe flows.

+ fact: Pahoehoe is common on the Hawaiian Islands.

Pahoehoe, unlike a'a, is a smooth and cohesive lava flow. It forms when the surface of molten lava cools into a thin, smooth skin. The skin is then pushed into folds by faster-moving lava below. Pahoehoe flows can eventually develop into lava tubes. When the flows harden, accumulations of the surface folds create wrinkles and ropelike masses. The surface of pahoehoe is relatively smooth, and footpaths on volcanic rock usually follow exposures of pahoehoe. Basaltic volcanoes often have both pahoehoe and a'a lava flows, depending on the viscosity and temperature of the lava as well as on the shape of the terrain.

Pillow Lava
Igneous terrain

Pillow lava is a type of lava flow that forms underwater. Outcrops look like stacks of elongate lobes and rounded blobs. Exposures of ancient pillow lavas can be highly deformed.

KEY FACTS

Formations consist of stacked lobes and protrusions, circular or elliptical in cross section.

+ fact: Outer layer is typically smooth and can be glassy.

+ fact: Pillow lavas in ancient rocks indicate the presence of water.

+ fact: Active pillow lavas are under Hawaiian waters; ancient pillow lavas are in uplifted oceanic rocks.

Pillow lava is formed when lava erupts underwater, along mid-ocean ridges and submarine flanks of seamounts and other volcanoes. Pillow lava is named for the globular shape of lava as it cools underwater. It develops because the surface of the lava rapidly cools and hardens to form a crust. As the molten lava continues to advance, it breaks through the crust in the shape of rounded blobs or tubes that are hard outside and molten inside. These blobs pinch off from the crust and roll down to the front of the flow. Pillow lavas in exposed ancient rocks are evidence of past oceanic environments.

Lava Tubes
Igneous terrain

Lava tubes are open tunnels made of shells of volcanic rock, typically basalt. They are conduits for molten lava. Some volcanoes have networks of lava tubes along their flanks.

KEY FACTS

Open tunnels are formed by hardened volcanic rock, commonly basalt.

+ fact: Lava tubes form in flows of low-viscosity lava.

+ fact: Lava tubes can create intricate networks of caves.

+ fact: Well-known lava tubes are found in New Mexico, Hawaii, California, Oregon, and Washington.

Lava tubes are large hollow tunnels through which lava travels away from its vent. Lava tubes tend to form in volcanoes that have relatively low eruption rates. High eruption rates can sustain open channels rather than closed tubes. Lava tubes are created when the surface of a moving lava flow cools and hardens into a crust. The crust insulates the lava flow below from the cool atmosphere, allowing it to remain molten. These streams of molten lava typically continue to drain through the tube until the magma supply is cut off. Usually the lava then drains away, leaving an empty tube in the volcanic landscape.

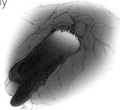

Cinder Cones
Igneous terrain

Cinder cones are small volcanoes that form out of hardened pieces of ejected lava that rain back down from the atmosphere after they are erupted.

KEY FACTS

These cone-shaped hills are made of chunks of hardened lava.

+ fact: Typically less than 1,000 ft (300 m) high

+ fact: They are built up over the course of multiple eruptions over many years.

+ fact: Examples are in Sunset Crater Volcano National Monument in Arizona and Lava Beds National Monument in California.

Cinder cones are a simple type of volcano in which most of the erupted magma is ejected as a fountain-like spray out of a central vent. The magma feeding cinder cones is charged with gasses. These gasses expand as they rise, causing the lava to spray out of the vent. The airborne lava hardens into volcanic bombs and smaller chunks of volcanic rock called cinders, lapilli, and ash. These particles, collectively known as tephra, fall back to Earth and accumulate around the vent, building up a cone-shaped landform. Most cinder cones have a rounded crater at their summit.

Caldera
Igneous terrain

A caldera is a large depression of rock formed in a volcano. Calderas form when volcanic eruptions empty underground magma chambers, causing the rock above to collapse.

KEY FACTS

These large volcanic depressions are often several miles wide.

+ fact: Common features of explosive volcanoes

+ fact: They're sometimes confused with craters, which are smaller, bowl-like formations at the summit of volcanoes.

+ fact: Much of Yellowstone National Park sits within an enormous caldera.

A caldera, named from the Spanish word for cauldron, is a large depression created by volcanic activity. It forms when a magma chamber beneath a volcano is emptied by eruption, causing the rocks above to collapse into a large depression. In very explosive volcanoes, calderas can form by a single eruption. In less explosive volcanoes, magma chambers may take numerous small eruptions to empty; thus, caldera formation is a slow, progressive process. Continued volcanic activity after formation of a caldera may cause the magma chamber to recharge, and then the center of the caldera will rise into a feature called a resurgent dome.

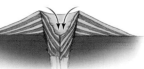

Volcanic Neck or Plug
Igneous terrain

A volcanic neck is the resistant core of a volcano that is exposed by erosion. Volcanic necks, also called plugs, are typically tall, somewhat cylindrical-shaped landforms.

KEY FACTS

A neck is an eroded cylindrical landform made of volcanic rock.

+ **fact:** Often found with eroded dikes; some surround the neck radially.

+ **fact:** Volcanic necks sometimes contain pieces of rock pulled from deep in the Earth during eruption.

+ **fact:** Famous necks include Devils Tower National Monument in Wyoming and Shiprock in New Mexico.

Most volcanoes are fed by a central vent that pipes magma from deep within the Earth. Cone-like outer structures of many volcanoes are created by buildup of lava flows and/or accumulation of ejected ash, cinder, and volcanic bombs. After volcanic activity ends, these are worn away by erosion. Outer flanks of volcanoes are sometimes easily eroded, especially if they consist of unconsolidated cinder and ash. Often the most resistant part of a volcano is the vent that becomes filled with hard volcanic rock at the end of eruption. Remnants of these vents projecting above the landscape are called volcanic necks or plugs.

Columnar Jointing
Igneous terrain

Symmetrical columns of volcanic rock form by a process called columnar jointing. As lava shrinks during cooling, fractures divide the rock into roughly hexagonal columns.

KEY FACTS

Polygonal columns consist of volcanic rock.

+ **fact:** Form in lava flows, ash flows, and shallow intrusions

+ **fact:** Columns can be hundreds of feet high.

+ **fact:** Famous examples are in Devils Postpile National Monument in California and Devils Tower National Monument in Wyoming.

Columnar jointing is a distinctive outcrop feature of some volcanic rocks, especially basalt flows in the West. As thick flows of lava cool, they shrink, causing the hardening lava to crack. In a homogeneous pool of lava without outside stresses, vertical cracks will develop with intersections of approximately 120 degrees, which is the most efficient orientation to accommodate shrinking. This orientation results in vertical columns with hexagonal cross sections. In the real world, however, columnar joints can have variable angles, and the resultant basalt columns take on various polygonal cross sections.

Bedding
Sedimentary terrain

Sedimentary beds are layers of rock with distinguishable tops and bases. Bed boundaries form by changes during deposition of sediment, before the sediment hardens into a rock formation.

KEY FACTS

Layers can be seen in sedimentary rock.

+ fact: Range from the millimeter to tens of meters scale

+ fact: Deposited horizontally with older layers below younger ones

+ fact: Dramatic horizontal patterns in walls of the Grand Canyon are created by bedding planes.

Sedimentary rocks are easily recognized by their layered appearance. The layers, called beds or strata, are made of accumulated sediments compacted and/or cemented into stone. Sediments are transported to their resting place by a mechanism such as wind or water, or they are deposited in place by living organisms such as coral, or by chemical precipitation out of a solution. Changes in this process are caused by changes in the environment or in the sediment source. The changes create natural breaks in the pile of sediment, which later define bed boundaries or bedding planes in the sedimentary rock.

Unconformity
Sedimentary terrain

An unconformity is a break in the sequence of sedimentary beds along which erosion occurred. There are three types: angular unconformities, disconformities, and nonconformities.

KEY FACTS

This feature is caused by disruption in the layering of sedimentary rocks.

+ fact: An unconformity represents a gap in the rock record.

+ fact: Unconformities can indicate deformation, erosion, or changes in sea level.

+ fact: John Wesley Powell described the Great Unconformity in the Grand Canyon in 1869.

In an angular unconformity, sedimentary layers below the boundary are at a different angle from layers above. These occur when the lower rocks are folded and/or faulted so that the layers are no longer horizontal. Later, new sediment is deposited, creating a new horizontal sequence of layers. Disconformities occur when there is a break in sediment deposition and erosion of a surface before the upper layers are deposited. These boundaries often show irregular surfaces between different rock types. A nonconformity is an erosive surface between a nonlayered rock such as granite and a layered sedimentary sequence.

Cross Bedding
Sedimentary terrain

"Cross bedding" is a term for layering that forms within a sedimentary bed and is oriented at an angle to the original horizontal bedding planes of the rock.

KEY FACTS

Angled layers occur within a sedimentary bed.

+ fact: Found in sedimentary rocks, primarily sandstones

+ fact: Some cross beds are a mark of ancient sand dunes.

+ fact: Geologists use cross beds to interpret previous environments and even the direction of water and wind currents.

Cross beds are sublayers in sedimentary rock formations that are oriented at an angle to the bedding planes. Cross beds form from particles moved by water or wind as they are being deposited. The grains accumulate into piles; in windblown sand, the piles can be large dunes. When they grow too high, the grains avalanche down the front and come to rest. Continued movement of particles rebuilds the piles. Repeated cycles create inclined layers or laminations that are preserved in rock. Cross bedding forms dramatic textures in sandstone outcrops in the Southwest, including famous exposures in Zion National Park in Utah.

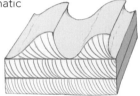

Ripple Marks
Sedimentary terrain

Ripples are low, elongate ridges that form when water or wind move sand particles. Ancient ripples are preserved in sandstones as features known as ripple marks or cross lamination.

KEY FACTS

Planar surfaces of sedimentary rocks contain sets of low, elongate ridges.

+ fact: Found in sedimentary rocks, primarily sandstones

+ fact: Look for ripple marks in outcrops of sedimentary rock in the Southwest.

+ fact: Geologists use ripple marks to interpret prior environments.

Hiking on sedimentary rocks, you may find exposed bedding planes in sandstones that are marked by low, elongate ridges—semiparallel sets of wrinkles protruding from the surface. These are ripple marks, which are essentially the preserved trace of sand ripples that formed during the deposition of the sand particles that make up the rock. Ripples are small, generally about an inch high, and form by the way sand particles aggregate and move as they are transported by currents or waves. Ripples can form from wind in desert environments or by the flow and/ or waves in rivers, lakes, and seas.

Mudcracks
Sedimentary terrain

Mudcracks are the cracks separating polygonal forms in clay and mud beds. Cracks form when wet, fine-grained sediments shrink as they dry.

KEY FACTS

Networks of cracks break dried mud or clay into plates.

+ fact: Ancient mudcracks are preserved in claystones and mudstones.

+ fact: Dinosaur tracks are commonly found in the same rock formations as mudcracks.

+ fact: Look for mudcracks in outcrops in Colorado, New Mexico, Utah, and Arizona.

When a bed of wet, fine-grained sediment such as a layer of clay or mud dries, the surface shrinks and contracts, resulting in a network of cracks. The cracks form plates of sediment with curled edges. The plates can form rough polygonal shapes, commonly hexagons. Environments creating mudcracks include marshes and lake beds that drain and become exposed to the atmosphere. Mudcracks can be preserved in sedimentary rocks if they are filled with sediment and then buried. Geologists use mudcracks in ancient rocks to understand "which way was up" in sedimentary rocks that have been disturbed by folding or faulting.

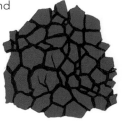

Karst
Sedimentary terrain

Karst is a distinctive type of terrain formed by the chemical dissolution of rocks. It develops in soluble rocks, primarily carbonates like limestone and dolomite.

KEY FACTS

Karst landforms include caves, springs, towers, rocks in fluted shapes, stalactites, and stalagmites.

+ fact: About 10 percent of the Earth's surface is karst.

+ fact: Much of the world's life depends on water from karst areas.

+ fact: Kentucky's Mammoth Cave is a famous karst, and karst is actively forming in Florida.

Carbonate rocks, especially limestone, dissolve in contact with groundwater or seawater that is undersaturated with calcium carbonate. Rock formations made of evaporite deposits such as gypsum and halite will also dissolve in contact with water. Landforms that develop by dissolution of bedrock are called karst. Karst surfaces are marked by springs, sinkholes, and caves. Dissolution along joints and faults can result in karst towers with fluted surfaces. Formations range from tiny chemical precipitates to entire landscapes. Water in underground conduits and caves, called karst aquifers, are important water resources.

Stalactites & Stalagmites
Sedimentary terrain

Stalactites and stalagmites are features of limestone caves. They are a type of cave deposit called dripstones that are formed by drops of water that precipitate calcium carbonate.

KEY FACTS

Stalactites point down from roofs of caves; stalagmites point up from floors.

+ fact: Stalactites and stalagmites are karst formations.

+ fact: Most are made of calcium carbonate.

+ fact: Spectacular examples include Natural Bridges Caverns in Texas, Carlsbad Caverns National Park in New Mexico, and California Caverns.

Stalactites and stalagmites are cave deposits called speleothems. The word "speleothem" is from the Greek *spelaion*, meaning "cave," and *thema*, meaning "deposit." Stalactites are cone or straw shaped, and protrude downward from the roof. These form from water trickling down from cracks in the cave roof. Stalagmites are cone shaped and protrude upward from the floor; they form from drops of water falling from stalactites. One way to remember the difference is to associate the letter *g* in stalagmite with "ground"—stalagmites point up from the ground. Stalactites and stalagmites can grow together forming columns.

Sinkholes
Sedimentary terrain

Sinkholes are areas where part of the bedrock below the land surface has dissolved and the land surface has collapsed into the space below, sometimes catastrophically.

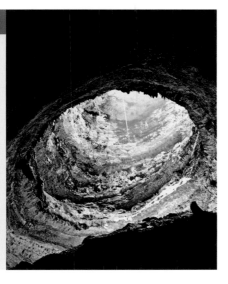

KEY FACTS

Bowl-shaped depressions in the land surface come in various sizes, some as large as hundreds of feet in diameter.

+ fact: Features of karst landscape

+ fact: Can develop rapidly, especially after a heavy rain

+ fact: Common in Florida, Texas, Alabama, Missouri, Kentucky, Tennessee, and Pennsylvania

Sinkholes are landforms that result from a collapse of the land surface. They occur in regions underlain by bedrock that is easily dissolved in groundwater. Examples are carbonate rocks and evaporite deposits such as salt and gypsum. The movement of groundwater causes the rock to dissolve over time, slowly enough that the ground above may stay intact. Eventually, however, dissolution leaves the bedrock unable to support the surface, and the land collapses into the hole. This can happen on various scales and can be catastrophic if buildings or roads are built on the affected surface.

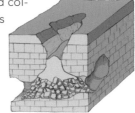

Desert Pavement
Desert environments

Desert pavement is a hard ground surface made of tightly packed pebbles and cobbles overlying a layer of fine sediment. Desert pavements are found in arid and semiarid environments.

KEY FACTS

A thin layer of packed rock fragments and clasts covers the ground.

+ fact: Many clasts are coated with a dark-colored patina called desert varnish.

+ fact: Exposures range from small patches to areas as large as hundreds of square miles.

+ fact: Newspaper Rock in Utah is a famous example.

Desert pavement is a ground surface common in some windblown deserts. It is formed by a thin layer of tightly packed clasts, pebble to cobble size, and generally of various shapes, sizes, and types. Ventifacts—wind-shaped particles—are common among these clasts. Desert pavements are found worldwide, even in Antarctica. The manner of their formation is controversial. Various ideas proposed to explain the features include deposition of the clasts during catastrophic rain events, uplift of the clasts due to wetting and drying or freezing and thawing processes, or creation of the pavement in place by wind erosion.

Ventifacts
Desert environments

Ventifacts are rocks shaped, polished, and etched by windblown sand. They come in all shapes and sizes, from small, polished, aerodynamic pebbles to large mushroom-shaped boulders.

KEY FACTS

These rocks in desert environments have smoothed, polished surfaces and/or pitted or fluted surfaces created by wind erosion.

+ fact: Look for ventifacts in exposures of desert pavement.

+ fact: The shape can indicate wind direction.

+ fact: Ventifacts have been identified on Mars and in Antarctica.

Ventifacts are formed in arid environments by the abrasive power of sand and smaller particles carried by the wind. They have various forms, depending on original characteristics of the rocks and energy and direction of the wind. Some ventifacts have smooth, wind-blasted faces. Others contain surfaces with networks of grooves and pits. Some ventifacts contain several wind-smoothed faces, suggesting that either the rock moved during wind abrasion or the wind direction changed during the ventifact's formation. Ancient ventifacts preserved in beds of sedimentary rock are indications of a stony desert in the past.

Alluvial Fans
Desert environments

Alluvial fans are cones of sediment similar to deltas, which form where there is an abrupt change in topography from steep uplands to a flat plain.

KEY FACTS

These are fan- or cone-shaped accumulations of gravel, sand, and smaller materials.

+ fact: Made of unconsolidated sediment known as alluvium

+ fact: Identified on Mars

+ fact: Spectacular alluvial fans are found in Death Valley National Park.

Alluvial fans are characteristic of arid and semi-arid environments in which mountains or hills abut a wide basin or plain. The high ground is drained along valley systems to the basin, where there is an abrupt change in topography. Here, flows of water and sediment rapidly lose energy and the sediment load is deposited. Repeated deposition builds up cones of sediment at the mouth of drainages. Alluvial fans are similar to deltas but occur above water in arid environments. Alluvial fans from neighboring valleys can converge when built up with enough sediment.

Sand Dunes
Desert environments

Sand dunes are distinctive features of sandy deserts, formed by wind-powered transport and deposition of sand grains. Dune shapes depend on wind direction and sand abundance.

KEY FACTS

Dunes are accumulations of sand grains in piles with straight or curved crests.

+ fact: Some cross beds in sandstones are the mark of ancient sand dunes.

+ fact: Colorado's Great Sand Dunes National Park contains the tallest sand dunes in North America.

+ fact: Sand dunes are features of expansive sandy deserts called ergs.

Arid deserts are characterized by landforms built from wind energy. As wind blows across sand, sand grains are progressively skipped forward by a process known as saltation. Saltating sand grains accumulate until the crest of the pile is unstable and the grains avalanche down the lee side, forming the shape of a dune. Straight-crested dunes called "transverse dunes" form perpendicular to the prevailing wind direction. When there is limited sand supply, individual half-moon shaped dunes called "barchan dunes" form. When there are two or more prominent wind directions, dunes with different orientations and shapes form.

Barrier Islands
Coastal environments

Barrier islands are sandbars or sandspits not connected to a coastline. They are separated from the coast by a sound or a lagoon, and they are high enough for dunes to form.

KEY FACTS

Sandy islands are separated from the continent by a sound, bay, or lagoon.

+ fact: Barrier islands move and change shape with ocean currents and storm events.

+ fact: Vegetation helps stabilize barrier islands.

+ fact: The Outer Banks of North Carolina are barrier islands.

Barrier islands are linear deposits of sand offshore, separated from the mainland by a sound, bay, or lagoon. They typically have a sandy beach on the seaward side, then an area of dunes, and then a marsh leading to a shallow lagoon on the inland side. Barrier islands are somewhat stabilized by dune vegetation, but they can change dramatically from season to season and storm to storm. Barrier islands "migrate" in two ways: Alongshore ocean currents move sediment from one tip of the island to the other, or storms remove beach sand from the ocean side and deposit it on the marsh side as overwash.

Shoals/Sandspits/Sandbars
Coastal environments

Shoals, sandspits, and sandbars are coastal landforms that develop where streams and alongshore currents create linear deposits of sand and mud.

KEY FACTS

These landforms consist of linear deposits of sand and mud.

+ fact: Shoals create local areas of shallow water and are navigation hazards for boats.

+ fact: A nearly continuous chain of sandspits and barrier islands extends from Long Island, New York, to Florida.

+ fact: Cape Cod, Massachusetts, is a sandspit.

Shoals are elongate deposits of sand and mud that extend into water in coastal environments. Shoals that are attached to a headland but that extend into the water past a headland bend are called spits or sandspits. These form via alongshore drift. When waves meet the beach at an oblique angle but recede perpendicularly, the change in current direction transports sand down the shore. When the angle of a headland changes, alongshore drift transports sand past the bend, extending the sand deposit into the sea until the current loses energy. Sandbars are isolated shoals detached from headlands and surrounded by water.

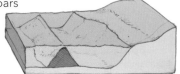

Marine Terraces or Platforms
Coastal environments

Marine terraces or platforms are bench-like coastal landforms that usually indicate a lowering of sea level or a rise of the land. These terraces are cut by wave erosion before they are exposed.

KEY FACTS

Broad "benches" are found along coasts.

+ fact: Some terraces along the California coast are uplifted by movement along the San Andreas Fault.

+ fact: Dramatic stacked terraces are exposed in the Palos Verdes Hills in Los Angeles County, California.

+ fact: Common on the Pacific coast of North America

Marine terraces are horizontal platforms separated by steep cliff faces on exposed coastlines. Ocean waves are one of nature's most powerful forces. Wave energy becomes concentrated on exposed headlands, cutting into the rock like a horizontal saw, causing overlying rock to collapse and retreat landward. This process cuts a nearly horizontal platform adjacent to a steep scarp. If the land rises with respect to sea level, this platform is exposed above the water. Periodic uplift (or sea level decline) will expose marine terraces in a steplike pattern, as along the coast of California.

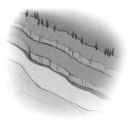

Sea Stacks
Coastal environments

A sea stack is a remnant of a coastal headland, carved by the powerful erosive force of ocean waves. Sea stacks generally form where resistant bedrock is exposed along a high-energy coast.

KEY FACTS

Isolated rock towers rise out of the ocean along a coastline.

+ fact: Can form from many different types of rock, including limestone, sandstone, and volcanic tuff

+ fact: Sea stacks are important bird habitats.

+ fact: Common along the West Coast; famous along Harris Beach, Oregon

Sea stacks are isolated towers or monoliths that rise steeply out of the sea. They are generally clustered along steep coastlines marked by resistant bedrock. Sea stacks often form when waves attack jointed or faulted rocks, preferentially eroding along these weak planes, leaving pillars or towers surrounded by seawater. Eventually, wave attack will continue to erode the stacks, and they will topple into the ocean and become pulverized to smaller and smaller particles. Sea stacks can also form in limestone bedrock by the combined effects of wave attack and limestone dissolution.

Cirques
Glaciated terrain

Cirques are bowl-shaped scoops on mountainsides, which are formed by glaciers. Glaciers carve cirques backward into the mountainside by breaking down and removing rock beneath the ice.

KEY FACTS

Bowl-shaped depressions appear in the sides of mountains.

+ fact: Cirques are named from the French word for "amphitheater."

+ fact: Those found in nonglaciated regions are important evidence for previous ice ages.

+ fact: Cirques filled with lakes are called tarns.

Mountain glaciers form along the flanks of high peaks and are relatively small compared with continental glaciers or ice sheets. The bowl-shaped depressions at the head of a mountain glacier are called cirques. Cirques are easily recognizable by their distinctive rounded, hollowed-out form, like the path of a gigantic ice cream scoop on the side of a mountain. Cirques form as the glacial ice and meltwater below the ice gouge out the rock behind the glacier. When two cirques on different sides of a mountain erode deeply enough to intersect, they form a sharp ridge between them called an arête.

Glacial Erratics
Glaciated terrain

Glacial erratics are rocks that have been deposited by a glacier, usually after being transported great distances from their bedrock source.

KEY FACTS

Erratics are out of place in their current environment due to transport by a glacier.

+ fact: Can range in size; boulders are well known.

+ fact: The largest known erratic in the U.S. and Canada is New Hampshire's Madison Boulder, weighing almost 6,000 tons.

+ fact: Some boulders are perched precariously on top of smaller stones.

Glacial erratics are clasts of rock, including large boulders, transported and deposited by glaciers. They are called erratics because they are incongruous with the surrounding bedrock, having been moved sometimes great distances by a glacier. Scientists use erratics to map the extent of the continental ice sheet that covered much of Canada and the U.S. during the last ice age, which reached its maximum extent approximately 21,000 years ago. Glacial erratics originate when glaciers pluck rocks as they scour over topographic highlands. When the glacier melts, it drops its sediment load into an unconsolidated pile.

Kettle Ponds
Glaciated terrain

Kettle holes or ponds are depressions formed by melting chunks of glacial ice that are no longer connected to the glacier. After the glacier has receded, these holes often fill with water.

KEY FACTS

Kettle ponds are depressions formed in glacial outwash plains that have filled in with water.

+ fact: The ponds are recharged by groundwater and/or rainwater.

+ fact: Thoreau's Walden Pond in Massachusetts is a famous kettle pond.

+ fact: The ponds of Cape Cod, Massachusetts, are kettle ponds.

Kettle ponds are important markers of glaciated terrain. These variably sized ponds are found in regions that were previously glaciated. They form when chunks of ice calve off the front of a receding glacier and are buried by glacial sediment called till. When the chunks of ice melt away, the till collapses into the space, creating a hole. Over time, these holes fill with water and become ponds. Kettle ponds are common landforms across the northern region of North America because of the continental ice sheet that reached its maximum extent approximately 21,000 years ago during the last glacial period.

U-shaped Valleys
Glaciated terrain

Valleys carved by glaciers develop a characteristic U-shaped cross section. U-shaped valleys are key indicators of glacial erosion, whereas V-shaped valleys are carved by rivers.

KEY FACTS

Broad troughs in the landscape have U-shaped cross sections.

+ fact: U-shaped valleys filled with seawater are called fjords.

+ fact: U-shaped valleys are important indicators of past glacial periods.

+ fact: Fantastic examples can be seen in Glacier National Park in Montana.

Distinctive markers of glaciated terrain are steep-walled valleys with broad floors that resemble the shape of a U in cross section. U-shaped valleys are formed by the downhill flow of glaciers. The ice carves out steep walls and a relatively flat base. These valleys are also often straighter in course than river-cut valleys. Some U-shaped valleys can be notched by V-shaped river channels from the flow of glacial meltwater. The floors of tributary glaciers can be significantly higher than the main valley floor, leading to waterfalls and hanging valleys after the retreat of the glaciers.

A mountain lake reflects a rainbow at Mount Robson Provincial Park in British Columbia.

4 | Weather

Nature You Can't Ignore

Most of the time, people go about their daily lives without thinking much about the weather. Before they leave home for the day, a brief radio forecast or a glance at a website tells them all they need to know: How should I dress to go out for the day? Many people, however, are fascinated by the ever-changing sky and weather, and they want to keep on learning about it.

■ A Quick Guide to Weather Science

Our day-to-day weather is the result of the sun's unequal heating of Earth, with the tropics—the belt circling Earth 1,600 miles (2,575 km) north and south of the Equator—receiving much more solar energy than the regions around the North and South Poles. Storms and ocean currents along with calmer winds move warm and cold air and water, redistributing heat, balancing Earth's heat budget. These movements of warm and cold air create our day-to-day weather.

While Earth sees relatively great temperature extremes, the planet's temperatures allow some forms of life over almost all of the planet. Earth's monthly average temperatures range from as cold as -80°F (-62°C) on Antarctica's polar plateau to 100°F (38°C) on tropical deserts. The world's single-day lowest temperature ever officially recorded was -129°F (-89.2°C) at the Russian Vostok Station in Antarctica on July 21, 1983, during the coldest part of the Southern Hemisphere winter. Earth's single hottest daily temperature on record is 134°F (56.7°C) at Furnace Creek, California, on July 10, 1913, at what is now the headquarters of Death Valley National Park. By the way, on January 8, 1913, Furnace Creek set its cold temperature record, 15°F (-10°C). Furnace Creek has tied this temperature since then, but has never been colder. This tells us something about desert climates: Temperatures drop rapidly after sunset because such dry climates have little water vapor in the air to absorb heat radiating away from the ground and radiate it back down.

■ Atmosphere & Water

If Earth didn't have an atmosphere and oceans, the planet's daily temperatures would probably be much like those on the moon, which range from daytime highs of approximately 200°F (93°C) and overnight lows of -280°F (-173°C). Copious amounts of water are important for Earth's weather, and not just because water fills the oceans. Water is the only natural substance that exists in all three states of matter at temperatures and air pressures found near Earth's surface. Water is the raw material for clouds, and all kinds of precipitation. Without water, Earth's weather would be like that on Mars: winds blowing dust around and wide temperature changes.

The air's invisible water vapor contains energy to power violent weather such as this summer storm building over Florida's Everglades.

■ Water Supplies Storm Energy

In addition to supplying raw material for clouds and precipitation, water supplies some of the energy that powers the weather. Water is the only natural substance that exists in all three states of matter—solid, liquid, and gas—at temperatures and pressures found near Earth's surface. When it changes among its phases, water either takes in energy as heat from the surrounding air, which cools the air, or adds heat to the surrounding air, warming it.

Heat is taken from the surroundings when water evaporates into vapor, when ice melts into water, or when ice sublimates directly to water vapor without first melting. This is why evaporation of perspiration from our skin cools us in hot weather. If the air is too humid for perspiration to evaporate easily, we become sticky and hot.

Heat taken from the surroundings is called latent heat. When water vapor condenses into liquid water or deposits directly to ice, it returns its latent heat to the surroundings as does water when it freezes into ice. Condensation, deposition, and freezing all add energy to the surrounding air,

BECOMING INVOLVED WITH WEATHER

The books, websites, and magazines in "Further Resources" will help you to learn more about weather and climate. If you want to look into becoming a meteorologist, or otherwise become directly involved with weather, you have various possibilities.

✛ If you like to watch storms and want to learn more about them and help the U.S. National Weather Service, you should consider becoming one of more than 290,000 trained volunteers who send reports of severe weather to a local NWS office. To become involved, contact the Warning Coordination Meteorologist at your local NWS office and arrange to take the training course these offices offer. To do this, go to *www.stormready.noaa.gov/contact .htm* and follow the directions. You will learn the basics of safety.

You can also check with your local weather office about becoming a volunteer cooperative weather observer. Since 1890, volunteers in this program have been sending data from NWS instruments on their property—mostly temperature and precipitation—to the NWS. These data are used to supplement climate reports from places away from NWS and other official weather stations.

✛ The Community Collaborative Rain, Hail and Snow Network is looking for volunteer observers to add to its thousands of volunteers. It is now the largest provider of daily precipitation observations in the United States, and is moving into Canada. Information is available at *www.cocorahs .org*. Becoming an observer is a good way to learn about weather.

✛ Environment Canada's Meteorological Service has a storm spotter program called the Canadian Weather Amateur Radio Network, which began as a program for amateur radio operators but now is open to anyone who wants to be a storm spotter. Contact your nearest Environment Canada weather office for information.

✛ The U.S. National Oceanic and Atmospheric Administration (NOAA)—the parent agency of the National Weather Service—offers a free smartphone app that enables anyone to report precipitation anonymously. Anyone can use the app to report storms to researchers at NOAA's Severe Storms Laboratory and the University of Oklahoma. NOAA and the university are using the reports to build a database. The reports are not used for warnings.

A thundercloud builds before a storm at Fort Whyte Centre in Winnipeg, Manitoba.

warming it. Energy added to rising air by condensation, deposition, and freezing adds heat to the air, making the heat rise faster and farther. Such latent heat releases the supply of energy for thunderstorms and hurricanes.

As air descends, as when falling rain drags it down, it warms, causing water to evaporate and ice to melt into water or sublimate directly into water vapor. These take energy from the falling air, making it colder and thus heavier. Normally, sinking air warms, but the heat taken away by water's phase changes more than offsets this warming. Under the right conditions, the air grows cool and heavy enough to smash into the ground as a camaging microburst.

■ Stable & Unstable Atmospheres

Meteorologists use the words "stable" and "unstable" in a technical sense to describe what kind of

*Sunshine is delicious,
rain is refreshing, wind braces us up,
snow is exhilarating; there is really
no such thing as bad weather,
only different kinds of good weather.*
—JOHN RUSKIN

weather to expect at particular times and places. Nevertheless, the words carry similar connotations when used to describe people. A "stable" person is one who stays calm even under pressure. When the atmosphere is stable, you might have rain or snow, but it will fall over a wide area without thunderstorms. An "unstable" person reacts with road rage when another driver gets in the way. When the atmosphere is unstable, the weather can produce strong thunderstorms, maybe even tornadoes.

When the atmosphere is stable, air that is given an upward shove (as by wind blowing upward on a mountain) will stop rising and begin sinking when the shove stops. In an unstable atmosphere, the air continues rising after the shove ceases. The temperature profile of the atmosphere makes the difference. Rising air cools by 3.5°F for each 1,000 feet (6.4°C/1,000 m) of altitude gained. When water vapor begins evaporating, it releases latent heat, warming the air, which

offsets the cooling to some degree. This means humid air can rise farther and faster than dry air.

Here's how this works. If the temperature is 60°F (15.5°C) at the surface, a bubble of air will cool to 54.5°F (12.5°C) when it rises 1,000 feet. If the temperature of the surrounding air here is 53°F (11.7°C), the bubble will be warmer. It will continue rising. The atmosphere here is unstable. When the bubble becomes colder than the surrounding air, it stops rising; the atmosphere at this height is stable. Forecasters need to know the stability to make predictions.

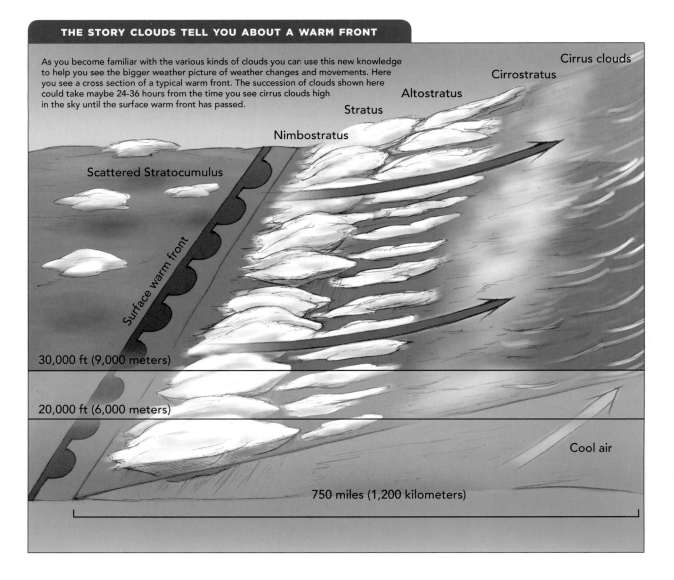

THE STORY CLOUDS TELL YOU ABOUT A WARM FRONT

As you become familiar with the various kinds of clouds you can use this new knowledge to help you see the bigger weather picture of weather changes and movements. Here you see a cross section of a typical warm front. The succession of clouds shown here could take maybe 24-36 hours from the time you see cirrus clouds high in the sky until the surface warm front has passed.

Cirrus clouds

Cirrostratus

Altostratus

Stratus

Nimbostratus

Scattered Stratocumulus

Surface warm front

30,000 ft (9,000 meters)

20,000 ft (6,000 meters)

Cool air

750 miles (1,200 kilometers)

WEATHER

What Causes the Seasons?

Earth's North Pole–South Pole axis tilts 23.5° in relation to its yearly path around the sun. This tilt causes the seasons, by changing the distribution of sunlight throughout the year.

KEY FACTS

Seasons are formed by the tilt of Earth on its axis, which causes unequal sunshine in various regions of Earth.

+ fact: On the equinoxes—March 21–22, and Sept. 22–23—the sun is directly above the Equator.

+ fact: On the Dec. 20–21 solstice—Northern Hemisphere winter—the sun is directly above the Tropic of Capricorn, latitude 23.5° S.

On September 22 or 23, the sun shines equally on all of Earth. Until December 21 or 22, days grow shorter in the Northern Hemisphere as the sun moves higher in the sky and days grow longer south of the Equator. On December 21 or 22, the winter solstice, the sun never rises north of latitude 66.5° N (the Arctic Circle); it never sets south of latitude 66.5° S (the Antarctic Circle). This process reverses until March 21 or 22, when the sun shines equally on all of Earth. Until June 20 or 21 the Southern Hemisphere receives less sun, turning colder as the Northern Hemisphere receives more sun and warms up.

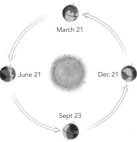

Climate Zones

The amount of solar energy any part of Earth receives determines whether it will have a tropical, temperate, or polar climate. A location's elevation and nearness to an ocean also affect climate.

KEY FACTS

A region's climate depends on the latitude.

+ fact: Tropical climates have average temperatures above 64°F (18°C).

+ fact: Temperate climates have four seasons without extreme temperatures.

+ fact: Polar climates have no monthly averages higher than 50°F (10°C). On some days the sun never sets; on others it never rises.

Climate is the long-term average weather of an area, including temperatures and precipitation. The tropics, immediately north and south of the Equator, receive the most solar energy: The sun is almost directly overhead all year, and so on average the tropics are Earth's warmest region. The polar regions, centered on the North and South Poles, receive the least amount of solar energy because the sun is always low in the sky, and it doesn't come up at all for part of the year in some regions. On average, these are Earth's coldest regions.

In between, the temperate zones have variable climates without widespread tropical or polar extremes.

Heat Balance

To a large degree, atmospheric winds and oceanic currents even out the temperature contrasts caused by day and night, and the seasonal variations caused by Earth's tilted axis.

KEY FACTS

Without oceans and atmosphere, Earth's temperatures would be like the moon's: -387°F (-233°C) at night, 253°F (123°C) during the day.

+ fact: Winds aloft and at Earth's surface move warm air toward the poles, cool air toward the Equator.

+ fact: Ocean-top currents carry warm water toward the poles; underwater currents haul cold water toward the Equator.

In the tropics, warm, humid air rises, forming towering Intertropical Convergence Zone thunderstorms. Air flowing in to replace the rising air creates tropical trade winds above the oceans. Air from the storms moves toward the south or north, but Earth's rotation upsets this simple, theoretical pattern. Earth's complex wind patterns include high-altitude jet streams and air sinking at some places, rising at others. These winds and the water vapor they carry are major drivers of storms, warm and cold outbreaks, and other weather events. We can wonder when we'll start seeing other expected changes.

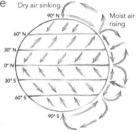

How Humans Affect the Weather

Earth's average overall temperature changes slowly and has warmed and cooled in the past. It is now warming, with greenhouse gases added to the air by humans playing an important role.

KEY FACTS

Without the natural greenhouse effect, Earth's average temperature would be 0°F (-19°C) instead of today's 60°F (15°C).

+ fact: Many prefer the phrase "climate change" because warming is only one phenomenon now occurring.

+ fact: Greenhouse gas emissions are falling in developed countries but are rising in developing nations.

Climate scientists are unwilling to say that climate change caused any particular event, such as a hurricane or heat wave, but evidence is strong that the Earth is warming faster than it would without human-related causes, especially by adding greenhouse gases, which trap heat that would otherwise escape. The number of local record high temperatures is increasing while record low temperatures are decreasing. Globally, glaciers are shrinking, and ice on lakes and rivers is breaking up earlier in the spring. Some of the biggest changes are happening in the Arctic where sea ice is shrinking. Scientists in the past had frequently predicted that many of the changes we are now experiencing would eventually occur.

Cloud Composition and Colors

Clouds are made of tiny water drops or ice crystals held up by rising air. Rain or snow falls when the drops or crystals grow large enough to overcome the resistance of rising air.

KEY FACTS

The faster air rises into clouds the bigger the drops or crystals can grow before they begin falling.

+ fact: Air rises as slowly as inches an hour in some clouds, faster than 100 mi (160 km) an hour in others.

+ fact: The lowest clouds touch the ground as fog; the highest can reach above 60,000 ft (18,000 m).

The water drops and ice crystals that make up clouds are large enough to scatter all wavelengths of sunlight, which makes the tops and sunny sides of clouds white. Bottoms of thin clouds are also white. Dark cloud bottoms don't necessarily mean rain. At least half of the sunlight hitting a cloud less than 3,000 feet (900 m) deep makes it through the cloud. This causes many cloud bottoms to be gray while the tops and sides facing the sun are white. Little sunlight passes through clouds more than 3,000 feet deep, which makes their bottoms dark. Shadows of other clouds also darken clouds. Sunrise and sunset turn clouds yellow, orange, or red.

Creating Clouds

Clouds form when rising air becomes cold enough for the air's humidity to begin turning into cloud drops or ice crystals.

KEY FACTS

The temperature of the surrounding air does not affect the cooling rate of rising air.

+ fact: When condensation begins, water vapor releases heat, which reduces rising air's cooling rate to less than 5.4°F per 1,000 ft (10°C per 1,000 m).

+ fact: Air is rising in all clouds at speeds ranging from inches an hour in many clouds to 100 mi (160 km) an hour in fierce thunderstorms.

Even when rising air is cool enough for water vapor to begin condensing into cloud drops, the vapor needs help. Condensation begins with water vapor molecules attaching to tiny particles in the air known as cloud condensation nuclei. These nuclei include dust and a variety of other natural substances as well as some kinds of pollution. Satellite images show long bright clouds created by nuclei from ship exhaust gases. These illustrate how added nuclei create many tiny cloud drops that reflect more light than nearby ordinary clouds with larger drops. This is one illustration of the complexities of clouds and why scientists continue working diligently to learn more about these important phenomena.

Cloud Names

Today, with a few additions and enhancements, we still use a system of cloud names devised in the early 1800s.

Scientist Luke Howard organized clouds much as the Swedish biologist Carl Nilsson Linnaeus had devised the system of "genus" and "species" for plants and animals. A British chemist whose company manufactured medications, Howard practiced meteorology on his own. He first presented his cloud system during an 1802 lecture in London, which was widely publicized and accepted by those developing the science of meteorology. Howard's decision to incorporate Latin into his naming system, as Linnaeus had done with biology, helped make the names international. At the time, scientists were beginning to figure out how clouds form and their importance in understanding weather. Howard's focus on clouds as visible indications of atmospheric changes helped advance the science.

Describing Cloud Cover

Meteorologists use specific terms for the general public and different ones for pilots to describe how much of the sky clouds cover currently or are forecast to cover.

Although the terms "scattered" and "partly cloudy" mean the same thing, as shown in Key Facts, the National Weather Service (NWS) uses "scattered" for pilots and "partly cloudy" for the public, and you might hear both. Pilots and controllers also need to know how far above the ground the bottoms of the clouds are. Weather stations have automated ceilometers, which send pulses of infrared light straight up and measure the time needed to reflect back to the instrument to calculate cloud heights. Each minute, the instrument uses the last 30 minutes of data on when clouds were above it to calculate cloud cover. For pilots and the public, "overcast" refers to clouds covering more than seven-eighths of the sky. Gradually increasing cloudiness can show that rain or snow is coming.

Cumuliform Clouds

Puffy cumulus clouds form in an unstable atmosphere humid enough for its water to condense into cloud drops in bubbles of warm air rising as narrow streams called thermals.

KEY FACTS

Water vapor begins condensing when air cools to its dew point temperature. Humid air has a higher dew point.

+ fact: Condensing water vapor releases the heat it gained when it evaporated, which warms the air, making it rise farther and faster.

+ fact: Growing cumulus clouds have flat bottoms at the height where condensation begins.

Meteorologists say the atmosphere at a particular time and place is unstable when rising air, which always cools at a steady rate, stays warmer—thus lighter—than the surrounding air. It continues in this direction—at times to great heights—as long as the rising air remains warmer than the surrounding air. Relatively warm air at the surface and cold air aloft creates instability. As air is rising to form cumulus clouds, some air aloft is sinking around the clouds, keeping the air between clouds clear. These up-and-down air movements—termed "convection"—form individual cumulus clouds of various sizes instead of a single, widespread cloud covering a large area.

Stratiform Clouds

Stratiform clouds form in a stable atmosphere that doesn't support cumuliform thermals because air doesn't continue rising after the initial shove upward.

KEY FACTS

Stratiform clouds are smooth, gray, and can cover most of the sky at a given location.

+ fact: Sometimes, stratiform clouds can cover large ocean areas, including the Pacific off the U.S. West Coast.

+ fact: Because stratiform clouds cool Earth, scientists are studying how climate change could affect them.

Meteorologists say the atmosphere at a particular time and place is stable when rising air, which always cools at a steady rate, grows colder—thus heavier—than the surrounding air. Once the air becomes heavier than the surrounding air, it will stop rising, which means cumulus clouds cannot form. Stratiform clouds form when all of the air in a particular area is pushed up, such as when wind carries warm air up and over heavier cold air at the surface. As stable air is pushed up, it spreads out to form a layer or layers of stratiform clouds that are thinner but wider than cumulus clouds.

Orographic Clouds

Rain or snow falling from orographic clouds are important sources of water for many regions, including much of the western United States.

KEY FACTS

When the atmosphere is stable, air continues rising as it reaches mountaintops and can create thunderstorms.

+ fact: When the atmosphere is unstable, air flowing over mountains forms turbulent rising and sinking that can stretch 100 mi (160 km) or more.

+ fact: Arid regions downstream of large mountains are called rain shadows.

When humid winds blow over hills or mountains, orographic clouds form. In some locations, these clouds bring most of the year's snow and rain. Air flowing up mountains cools enough for its humidity to condense into raindrops or to form snow. This orographic precipitation waters trees and other plants that cover mountains, even while nearby lower elevations remain arid. Spring and summer melting mountain snow feeds rivers, fills reservoirs, and becomes the major source of water for places far from the mountains. In the Sierra and Cascades ranges of the West, for example, flowing air loses moisture over the mountains, creating the dry Great Basin east of the mountains.

Cirrus Clouds

Meteorologists use specific terms for the general public and different ones for pilots to describe how much of the sky clouds cover currently or are forecast to cover.

KEY FACTS

Cirrus clouds block little sunlight but absorb infrared radiation from Earth, contributing to the greenhouse effect.

+ fact: Cirrus clouds are made mostly of ice crystals—they are located in -50°F (-46°C) air.

+ fact: At any one time, cirrus clouds cover approximately 25 percent of the Earth.

Cirrus clouds form when water vapor that has been pumped high into the air, often by a storm, turns directly into ice crystals—a process called deposition. Cirrus clouds can be a sign that rain or snow is on the way, maybe in a day or so. The only precipitation these clouds produce is light snow, which evaporates long before reaching the ground. Meteorologists identify these wisps of snow as "fall streaks," but they are commonly called "mares' tails." Some cirrus clouds are as thin as roughly 300 feet (100 m); others are as thick as 5,000 feet (1,500 m). Cirrus clouds are white—except around sunrise and sunset— and are generally transparent.

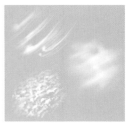

Nacreous Clouds

Nacreous clouds form in the stratosphere high above the Arctic and Antarctic, where they help to destroy stratospheric ozone. They are sometimes, but very rarely, seen at lower latitudes.

KEY FACTS

Destruction of stratospheric ozone is a concern because it blocks harmful ultraviolet radiation.

+ fact: Nacreous clouds are 49,000–82,000 ft (5,000–25,000 m) high, where temperatures are below -108°F (-78°C).

+ fact: They are named for iridescent nacre, a material that some mollusks make.

Nacreous clouds can be seen from the northern regions of Europe, Asia, and North America and over Antarctica, but they are very rarely seen elsewhere. Little was known about them until 1982, when NASA scientists using satellite data described them in detail, including the fact that they form regularly over Antarctica in the ozone layer and sometimes over the Arctic. The researchers named them "polar stratospheric clouds." In 1986 and 1987, scientists working in Antarctica found that these clouds are, in effect, natural laboratories that enable man-made substances including chlorofluorocarbons (such as Freon) to destroy stratospheric ozone. Temperatures over Antarctica are low enough for the clouds to form every winter, but they form only occasionally over the Arctic.

Cirrocumulus Clouds

Cirrocumulus clouds are white patches of high cloud without gray shadows and with lumps called cloudlets, often a sign that rain or snow could arrive in a day or sooner.

KEY FACTS

Cirrocumulus clouds are the least common high clouds.

+ fact: Cirrocumulus cloudlets appear to be the size of the tip of your little finger held at arm's length (or smaller).

+ fact: Cirrocumulus clouds are mostly ice crystals but often include liquid drops that have not frozen in well-below-freezing air.

Cirrocumulus clouds are usually found higher than 20,000 feet (6,100 m). If you watch one for a while, you might see it turn into a cirrostratus or cirrus cloud or dissipate because any particular cirrocumulus cloud tends to have a short life. The cloudlets show that convection—up-and-down air movement—is occurring, as it is in any cloud with "cumulus" as part of its name. These clouds are usually in relatively warm air that is moving over cooler, denser air at the surface, maybe 500 miles (800 km) or more ahead of a warm front, which is the surface boundary between warm and cold air.

Cirrostratus Clouds

Cirrostratus clouds are high clouds, above 20,000 feet (6,100 m). They are sometimes too thin to hide the sun or moon, even though they might be as much as 1,000 feet (305 m) thick.

KEY FACTS

Some very thin cirrostratus clouds appear only as a halo around the sun or moon.

+ fact: Forecasters use "hazy sunshine," to describe the milky look of the sky with cirrostratus clouds.

+ fact: Fibrous cirrostratus clouds with no halos are called "cirrostratus fibratus."

Cirrostratus clouds are made of ice crystals and are generally thin and uniform. They form when warm air moving over heavier cold air ahead of a warm front or rising air in the center of a surface low-pressure area lifts humid air into the upper atmosphere. If you see them replace cirrus clouds over a few hours and then grow thick enough to hide the sun or moon, you know snow or rain has a good chance of arriving within 24 hours—and maybe sooner. Just as other high clouds do, cirrostratus clouds frequently reflect red and yellow patterns that create spectacular sunrises and sunsets.

Contrails

Contrails are long, thin clouds that form behind high-flying airplanes as water vapor in the airplane's engine exhaust creates a narrow stream of air humid enough for a cloud to form.

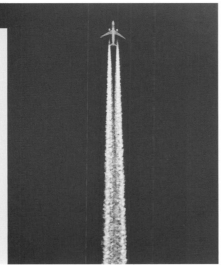

KEY FACTS

Both piston engines and jet engines exhaust water vapor, which can create contrails.

+ fact: During World War II, contrails helped enemy pilots and anti-aircraft gunners spot high-flying bombers and fighters.

+ fact: Ice crystals in contrails fall approximately 6.56 ft (2 m) a second.

"Contrail" is short for "condensation trails," which are narrow clouds of condensed water vapor that become visible behind airplanes flying higher than 26,000 feet (8,000 m), where the temperature is colder than approximately -40°F (-40°C). Water vapor in the airplane's exhaust added to the air's own water vapor transforms into ice crystals around nuclei of material including particles in the exhaust. Contrails grow long when the air at their altitude is humid, a phenomenon that can indicate rain or snow is on the way. Minimal contrails behind a high-flying airplane means the air at its altitude is dry, so no contrails form or they quickly evaporate. When contrails persist for more than a few minutes, winds aloft often push them into wavy paths.

Altocumulus Clouds

These puffy clouds, white and gray with darker patches, generally form about 6,500 to 20,000 feet (2,000 to 6,100 m) above the ground and can cover large areas of the sky.

KEY FACTS

Altocumulus clouds on warm, humid mornings indicate thunderstorms that day.

+ fact: Individual clouds appear as wide as your thumb with your hand at arm's length.

+ fact: A sky with wavy, mixed altocumulus and cirrocumulus clouds and blue-sky gaps is called a "mackerel sky" because it resembles fish scales.

Altocumulus clouds form in various ways. They might develop from cumulus clouds that grow to an altitude where rising air slows and spreads out, or they might be transformations of other clouds including altostratus, stratocumulus, or nimbostratus. The clouds are usually made of water drops, although ice sometimes is found. They are usually less than 3,000 feet (1,000 m) thick and sometimes produce virga—rain that evaporates on the way down—but they rarely produce rain that reaches the ground. They may form distinct layers or parallel bands of clouds, called cloud streets, with air rising into the clouds and sinking between the bands.

Undulatus Asperatus Clouds

Since 2009, news reports in the United States and the United Kingdom have suggested that this is a new variety of cloud. Although such clouds are rare, there is no reason to think they are new.

KEY FACTS

Billow clouds are the undulatus variety you are most likely to see, often with a blue-sky background.

+ fact: Gavin Pretor-Pinney, founder of the Cloud Appreciation Society in the U.K., suggested the cloud's name.

+ fact: Undulatus clouds highlight the atmosphere's many natural wavy motions

An "undulatus" cloud, according to the American Meteorological Society glossary, is composed of long, parallel elements, merged or separate, that look like undulating ocean waves. The term "asperatus" comes from the Latin verb for "rough or difficult." Peggy LeMone, a scientist at the National Center for Atmospheric Research, recalls taking her first photo of such a cloud "on a wintry day in Columbia, Missouri, probably in the 1970s," and says she has seen others since then. Is it a new type? Maybe, instead, this cloud reminds us that if you look at enough clouds, you're bound to see some that are hard to classify. If you continue closely observing clouds, you'll learn that such ambiguity is common. This is a reason why many people find the sky and weather fascinating.

Lenticular Clouds

You're a few miles from a mountain or mountain range and see something like a flying saucer. You are almost surely looking at a lenticular cloud formed by wind blowing over mountains.

When the atmosphere is stable, wind goes down and up in a wavy pattern after it has crossed mountains. (When it is unstable, the air keeps rising.) Unless the air is very dry, a lens-shaped cloud or line of clouds forms atop the wave or waves. Air rising toward the top of a wave cools, and its humidity condenses into cloud drops. Air descending from the top of a wave warms, which evaporates cloud drops. The combined processes create this lens-shaped cloud. Lenticular clouds change shape little and stay the same distance from the mountains instead of traveling downstream with the winds. These clouds remain stationary as air moves through them.

Mammatus Clouds

Mammatus clouds are pouches that hang from the bottoms of clouds, most commonly from thunderstorm anvils, a cloud spreading out from the storm's top.

Mammatus pouches, which can be transparent or opaque, form when blobs of cold air containing water drops, ice crystals, or both begin sinking into clear air below a cloud. They descend into increasing air pressure, which warms the falling drops and crystals. At the same time, the crystals or drops are evaporating into water vapor, which cools them, offsetting the warming to some extent and keeping them cooler, thus heavier, than the surrounding air. The blob of crystals or drops sinks below the bottom of the cloud and looks like a pouch. Individual pouches may last ten minutes, but a cluster can last for hours.

Altocumulus Castellanus Clouds

Altocumulus castellanus clouds look somewhat like the turrets of a castle. If you see these early in the day, it could signal rain or thunderstorms later that day.

KEY FACTS

Tall, narrow castellanus clouds are called turkey necks.

+ fact: These can show that rising air is breaking through a warm layer that has been suppressing thunderstorms.

+ fact: Castellanus towers are easier to see from the side rather than along the narrow cloud's length.

When rolls of altocumulus clouds begin sprouting towers, the atmosphere above the clouds is becoming unstable enough that they might grow into towering cumulus and possibly thunderstorms that day. Such instability means the atmosphere above altocumulus clouds is cold enough for rising air to stay warmer than the surrounding air and continue rising. The towers are not a guarantee of thunderstorms, just a possible predictor. If the air near the ground is humid, thunderstorms are more likely. If you are on an airplane about to take off with castellanus clouds overhead, expect some turbulence as the airplane flies through the clouds.

Stratocumulus Clouds

Like other kinds of stratiform clouds, stratccumulus clouds spread across large areas with small breaks between individual clouds. In addition, they display a rounded cumuliform shape.

KEY FACTS

Stratocumulus clouds cover an average of 23 percent of the oceans and 12 percent of land.

+ fact: Individual stratocumulus clouds appear roughly the size of your fist when you extend your arm full length toward the cloud.

+ fact: Stratocumulus clouds produce little precipitation—mostly drizzle, light rain, or snow.

Stratocumulus clouds are most common over chilly oceans in the subtropics where air is slowly sinking from aloft, warming it. The layer of warm air blocks humid air from rising far from cool oceans. Thus, stratocumulus clouds generally don't rise above 8,000 feet (2,400 m). Stratocumulus clouds also form over land where air is descending into a surface high-pressure area and warming. If cumulus clouds are forming in such an area, they stop growing taller at the warm layer and spread out. Because the tops of stratocumulus clouds reflect a good share of the sunlight that hits them, they tend to cool the Earth.

Stratus Clouds

Stratus clouds are featureless gray layers with generally uniform bases. They sometimes, but rarely produce drizzle, small ice crystals, and snow grains.

KEY FACTS

Instead of rain or snow, drizzle, snow grains, or tiny ice crystals usually fall from stratus clouds.

+ fact: A stratus fractus cloud has parts of different sizes and brightness that change rapidly.

+ fact: When you can see the sun through the clouds, its outline is usually clearly discernible.

Stratus clouds are common in coastal areas because the air contains abundant humidity at low levels and the atmosphere is likely to be stable, which favors their development. Fog moving in from the ocean at night can set the stage. Stratus sometimes forms when the bottom of a layer of fog evaporates, and the lower part of the fog rises off the ground. Stratocumulus clouds can also form into stratus when the bottom of the stratocumulus descends and spreads out under a layer of warm air aloft. Stratus clouds are most common at night and in the early morning, before the sun begins to evaporate them. Stratus clouds can form in air moving into thunderstorms.

Nimbostratus Clouds

When dark, gloomy nimbostratus clouds move in with their steady rain or snow, they leave no doubt that you are in for a period of wet weather and no sun.

KEY FACTS

An altostratus becomes a nimbostratus cloud if it totally blocks the sun or its precipitation reaches the ground.

+ fact: Unattached cloud fragments called "pannus" or "scud" clouds often form below nimbostratus, cumulus, and cumulonimbus clouds.

+ fact: Nimbostratus clouds do not bring lightning, thunder, or hail.

Nimbostratus clouds are low clouds that hang below 6,500 feet (2,000 m) above ground level. They can be a few thousand feet thick and are heavy with suspended water drops, possibly ice crystals, and falling rain or snow that is likely to last for several hours or even more than a day. Little or no sunlight makes it through the cloud. The rain or snow from a nimbostratus is steady, and it falls at a light or moderate rate, not the on-and-off but sometimes drenching rain showers that cumulonimbus clouds produce. The bottoms of nimbostratus clouds tend to be ragged instead of well defined.

Altostratus Clouds

Given their name because they are the highest (alto-) of the stratus or sheet-type clouds, the thin gray altostratus sky cover signifies a storm could be on the way.

KEY FACTS

Not enough sunlight passes through altostratus clouds to cast shadows on the ground.

+ fact: Altostratus clouds tend to be translucent; you can often see a watery sun or moon through them.

+ fact: A watery moon appearing and disappearing through altostratus clouds is a classic horror movie scene.

If the sky is covered as far as you can see by gray or blue-gray clouds that are 6,000 to 20,000 feet (1,800 to 6,100 m) above the ground, they are altostratus clouds. If the clouds are white or have areas of white, you are instead looking at cirrostratus clouds. Altostratus clouds are made of ice crystals, usually near the top of the cloud, and water drops, near the bottom. They usually arrive ahead of storms that carry widespread, steady rain or snow. As rain or snow continues falling from the altostratus, the cloud's bottom can sink below 6,000 feet, and it becomes a nimbostratus cloud.

Fair-weather Cumulus Clouds

These small, puffy, white clouds portend calm, dry weather for the immediate future, at least early in the day. As the day goes on, they can grow into thunderstorm clouds.

KEY FACTS

The scientific name for fair-weather cumulus clouds is *cumulus humilis*, from the Latin word for "low, lowly, small, or shallow."

+ fact: If you're on an airplane taking off under fair-weather cumulus clouds, expect mild turbulence until you're above the clouds.

+ fact: Fair-weather cumulus commonly form under cirrostratus clouds.

Fair-weather cumulus develop as the sun warms the ground, creating thermals that rise until water vapor begins condensing. Warm air aloft, or air aloft that cools too slowly with height, blocks air from rising higher than the cloud tops. If this continues, the day will remain calm. If you see one or a few clouds growing higher than the others, you know that either air is rising into them with enough force to break through the warm layer, or that the atmosphere above the clouds has cooled, making it more unstable. When this happens, one or a few of the smaller cumulus clouds can grow into cumulus congestus (opposite) and even thunderstorms.

Cumulus Congestus Clouds

These are impressive, hard-to-ignore clouds that tower high in the sky with solid-looking, cauliflower-like towers. Some of these clouds grow into fierce thunderstorms.

KEY FACTS

Cumulus congestus clouds are also called "towering cumulus" because they are usually taller than wide.

+ fact: Congestus clouds can produce heavy and prolonged rain or snow showers without growing into cumulonimbus clouds.

+ fact: Cumulus congestus clouds can form as individual clouds or as a "wall" of clouds.

Cumulus congestus clouds form on days when the atmosphere up to great heights is unstable. Thermals of warm air begin rising from the ground at about 50 mph (80 kph) or faster. As the rising air cools, water vapor begins condensing into tiny cloud drops. The cauliflower-like parts of the cloud are made of tiny water drops that reflect more light than the softer, fibrous parts of the cloud, which are made of ice crystals. When parts of the cloud begin to take on the softer look, it shows the cloud is "glaciating"—ice crystals are forming, which is the beginning of its transformation into a cumulonimbus or thunderstorm.

Cumulonimbus Clouds

Cumulonimbus, commonly called a thunderstorm, is the only cloud requiring caution when nearby. Its potential dangers include lightning, tornadoes, and dangerous straight-line winds.

KEY FACTS

Thunderstorm tops are typically 20,000 ft (6,000 m) above the ground and at times 75,000 ft (23,000 m) high.

+ fact: Cumulonimbus clouds usually reach their peak strength in late afternoon.

+ fact: Thunderstorms occur as individual storms, in large and small clusters, and in lines 200 mi (300 km) or more long.

Meteorologists consider a cumulus congestus to have become a cumulonimbus when at least the top part of the cloud has taken on the smooth, fibrous, glaciated appearance that comes when this part of the cloud is mostly ice. A sure sign is that the top of the cloud begins to flatten out into a characteristic anvil shape. Thunderstorm precipitation is showery. It starts and stops suddenly, and it can be quite heavy over a relatively small area while nearby areas are dry. The general winds in an area push thunderstorms across the countryside, as they do other clouds, sometimes as fast at 50 mph (80 kph) but usually more slowly. Thunderstorms are quite turbulent inside the cloud.

Pyrocumulus Clouds

Pyrocumulus or "fire cumulus" clouds form over large fires, usually wildfires, under certain atmospheric conditions. They can threaten wildfire-fighters with hard-to-forecast wind shifts.

KEY FACTS

Fires sometimes create tornado-like "fire whirls"; most last a few minutes but some persist for 20 minutes and spread the fire.

+ fact: Pyrocumulus clouds sometimes grow into thunderstorms—pyrocumulonimbus.

+ fact: Volcanic eruptions create pyrocumulus clouds. The mushroom cloud of a nuclear bomb is a pyrocumulus.

Wildfires are dangerous enough, but if they burn during days when the atmosphere is unstable and dry with light upper air winds, they become especially dangerous for firefighters. Under those conditions, a fire's hot air will rise straight up, 20,000 feet (6,000 m) and more, to create a pyrocumulus cloud. Air rushing in from around the fire to replace the rising air fans the flames. These winds can also change direction with little warning, especially on mountains and in hills, sending fire in new directions and potentially trapping firefighters. While the bottom of a pyrocumulus might be brown or gray with smoke, the top is bright white, like the tops of other cumulus clouds. Pyrocumulus clouds don't have enough water to produce the rain that would put out a fire.

Billow Clouds

The lovely wave formations atop billow clouds are spawned by forces that are common in the atmosphere and create other phenomena, such as ocean waves.

KEY FACTS

Lord Kelvin, a Scottish physicist, and Hermann von Helmholtz, a German scientist, mathematically analyzed waves in the 19th century.

+ fact: Such waves form at the boundary between fluids with different densities.

+ fact: The distance between each cloud wave is usually between 3,200 and 6,500 ft (1,000 and 2,000 m).

Billow clouds, also called Kelvin-Helmholtz clouds or waves, usually occur at a layer of warm air high above the ground. Up there, with the warm air above and cold air below, winds are blowing in opposing directions, and the curving pattern develops. To envision what happens, imagine holding a ball between the palms of your hands and moving your hands back and forth in opposite directions. To form billow clouds, the air rolls between two outside layers of air moving in opposite directions, just like the ball rolls between your hands. The rolling air might not complete the circle, though, creating forms that look like breaking ocean waves.

Supercell Thunderstorm Clouds

Supercells are the most dangerous but the least common thunderstorms. They produce almost all of the deadliest tornadoes. Distinct features make supercells stand out.

KEY FACTS

Haze and hills in the U.S. Southeast and East often make it difficult to see enough of a thunderstorm to determine if it is a supercell.

+ fact: Strong supercells have a dome on the top, called an "overshooting top."

+ fact: Low-precipitation supercells on the arid High Plains produce little rain or hail.

A supercell is a long-lasting kind of thunderstorm that produces strong tornadoes and other dangerous weather. A mesocyclone—a one- to ten-mile-wide rotating updraft that carries air from the ground to the storm's top—distinguishes supercells from all other thunderstorms. You can often see its barber-pole striations. Supercells develop a wall cloud between the area of precipitation from the thunderstorm and a precipitation-free base. The wall cloud forms where wet, cool air is being drawn into the thunderstorm's updraft. A wall cloud that lasts more than ten minutes and moves violently is most likely to produce a tornado.

Shelf and Roll Clouds

Shelf and roll clouds form on top of air that descended in a thunderstorm and moves away from the storm. A shelf cloud is attached to its parent thunderstorm; a roll cloud has broken away.

KEY FACTS

Sometimes a shelf cloud spins out a "gustnado," a small, weak tornado that rarely causes damage.

+ fact: All shelf clouds come from thunderstorms; some roll clouds come from cold fronts and sea breeze fronts.

+ fact: Roll clouds spin around a horizontal axis and are not as common as shelf clouds.

Air descending from a thunderstorm travels away from the storm, becoming a dome of cool air that is called a gust front. As its leading edge meets and plows into warm, humid air, it pushes the air up. The warm air cools, and its humidity condenses to form a shelf cloud atop the cool downdraft air. Such a gust front can last more than a day and travel hundreds of miles to help trigger new thunderstorms by pushing up warm, humid air. If the front of the shelf cloud is ragged, pushing rising, small, ragged clouds in front of it, damaging wind squalls and shifts in wind direction will likely result.

Dew

Dew, the water that you often find on the grass and your car on some early mornings, is one of the many ways water moves out of and back into the atmosphere.

KEY FACTS

The term "dew point" refers to the temperature at which condensation begins on the ground or in the air.

+ fact: Dew on grass mostly evaporated from the grass and stayed in the air nearby until condensing back onto the grass.

+ fact: Dew often forms first on car roofs because they radiate heat directly away.

Dew begins condensing when the air cools to the dew point—a temperature point that varies depending on how much water vapor is in the air. Because warm air can hold more water than cold air, water vapor will begin condensing from very humid air at a higher temperature than it would condense from air with very little water vapor. Dew begins forming on grass and other low-lying plants, because air sinks as it cools and air along the ground is usually colder than air just a little higher up. Calm, clear, still nights encourage dew. These are nights without wind that mixes the air, and with heat radiating directly to space. You can't rely on the old folklore that says a dewy morning means the day will be clear, because a new weather system can move in with clouds and rain.

Frost

Frost forms when water vapor in the air becomes ice without first condensing into water through a process called deposition. It forms on clear, calm nights with temperatures below 32°F (0°C).

KEY FACTS

Frost can form on the ground when the official temperature—measured above ground level—is above freezing.

+ fact: "Black frost" occurs when frigid air kills plants without visible frost.

+ fact: The growing season runs from the average date of spring's last frost to the average date of fall's first frost.

Frost forms overnight as white crystals on grass and other objects. With enough humidity in the air, these begin growing new crystals, which are called hoarfrost. Snowbanks are a good place to look for hoarfrost because some of the snow sublimates directly into water vapor during the day, and the vapor stays near the snow on calm days. At night, the vapor deposits into hoarfrost crystals, which cause the snow to sparkle. Frost forms on the inside of windowpanes that are not well insulated when it is moderately humid inside and very cold outside. Imperfections or scratches help shape the patterns that form. Frost can form on double-pane windows.

Rain

Clouds are made of water in the atmosphere, but they do not always cause precipitation. All rain falls from clouds, but most clouds do not produce rain.

KEY FACTS

Meteorologists have agreed on standard descriptions of falling rain.

+ **fact:** Light rain falls at up to 0.10 in (2.54 mm) an hour; scattered drops are seen.

+ **fact:** Moderate rain falls from 0.11–0.30 in (2.79–7.62 mm) an hour; drops aren't clearly seen.

+ **fact:** Heavy rain falls at more than 0.30 in an hour in sheets rather than drops.

A freezing cold rain and a warm rain both involve precipitation, but clouds that are colder than 32°F (0°C) and those that are warmer produce rain in different ways. In both, roughly a million cloud drops come together to produce a raindrop 0.08 inch (2 mm) in size. But in warm clouds, a few slightly enlarged cloud drops fall and sweep up others to grow into raindrops. In freezing cold, when ice and water mix, water vapor migrates into ice crystals that grow heavy enough to fall. If the air below is warmer, the ice crystals melt into rain; if not, they come down as freezing rain. See shapes of small and large drops at right.

Drizzle

Drizzle consists of water drops less than 0.02 in (0.5 mm) in diameter. Ordinary drizzle causes few problems to everyday life, but freezing drizzle turns to ice on impact and can pose a danger.

KEY FACTS

Visibility determines drizzle's intensity. More than a half mile (0.8 km) is light; between a quarter (0.4 km) and a half mile is moderate; less than a quarter mile is heavy.

+ **fact:** Drizzle drops fall close together and float in air currents.

+ **fact:** Drizzle is most likely in November and least likely in July in North America.

Drizzle falls mostly from stratiform or stratocumulus clouds. Climate scientists are especially interested in the drizzle from the shallow stratocumulus clouds that cover large areas of subtropical oceans—the areas just north and south of the tropics. These clouds are important for cooling the Earth, and researchers are investigating the role oceanic drizzle plays on these clouds and their effects. For most of us, drizzle means nothing more than a gloomy day—except for the ice that freezing drizzle leaves on roads and sidewalks. Drizzle-size drops that freeze on impact are especially dangerous for aircraft encountering them in the clouds or as they fall. Drizzle helps illustrate how even tiny atmospheric phenomena can be important.

Thunderstorms

Cumulus clouds grow and proceed through predictable stages as they turn into thunderstorms. Learning about thunderstorm stages is a first step toward understanding them.

KEY FACTS

A thunderstorm's life cycle has three stages.

+ fact: First is a towering cumulus stage as rising air forms a growing cloud.

+ fact: The mature stage, the longest, begins when rain starts falling, dragging down cold air.

+ fact: The dissipating stage begins when updrafts end, leaving only downdrafts until the storm dies.

As an ordinary cumulus cloud grows into a cumulus congestus and then a cumulonimbus, or thunderstorm, it's an indication of surging updrafts rising to form the cauliflower-like clouds you see as water vapor becomes cloud drops, which grow into raindrops. When parts of the cloud become smooth, ice crystals are forming. This process of glaciation releases more energy and generates more updraft, which can help spur the formation of the telltale anvil shape at the top of a thunderstorm cloud. In the mature stage, updrafts can be faster than 100 mph (160 kph); accompanying downdrafts measure half that speed. In fierce thunderstorms, especially supercells, a storm's mature stage can last for hours.

Multicell Cluster Thunderstorms

A large cumulonimbus cloud that is producing lightning with more than one dome on top, maybe an anvil on one side, is a multicell cluster of thunderstorms.

KEY FACTS

Multicell cluster thunderstorms line up in the direction the winds are blowing.

+ fact: New cells usually form at the side from which the wind is blowing; cells mature in the cluster's center.

+ fact: Each cell of a cluster lasts approximately 20 minutes, but the cluster itself can keep going for hours.

Although many thunderstorms go though three distinct stages as separate, single-cell storms, groups of related storms either in clusters or lines are more common. A multicell cluster forms when the downdrafts from one thunderstorm push air up to trigger an adjacent thunderstorm. This storm in turn can trigger a new one as the original storm is dissipating. A cluster will often have at least one storm in the dissipating stage, one mature storm, and one still in the towering cumulus stage, with the clouds of all of them blending. The wind pushes thunderstorms in a cluster in the same direction.

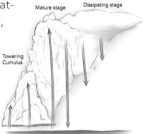

Mature stage   Dissipating stage

Towering Cumulus

Supercells

These thunderstorms are "super" in terms of their size, how long they last, and their potential for causing death and destruction. Supercells produce almost all of the most deadly tornadoes.

Although most single-cell thunderstorms last less than a half hour, supercells last for hours. Winds from different directions at various altitudes cause the main updraft to lean instead of traveling straight up and down, which means that rain falling from the top of the storm doesn't cool the cloud's rising warm air, and so the updraft keeps going. Many supercells produce only weak tornadoes, but some can be deadly. Researchers are looking for ways to distinguish well in advance which supercells will be the most dangerous and which will bring no major threats. When you see a supercell, you should stay alert for a tornado.

Squall Line Thunderstorms

If you see a long line of approaching thunderstorms preceded by a shelf cloud—a low-hanging, horizontal, wedge-shaped cloud—prepare for strong winds and lightning.

Steady winds can push a line of thunderstorms together in the same direction. The advancing storms scoop up warm, humid air, which feeds the storms even further. As individual storms die, new ones take their place in the line. Tornadoes sometimes occur, but a squall line's major danger is fierce straight-line winds that blast down in the direction the storms are moving. Individual storms in a line can reach more than 40,000 feet (12,000 m) into the air, too high for airliners to fly over. Squall lines can occur right along cold fronts, but the strongest are usually those several miles ahead of a cold front.

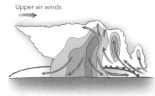

Upper air winds

Derecho

A derecho is an extremely long-lasting, fast-moving thunderstorm squall line that produces winds of at least 57 mph (92 kph) along a path at least 240 miles (386 km) long.

KEY FACTS

The term "derecho" was first used by Gustavus Hinrichs, a University of Iowa professor, in an 1888 scientific journal article.

+ fact: Hinrichs chose *derecho*, Spanish for "straight ahead," to distinguish its straight-line winds from a tornado's rotating winds.

+ fact: The National Weather Service began using "derecho" in 1987.

In the late spring and early summer, especially strong, long-lasting squall lines called derechos move across the Great Plains from the Rocky Mountains and sometimes all the way to the Atlantic coast. As in any squall line, the individual thunderstorms making up a derecho weaken and die, soon replaced by new ones, and so the damage along a derecho's path is not consistent. Winds build up to extreme speeds because the downburst force from individual thunderstorms is added to the speed of the wind pushing the squall line to the east. A derecho dies when it runs into dry air in the upper atmosphere or when the winds pushing it die down. Most derechos occur in the summer, most likely during heat waves, and mostly east of the Rocky Mountains.

Bow Echoes

A bow echo storm is an especially dangerous curved line of thunderstorms. The most dangerous winds occur at the crest or center of the curving bow-shaped formation of clouds.

KEY FACTS

A bow echo can range in size from 12 to 125 mi (20 to 200 km) across and last from 3 to 6 hours.

+ fact: Tornadoes often form on each end of a bow echo, but these are usually weak, doing little damage.

+ fact: On November 2, 1995, a bow echo hit the Hawaiian island of Kauai with winds of 90 mph (145 kph).

At times, winds blast down from part of a squall line or an isolated supercell and race ahead of the line or the supercell right above the ground. This soft, fast-moving wind, called a rear-inflow jet, pushes warm, humid air up and triggers a line of thunderstorms that form in a bow shape. T. Theodore Fujita, the famed 20th-century Japanese American tornado researcher, discovered and named bow echo damage patterns and radar images while investigating a derecho that struck from Michigan to Minnesota on July 4, 1977. Almost all derechos produce bow echoes, which often cause the derecho's most destructive winds.

Mesoscale Convective Complex

A mesoscale convective complex (MCC) is experienced as a long summer night with constant thunderstorm downpours and frequent lightning, usually on the Great Plains.

KEY FACTS

Each word in "meso-scale convective complex" describes an element of this sort of storm.

+ fact: Mesoscale: midsize weather phe-nomena, from a few miles to a few hundred miles across

+ fact: Convective: movement of air up and down

+ fact: As a complex, these storms are inter-related, not just hap-pening to be together.

By definition, a mesoscale convective complex (MCC) is a persistent, nearly circular area of clouds measuring a temperature of -25°F (-32°C). Generally thunderstorm anvils, they cover a huge area—at least 38,500 square miles (99,700 sq km), roughly the size of Iowa. The system begins in the late afternoon and early evening with heavy rain and sometimes strong winds. By early morn-ing, the thunderstorms die and the complex's rotating vortex, taller than 10,000 feet (3,000 m), continues traveling to the east. It can reignite the MCC the next evening. Meteorologists didn't real-ize MCCs were organized systems until weather satellite images showing cloud-top temperatures, thus their heights, became available in the 1970s. MCCs mostly affect the middle of the U.S.

Lightning

Lightning is a huge electrical spark flashing between areas of opposite electrical charge. It occurs inside clouds, from a cloud to the ground, from a cloud to another cloud, or into empty air.

KEY FACTS

Separate current strokes lasting a few tenths of a second cause the flickering lightning flash you see.

+ fact: Lightning's rapid heating and cooling of the air cre-ates the sound waves we hear as thunder.

+ fact: Thunder rum-bles as sounds from different parts of the flash arrive at slightly different times.

The violent churning of mixed ice crystals and water drops within a thunderstorm leaves areas of negative and positive charge in different parts of a cloud, usually with a strong negative charge near the cloud's bottom. Attraction between this and positive charge on the ground causes streams of negative charge to begin working their way down through the air as stepped leaders that zig one way, zag another. These create paths for stronger currents. When one of these connects with something on the ground, such as a tree or a lightning rod, a strong return stroke, which we see as lightning, flashes from the ground to the cloud.

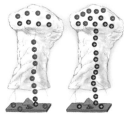

Lightning to Worry About

About 100,000 thunderstorms occur in the United States each year, according to the National Weather Service. Lightning makes each one dangerous, no matter how weak it is.

KEY FACTS

A flash of lightning measures approximately 50,000°F (27,760°C), but it lasts so briefly that a victim does not suffer deep burns.

+ fact: Lightning can cause random neurological damage or stop a victim's heart.

+ fact: Lightning rods, which carry lightning into the ground, have changed little since Benjamin Franklin's invention.

Most lightning flashes are a negative charge attracted to the ground's strong positive charge. To avoid lightning damage, you need to provide lightning a low-resistance path to the ground such as a lightning rod kept in good condition with no breaks in its path into the ground. To prevent injury or death by a lightning strike outside, stay away from lightning's path to the ground by being inside a sturdy building or a hardtop vehicle. Because lightning goes into the ground, it can hit and damage underground utility lines. Water pipes can give lightning a path inside your home. Indoors during a thunderstorm you shouldn't be near plugged-in appliances or computers, take a shower or bath, or talk on a phone with a cord. Cordless and cell phones are safe to use indoors during a thunderstorm.

Upper Atmospheric Lightning

A few thunderstorms put on stunning but hard-to-see shows of phenomena called sprites, elves, and blue jets above the storms as lightning flashes in and below the clouds.

KEY FACTS

Now, thanks to aircraft and imaging technology, we have images of all three phenomena, caused by intense electrostatic currents above dying thunderstorm clouds.

+ fact: Sprites are reddish orange or greenish blue with tendrils hanging down.

+ fact: Blue jets are narrow cones that fan out above the top of a storm cloud.

Going back to at least World War II, airplane pilots reported unusual lights above thunderstorms, causing scientists to wonder what happens above as well as below thunderstorms. Finally, scientists testing a low-light video in 1989 saw the first images of a sprite, a large discharge of energy above a thunderstorm. Unless you're an airline pilot, you will have to make a special effort to see these upper atmospheric lightning phenomena. The best way to try to see them is to be somewhere high, such as the Rocky Mountain Front Range. Choose a clear night and pick a location that is far from city lights. Wait until your eyes become adapted to the darkness, and then look out over the tops of thunderstorms on the Great Plains.

When Lightning Hits Airplanes

On average, lightning hits each airliner flying over North America once a year, but today's aircraft are built to shrug off the charge. Lightning last caused a U.S. airline crash in 1962.

KEY FACTS

Lightning frequently hits aircraft today.

+ fact: Airplanes and rockets can trigger lightning by flying into strong electrical fields.

+ fact: Lightning hit NASA's lightning research plane 714 times.

+ fact: After being hit by lightning 26.5 seconds after blastoff in 1969, Apollo 12 continued on to make the second manned moon landing.

When lightning hits an airplane, the electricity spreads out and flows through the aluminum skin and then back out into the air. The only damage caused includes small burn marks where the lightning entered and left the airplane. Fuel tanks have extra metal around them to keep lightning hits from burning through. Shielding protects electrical systems from the direct flow of lightning currents and from currents lightning can induce in wires. Advanced composite aircraft such as Boeing's 787 Dreamliner have conductive material embedded in the skin. Even if lightning hits, no one aboard a commercial aircraft today is in danger of being shocked. Lightning's threat to those servicing airplanes on the ground can delay flights.

Saint Elmo's Fire

Strong electrical fields rip electrons from nitrogen and oxygen molecules in the air causing a harmless blue-violet glow called Saint Elmo's fire, usually on tapered objects.

KEY FACTS

Saint Elmo's fire is named for Saint Erasmus of Formiae—Saint Elmo in English—the patron saint of sailors.

+ fact: Neon gas in neon and mercury vapor in fluorescent tubes creates glow discharges like Saint Elmo's fire.

+ fact: Flying through volcanic ash causes bright Saint Elmo's fire on wings and passenger windows.

Saint Elmo's fire is a natural blue-violet glow caused by a strong electrical field tearing apart the air's molecules of nitrogen and oxygen—a process called ionization. The resulting Saint Elmo's fire is a plasma. Curves and sharp points concentrate electronic fields, making the glow stronger. In past centuries, people saw Saint Elmo's fire at the top of ship masts and church steeples. The glow seemed miraculous because, though it looks like fire, it produces no heat and does not burn ships' wooden masts. Today, you are most likely to see Saint Elmo's fire on the tips of airplane wings. Pilots regularly see it around their windshields. In 1749 Benjamin Franklin became the first scientist to describe Saint Elmo's fire as an atmospheric electrical phenomenon. He sometimes saw it on the tips of lightning rods before lightning hit the rods.

Ball Lightning

Since the time of the ancient Greeks and Romans, people have reported glowing, floating balls in the air, associating them with lightning, yet scientists are still unable to explain what's going on.

KEY FACTS

Most reports of ball lightning indicate that it is harmless. Scientists can't explain it.

+ fact: Ball lightning is roughly the size of a grapefruit and as bright as a 60-watt bulb.

+ fact: Laboratory experiments have created objects with aspects of ball lightning, but have not entirely matched reports of it.

This unusual phenomenon is called ball lightning because it often resembles a floating ball. Almost everyone who has viewed a ball says it appeared either just after a lightning strike or when lightning was striking nearby. Researchers have compiled at least 10,000 reports of ball lightning in recent decades. Some observers report ball lightning passing through windows without causing damage; others report seeing it appear and disappear inside airplanes in flight, again with no damage. Mysteriously, it glows without giving off heat and does not appear to generate power. Individuals and groups of people have reported ball lightning since the time of ancient Greece. Glowing balls approximately six inches in diameter floating in the air have persuaded many scientists that ball lightning is real.

Hail

Hailstones are pieces of ice that form in thunderstorm updrafts, which keep them from falling while more water freezes onto them. Updraft speed determines the size of hailstones.

KEY FACTS

A hailstone ½ inch in diameter needs a 20 mph (32 kph) updraft to form; a ¾-inch stone needs a 64 mph (103 kph) updraft.

+ fact: More hail falls yearly on northeastern Colorado and southeastern Wyoming than elsewhere in the U.S.

+ fact: The National Oceanic and Atmospheric Administration says hail injures 24 people in the U.S. each year, deaths are very rare.

Most hailstones form in multicell, supercell, or cold-front squall line thunderstorms, usually near the center of a storm. Hail begins forming as tiny ice pellets collide with supercooled water droplets that freeze on contact with the ice. As they grow, hailstones may make several up-and-down trips within the center of the storm before the updraft weakens or when they become heavy enough to fall. By definition, a thunderstorm that produces ¾-inch (19 mm) hail is severe because of the strong updrafts and other winds in the system. Very strong updrafts can carry hailstones high into a storm and then sweep to one side, so that they fall outside the storm itself.

Microbursts

A microburst is an intense, concentrated wind that blasts down from a shower or thunderstorm, affecting an area no longer than 2.5 miles (4 km) on a side with strong, gusty winds.

KEY FACTS

Wind damage often shows whether a microburst or a tornado caused it.

+ fact: The National Weather Service has 48 Terminal Doppler Weather Radars to detect airport microbursts.

+ fact: In 1983, a 138-mph (220 kph) microburst hit Andrews Air Force Base three minutes after *Air Force One* carrying President Ronald Reagan landed.

Microbursts can blow down big trees, so people should avoid standing near trees during a windy thunderstorm. The greater danger of microbursts is to flying aircraft. Meteorologists did not have a name for microbursts until the 1970s, when researcher T. Theodore Fujita coined the term "microburst" to describe the concentrated winds that had caused several airplane crashes. Once the term was in use, pilots and air traffic controllers were better able to observe and avoid them. Through 1985, the U.S. air industry suffered roughly one fatal microburst airliner crash every 18 months, but since then, only one has occurred, thanks to better warnings.

Gust Front

Air coming down from a thunderstorm moves over the ground, as a miniature cold front. Effects include bringing slightly cooler temperatures and triggering new thunderstorms.

KEY FACTS

You may feel a distant thunderstorm's gust front as a cool breeze on a hot day.

+ fact: Outflow boundaries interacting with supercell thunderstorms increase the odds that the supercell will produce tornadoes.

+ fact: A gust front hitting an airport can endanger takeoffs and landings with unexpected wind shifts.

Meteorologists call gust fronts "outflow boundaries" because they do more than bring a quick shot of refreshing cool air. They can persist longer than 24 hours with small temperature differences in the air across the front. Even small changes can help trigger new thunderstorms and affect existing storms. Air converges along gust fronts, bringing dust, insects, and other small objects that Doppler radar can see long after the gust front forms. Meteorologists use such gust front images to forecast where new thunderstorms are likely to begin. For instance, the meeting point of two gust fronts, each of which may be a day old, is a prime location for where storms may be expected to originate and then develop.

Funnel Cloud

A funnel-shaped cloud full of condensed water stretching down from a towering cumulus cloud or cumulonimbus might be called a tornado that is not touching the ground.

KEY FACTS

When the air is too dry for a condensation funnel to form, the first sign of a tornado can be spinning debris on the ground.

+ fact: A tornado's spinning winds extend wider than the visible funnel.

+ fact: Funnels turn from white to black or other colors when they pick up dirt or dust from the ground.

A funnel dipping down out of a storm cloud is a funnel cloud as long as it neither touches the ground nor kicks up dust or debris from the ground. Once either of those conditions occur, it has become a tornado. This type of condensation funnel forms when falling atmospheric pressure in the vortex, or spinning funnel, cools the air enough for water vapor to condense into a swirling cloud. Almost all tornadoes begin as funnel clouds, but many funnel clouds never become tornadoes. Meteorologists sometimes talk of cold air funnels: They don't form in thunderstorms, they are generally weak, and they don't last long, although a few might touch down briefly as weak tornadoes or waterspouts.

Tornadoes

Tornadoes are rotating columns of air hanging from a cumulonimbus or sometimes a cumulus congestus cloud that have developed long enough that they come in contact with the ground.

KEY FACTS

Tornadoes' strength is ranked from F0 to F5 on the Fujita scale.

+ fact: Weak tornadoes: F0, 65–85 mph (105–137 kph); F1, 86–110 mph (138–177 kph)

+ fact: Strong tornadoes: F2, 111–135 mph (179–218 kph); F3, 135–165 mph (219–266 kph)

+ fact: Violent tornadoes: F4, 166–200 mph (267–322 kph); F5, faster than 200 mph (322 kph)

The wind speed of tornadoes can range from 40 mph (64 kph) to more than 300 mph (483 kph). Tornadoes are the most destructive of all local weather phenomena. Those clocking faster than 166 mph (267 mph) account for more than 70 percent of all tornado deaths and the greatest destruction. Fortunately, about three-fourths of all tornadoes are weaker, with winds no faster than 110 mph (179 kph). Because strong tornadoes destroy wind instruments, meteorologists must often examine damage to determine the wind speed. The U.S. averages more than 1,000 tornadoes a year, more than any other nation; Canada ranks second.

Multiple Vortex Tornadoes

Many tornadoes, especially the larger ones, are packages of twisters with smaller vortices circling around the central tornado's perimeter, sometimes seen but often hidden.

KEY FACTS

Three seems the most common number of subvortices in multiple vortex tornadoes, but spotters have seen as many as seven.

+ fact: Some hurricane eye walls have small vortices that look and act like tornado subvortices.

+ fact: Spotters reported multiple vortices before the May 22, 2011, tornado hit Joplin, Missouri, killing 161 people.

Why does it happen that a tornado can destroy a well-built house while a similar house, right next door, escapes with minor damage? The existence of multiple vortex tornadoes explains this phenomenon. The heavy damage occurs where the winds of the main tornado combine with the winds of a subvortex moving in the same direction, thus multiplying the impact. At times, this combination of forces can add 100 mph (161 kph) to the wind, spotlighting small areas in the tornado's larger path. Tornado spotters most likely see multiple vortices as a tornado is beginning, before dust and debris darken the main funnel. Damage patterns can confirm the vortices.

Waterspouts

A common definition of a waterspout is "a tornado over water," but meteorologists prefer to use the term for weak, non-supercell vortices that form below cumulus congestus clouds.

KEY FACTS

A waterspout's condensation funnel is not water sucked up from below; it is condensed from water vapor in the air above.

+ fact: A waterspout's life cycle is usually less than 20 minutes; winds reach speeds no greater than 85 mph (137 kph).

+ fact: A fall waterspout season often occurs on the Great Lakes.

Waterspouts most often form from the bottoms of dark, growing cumulus clouds. The first sign of a waterspout forming is a dark spot on the water's surface where the still invisible funnel has touched down; it may be kicking up water spray as well. Many waterspouts die out in this early stage, but sometimes they develop to the stage that a condensation funnel is visible, coming down from the cloud to the dark spot. The oceans around southern Florida, especially in the Keys, have more waterspouts than any other part of the U.S. Non-supercell waterspouts are generally weak, which means that they aren't as dangerous as most ordinary tornadoes, but boaters should not ignore them. Waterspouts can be stronger than they appear.

Dust Devils

A dust devil is a small whirlwind that, unlike a tornado, is not attached to a cloud. Dust devils usually occur in hot, dry locations where dust or sand makes them visible.

KEY FACTS

The strongest dust devils occur in the hottest deserts, but small ones can occasionally occur in cities as well.

+ fact: Dust devil winds sometimes build to the speed of an F1 tornado, up to 110 mph (177 kph).

+ fact: The Viking orbiters NASA sent to Mars in the 1970s discovered that dust devils occur on the red planet.

Dust devils form in thermals of rising air and often develop a funnel-like shape as air flows out of the top. The air cools as it rises and then eventually sinks back to the ground, where it can create the light breezes that spin thermals into dust devils. Dust devils can develop to a height of 3,000 feet (900 m) or more on some of the hottest deserts. They are usually benign, but they occasionally cause damage and injuries. On September 14, 2000, for example, a dust devil hit the Coconino County Fairgrounds in Flagstaff, Arizona, causing minor injuries and damaging booths and tents.

Acid Rain

Acid rain describes precipitation made slightly acidic by air pollution often from far away. It wreaks damage on both the natural environment and on limestone and marble buildings.

KEY FACTS

Acid rain's most damaging effects occur in locations where the soils do not naturally neutralize acids.

+ fact: Acid rain leaches calcium from soil, depriving plants of an important nutrient.

+ fact: Robert Angus Smith, a Scottish chemist, coined the term "acid rain" in 1852 and linked it to damage it causes.

Although the phenomenon likely dates back a century or two, only in the mid-20th century did people begin to take responsibility for acid rain and its consequences. Most acid rain in North America develops from sulfur dioxide and nitrogen oxides produced by burning fossil fuels, mostly to generate electricity and run cars and trucks. The acid rain does not come directly from smokestacks or tailpipes. Instead, the emissions react with water in the atmosphere to create damaging compounds that can travel hundreds of miles from their sources. The pollutants can fall as dry materials as well as in rain or snow. The chemicals make soil, rivers, lakes, and ponds slightly acidic, potentially killing smaller animals and plants. Air quality rules are reducing acid rain, and some hard-hit areas in the Northeast are recovering.

Hurricanes

A hurricane is a tropical cyclone in the Atlantic, the eastern Pacific, the Caribbean Sea, or the Gulf of Mexico—in which winds reach a sustained speed of 74 mph (119 kph) for a minute or more.

KEY FACTS

A tropical cyclone becomes a tropical storm when its winds reach 39 mph (63 kph), at which time it is assigned a name from a pre-selected list.

+ fact: Sustained winds of 74 mph (119 kph) make it a hurricane.

+ fact: Hurricanes are classified on a scale from 1 to 5 in order of severity based on wind speed.

Tropical cyclones are born over ocean water that is warmer than 80°F (27°C) and in a condition of high humidity. Because they draw their energy from warm water, these storms weaken and die when they move over cool water or land. A mature hurricane system rotates counterclockwise with a diameter of hundreds of miles and an eye of mostly clear air and light winds in the center. The fastest winds are usually in the eyewall—the ring of thunderstorms encircling the eye. Rain bands of thunderstorms spiral around the storm and into the eyewall. A hurricane pushes water ahead, creating a storm surge when it comes ashore.

Steering current · Air leaving the storm's top · Low-level winds spiraling into the storm

Hurricane Forecasting

Radar, satellites, sea buoys, and reconnaissance flights contribute to improved methods of hurricane forecasting, so that meteorologists can often predict a hurricane's path a week ahead.

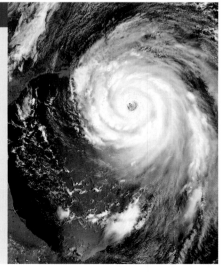

KEY FACTS

Hurricane researchers are focusing on ways to improve forecasts for when a storm will strengthen.

+ fact: Weather satellites ensure that no hurricane will hit land without warning.

+ fact: The National Hurricane Center regularly sends airplanes into hurricanes to gather data that are unavailable via satellite.

Hurricane forecasts are fast improving, but they are still far from perfect. From 1970 to 2012, the average error in track forecasts for three days into the future shrank from 518 miles to 138 miles (834 to 222 km). That is a significant improvement, but it means that if a strong hurricane is forecast to hit in three days, anyone located 200 miles (322 km) on each side of the forecast strike zone should be alert for changes in the forecast and ready to evacuate. Forecasters also assess climate variables to predict the number of hurricanes in a given season, still an inexact science as well.

Extratropical Cyclones

In brief, these are large storm systems that form over cold air or cold water away from the tropics. They account for almost all stormy weather in the middle and polar latitudes all year.

KEY FACTS

Fronts are defined by the interaction of air masses at their boundary.

+ fact: At a cold front relatively cold air is advancing to replace warmer air.

+ fact: At a warm front relatively warm air is advancing to replace cooler air.

+ fact: At a stationary front warm air and cold air meet, with neither advancing.

You're not likely to hear a television meteorologist describe the weather system or storm threatening to bring widespread rain or snow as an extratropical cyclone—but chances are it is one. The term applies to any storm that is not a tropical cyclone (or hurricane). Unlike tropical cyclones, these storms contain both warm and cold air masses with fronts as boundaries between them. The temperature contrast between the large masses of cold and warm air supplies the storm's energy. The larger the contrast, the stronger the storm. Extratropical cyclones range in diameter from 600 to 2,500 miles (1,000 to 4,000 km). The fronts usually run like spokes of a wheel from the storm's central area of low pressure.

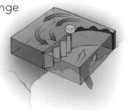

Winter Storms

Winter storms commonly contain ice, snow, and blizzards, and yet extratropical cyclones that cross North America in the winter can bring thunderstorms, tornadoes, and flooding downpours.

KEY FACTS

Humid winds blowing uphill bring heavy upslope snowstorms to mountains in the West and East.

+ fact: A nor'easter is a winter storm that moves northward along the East Coast of the U.S. and Canada.

+ fact: Alberta clippers are small winter storms that zip from Canada to the East Coast with sharp temperature drops.

From time to time, a strong extratropical cyclone slams into the West Coast from the Pacific Ocean with flooding rain in lower elevations and heavy snow in the mountains. Crossing the Rockies disrupts the storm's surface winds, but as the upper-level winds move over the Great Plains, they stir up surface winds reviving the winter storm with cold air from Canada and warm air from the Gulf of Mexico. Some storms move to the Gulf of Mexico and then northeastward along the Atlantic coast; others travel into the Midwest, bringing blizzard conditions. A winter storm's cold front can stretch into the South to produce severe thunderstorms and sometimes tornadoes. Snow, of course, is also the lifeblood of ski resorts.

Blizzards

A snowstorm becomes a blizzard when winds reach a speed of 35 mph (56 kph) accompanying falling or blowing snow that reduces visibility to less than a quarter mile (400 m).

The combination of low temperatures and poor visibility in blowing snow makes blizzard conditions the most dangerous winter weather. The blowing snow creates a "whiteout" when the horizon disappears, no shadows appear, and objects are hidden. Disoriented victims can become lost in places they know well, even between their house and barn. A more common hazard in today's automobile world is a chain-reaction collision often caused by a driver who stops suddenly when visibility drops to zero. In both cases, low temperatures can be fatal for victims who cannot reach warm shelter soon enough. Many blizzard victims die of carbon monoxide poisoning when they run a snowbound car to keep warm and exhaust gas leaks in, or when they use an unvented heat source such as a charcoal grill indoors.

Lake-Effect Snow

Bitter cold air flowing over much warmer water brings heavy snow to areas downwind of the Great Lakes and a few other bodies of water. The lakes are needed for this weather system.

Frigid air blowing across lakes that are at least 20°F (11°C) warmer than the air creates cumulus clouds as the warmer lake water evaporates into the cold air. These clouds dump heavy snow as they move inland and over hills. Places downwind of the Great Lakes, such as Buffalo, New York, often have 100 or more inches (2.5 m) of snow a year—double the amount that falls on places at the same latitude not downwind of the lakes. The greatest amounts of lake-effect snow fall early in the season, before the water in the lakes cools. A lake stops making snow if it freezes over.

River Floods

Rivers flood after prolonged heavy rain over a large area or when deep snow covering a large area melts. Hydrologists can predict river floods in time for those threatened to flee.

KEY FACTS

A floodplain is flat or nearly flat land adjacent to a river that floodwaters naturally cover.

+ fact: Sediments deposited by previous floods make floodplains prime farmland with rich soils.

+ fact: A river flood watch means water is close to rising above flood stage; a river flood warning means flooding may begin soon.

River floods are usually slow-motion disasters, with the highest part of the flood—the crest—moving downstream at less than 10 mph (16 kph). Hydrologists describe actual and forecast flood heights in relation to the "flood stage" at particular gauging stations. Flood stage is the water height at which flooding begins to cause damage at a location. Because flood stage is different at each station, you cannot rely on figures from upstream to indicate what your area should expect. Levees—long, earthen banks that hold back floodwater—sometimes break, which can quickly enlarge the area covered by floodwater. Because forecasting isn't perfect, if flooding is predicted for an area near your home you should be prepared for the flood to be higher than predicted.

Flash Floods

Flash floods, which inundate low-lying areas in less than six hours, are a leading cause of weather deaths in the United States. Intense rainfall or dam failures cause most flash floods.

KEY FACTS

Slowly moving water 2 ft (0.61 m) deep can carry away an SUV-size vehicle.

+ fact: More than half of flash-flood victims were in vehicles driven into water covering a road.

+ fact: A dam failure caused the worst flash flood in U.S. history, killing 2,209 people in Johnstown, Pennsylvania, on May 31, 1889.

Intense rain from thunderstorms or a dying hurricane can create floods that quickly turn usually quiet streams into death traps. In the spring, chunks break off from the ice that has covered a stream. They wash down and pile up into ice dams that can suddenly collapse, creating flash floods. Some of the worst flash floods occur in deserts, where water doesn't soak into the ground. Rain from a thunderstorm too far away to see or hear can race down dry streambeds and catch hikers unaware. Hurricanes and tropical storms that have moved far inland often cause flash floods as well. Those living near streams in hilly or mountainous areas need to be especially aware of the flash-flood danger. A weather radio that turns on and sounds an alarm when a warning is issued can save your life.

Dust Storms

Dust storms, which affect arid regions, fill the air with wind-driven, microscopic dust particles over an extensive area and reduce horizontal visibility to less than ⅝ of a mile (1 km).

KEY FACTS

The leading edge of a dust storm often looks like a knobby vertical or convex wall.

+ fact: The Arabic word "haboob" is often used for dust storms in the southwestern U.S.

+ fact: Sandstorms do not grow as tall as dust storms because sand particles are larger and heavier. Winds rarely lift sand particles above 50 ft (15 m).

Thunderstorm gust fronts pushing over dry, dusty ground can stir up dust storms. Unlike cold fronts moving into humid areas, which trigger thunderstorms, a cold front in an arid region often lifts dust high into the air. Dust storms can reach a height of roughly 3,000 feet (1,000 m). Cold fronts advancing into dry air create the largest dust storms. In the U.S. Southwest, the dust storm season is May through September. The most severe storms occur when the soil is driest, between April and June, depending on the year's weather. Winds in dust storms are rarely faster than 30 mph (48 kph), but they have been clocked as fast as 62 mph (100 kph). During the 1930s Dust Bowl drought, storms carried dust from the plains to the East Coast.

Volcanic Ash

Volcanic eruptions are not weather events, but can substantially affect the atmosphere. Volcanic ash shot high into the air can stop airplane engines. Some eruptions affect global weather.

KEY FACTS

Between 1982 and 1989, volcanic ash briefly stopped all engines of three Boeing 747s. All three landed safely.

+ fact: Sulfur from the 1991 Mount Pinatubo eruption in the Philippines cooled the Earth for a year.

+ fact: Eruptions in 2010 and 2011 forced *Air Force One* pilots to change President Obama's schedule on three overseas trips.

Both the ash and the gases shot into the air during a volcanic eruption have deleterious effects on weather and daily life. The ash, composed of tiny particles, can block the sun and cool the Earth temporarily. Sulfurous gas shot into the stratosphere during large eruptions forms a haze of sulfuric acid that can block the sunlight and combine with water to form acid rain. When a jet aircraft runs into a cloud of volcanic ash, the tiny particles invade the spaces between moving parts in the engine and drivetrain and can melt and fuse inside the works. This is why volcanic eruptions such as that of Eyjafjallajökull in Iceland in 2010 interrupt air travel until the atmosphere has cleared, which often takes days.

Air Pressure

Even though we hardly notice the air around us, its pressure is one of the most important forces driving the weather. Unequal air pressures in large masses of air cause the winds to blow.

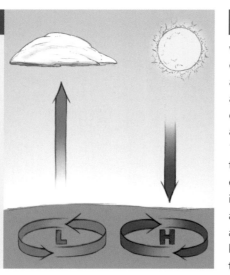

KEY FACTS

At sea level, the air's average pressure is 14.7 lb per in².

+ fact: At 18,000 ft (6,000 m) above sea level, air pressure averages 7.25 lb per in²; half of Earth's air is below that altitude.

+ fact: At 102,000 ft (31,090 m) above sea level, pressure averages 0.147 lb per in²; 99 percent of Earth's air is below.

Dry air consists of roughly 78 percent nitrogen molecules and 20 percent oxygen, with other gases making up the rest. Air is easily compressed, and the pressure at any altitude depends on the weight of all of the air above pressing down. (That is why air pressure decreases rapidly with increasing altitude.) The air's molecules are zipping around at roughly 1,000 mph (1,600 kph)—the higher the temperature, the faster they're going. Fast-moving molecules create pressure pushing in all directions, including up, to oppose the weight of molecules above. We experience the movement of the air as wind. Differences in air pressure at different locations and different altitudes cause winds to blow.

Measuring Air Pressure

Meteorologists measure atmospheric air pressure both at Earth's surface and aloft, because pressure differences between locations determine wind speeds, directions, and weather.

KEY FACTS

The height of mercury in a barometer tube—in inches or millimeters—was the original air pressure measurement.

+ fact: Today the U.S. National Weather Service uses millibars to describe upper air pressures and in surface reports for meteorologists.

+ fact: Canada, like most other nations, uses hectopascals for barometric measurements.

In the late 19th century, as meteorology was becoming a mathematical science, meteorologists began using what are now called hectopascals, a metric unit of pressure, like pounds per square inch, that can easily be used in mathematical formulas. In common parlance, one more likely hears about "inches of mercury," a unit the U.S. National Weather Service uses for surface atmospheric pressure in reports for the public. The phrase harkens back to the mercury barometer, invented by the Italian Evangelista Torricelli in the 1640s. Most weather observers today use electronic devices that sense air pressure rather than mercury barometers. These devices are at the heart of automated barometers, and hikers can easily carry them.

Why Winds Blow

Winds blow as air moves from areas of high atmospheric pressure toward areas of lower pressure. Because the Earth rotates underneath, the wind follows slightly curved paths.

KEY FACTS

The Coriolis force, named for Gaspard G. Coriolis, describes how Earth's rotation causes winds to follow curved paths.

+ fact: It causes counterclockwise winds around large Northern Hemisphere storms and clockwise winds in the Southern Hemisphere.

+ fact: It has no effect on water draining from a sink or down a toilet.

To see why winds blow, let the air out of a balloon or a bicycle tire and feel the escaping air create a mini-wind as it moves from the high-pressure air inside to the lower-pressure outside air. The same phenomenon happens in the atmosphere on a much larger scale as wind blows from an area of high pressure toward an area of lower pressure. Two factors determine the wind's speed: the difference in pressure and the distance between the two areas. Masses of air close together with distinctly different pressure induce strong winds; masses of air far apart with similar pressure induce little to no wind. Because friction with the ground slows winds near the ground, they are normally slower than winds aloft and also low winds above an ocean or large lake.

Measuring Winds

To describe and predict the weather, meteorologists use various anemometers to measure both wind speed and direction. A wind's direction is named for the direction from which it is blowing.

KEY FACTS

Official surface winds are measured by instruments mounted 33 ft (10 m) above the ground.

+ fact: Wind speeds are calculated as 2-minute averages.

+ fact: A squall is a wind 18 mph (30 kph) faster than the sustained (or steady, underlying) wind. A squall happens suddenly and lasts at least 2 minutes.

The observation of wind speed and direction is an ancient art and a modern science. For many years, most weather stations used cup-and-vane anemometers with spinning cups to measure wind speed and with a moveable vane to show the wind direction. Snow and ice could disable these, so measurements during severe weather were disrupted. Today, sonic anemometers are becoming standard. They send ultrasound waves between three arms 4 to 8 inches (10 to 20 cm) apart. Winds slow or speed the sound waves, and a processor inside the instrument determines the time it takes for the sound to travel between the arms, and uses those findings to calculate wind speed and direction. Sonic anemometers work well in turbulence.

Local Winds

Differences in air temperatures over relatively short distances, such as a couple of hundred miles, cause local winds—from gentle sea breezes to the roaring winds that whip up California wildfires.

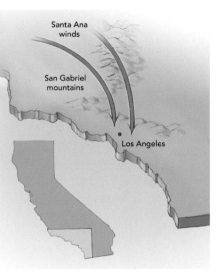

KEY FACTS

Cold air spilling down hills, mountains, or the sides of canyons is called a katabatic wind.

+ fact: Santa Ana winds race through dry southern California canyons, often fanning the flames of wildfires.

+ fact: Wildfire fighters need to be prepared for unexpected mountain wind changes that could endanger them.

Local winds occur apart from large air masses and weather systems, caused by the interaction of winds and geography. California's Santa Ana winds begin when dense, cold air builds up east of the Sierras and the southern coastal range. The cold air spills through the canyons, warming as it falls downhill—a local phenomenon that occurs in many places. Sea breezes begin when land warms faster than water. Air rises over the warmed land and air from over the water flows in to replace it. As mountaintops warm during the day, air begins rising, and air from valleys flows uphill to replace it. At night, as the mountaintop air cools, the air becomes heavier and flows down into valleys, making them colder than nearby elevations.

Regional Winds

Various kinds of winds other than parts of tropical or extratropical cyclones can have major regional effects on the weather. These include monsoon winds and chinook winds.

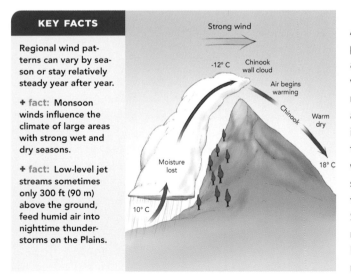

KEY FACTS

Regional wind patterns can vary by season or stay relatively steady year after year.

+ fact: Monsoon winds influence the climate of large areas with strong wet and dry seasons.

+ fact: Low-level jet streams sometimes only 300 ft (90 m) above the ground, feed humid air into nighttime thunderstorms on the Plains.

At the intersection of weather systems, landforms, and major bodies of water, regional weather patterns can be expected. Monsoon climates in Asia and in the southwestern U.S. and adjacent Mexico vary between very dry and very wet seasons. In these regions during the summer, warmed inland air rises and humid winds from the ocean bring humidity, feeding rainstorms. In winter, dry winds blow from inland to the oceans. On the east side of the Rockies, chinook winds warm up as they blow down and melt winter snow. In one case, chinook winds caused temperatures to rise from -54° to 48°F (-48° to 9°C) in 24 hours. Strictly speaking, "monsoon" refers to winds with pronounced seasonal shifts or climates with such shifts. It's also commonly used for heavy rain the humid summer winds bring, or even for any heavy rain.

Global Winds

Global-scale winds blow constantly above the Earth, moving warm air out of the tropics and cold air out of the polar regions, setting the stage for smaller weather events that directly affect us.

In the tropics or the polar regions, the winds blow from the east most of the time. In the middle latitudes, north and south, while the general flow is from the west, storms complicate the surface picture with changing wind directions. The winds high aloft, including jet streams—concentrated horizontal, high-altitude winds—move from west to east with deviations to the north and south. Extratropical cyclones travel generally west to east with diversions like those of the jet streams. When tropical cyclones such as hurricanes move into the middle latitudes, their paths begin curving toward a west-to-east direction. Weather forecasters focus a great deal of their attention on measuring and forecasting global-scale winds because they determine the paths and strengths of storms and their winds and precipitation.

Jet Streams

Jet stream paths follow the locations of cold and warm air at the surface and are intimately linked with the movements of cold air toward the Equator and warm air toward the poles.

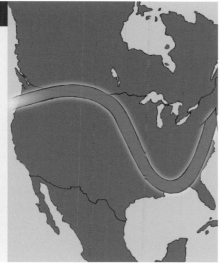

Meteorologists define a jet stream as "a relatively narrow river of very strong horizontal winds embedded in the winds that circle Earth aloft." Jet streams skirt the boundaries between deep layers of warm and cold air—fronts on Earth's surface, and often locations of potentially dangerous weather. A wavy jet stream shows that warm air is moving north and cold air south, possibly destined to mix it up in a new storm. Nevertheless, jet streams and other upper air winds steer storms and determine where areas of high and low pressure form at the surface. Jet streams and surface weather dance with one another, neither one always taking the lead. Fair jet stream winds that dip far over the South are a characteristic of strong winter storms.

The Polar Jet Stream

The northern polar jet stream is an upper-atmosphere band of high-speed winds circling the globe above the ever-shifting boundary between cold, dry polar air and warmer, moist mid-latitude air.

KEY FACTS

The polar jet is fastest and farthest south during the coldest parts of winter, farther north in summer.

+ fact: The polar jet stream helps extratropical cyclones form and grow and helps steer them.

+ fact: The Southern Hemisphere's polar jet stream usually circles over the continent of Antarctica all year long.

The polar front separates cold polar air and warm mid-latitude air, but often, polar and mid-latitude air blend with no sharp temperature differences. At such places, the polar jet fades and then forms again where the boundary has larger temperature contrasts. When polar air plunges south, the polar jet turns south and loops around to the north, staying above the warm–cold boundary as a trough. Fast jet streams characterize fierce storms. During the March 12–14, 1963, "superstorm" that paralyzed the East, the polar jet trough dipped all the way over the Gulf of Mexico, and jet stream winds were as fast as 224 mph (360 kph). The extratropical storm below this jet stream had surface winds faster than 74 mph (119 kph).

Atmospheric Rivers

Atmospheric rivers are narrow bands of strong low-level winds—5,000–8,000 feet (1,500–2,250 m) above oceans—that feed tropical water vapor to mid-latitude storms.

KEY FACTS

Atmospheric rivers supply between a third and a half of all U.S. West Coast precipitation.

+ fact: An atmospheric river from the Pacific Ocean crossed Central America to feed the February 2010 East Coast "Snowmageddon" blizzard.

+ fact: West Coast meteorologists often call atmospheric rivers the "Pineapple Express."

As far back as the 1930s, scientists hypothesized the existence of narrow bands of strong low-level winds that supplied moisture for middle-latitude storms. It took computers and weather satellites to confirm it. In the 1990s, researchers suggested that three to five narrow rivers of air supply 90 percent of the tropical water vapor that reaches the middle latitudes. In 2004, using data collected during airplane flights into these atmospheric rivers and other sources, NOAA scientists confirmed the hypothesis. West Coast forecasters now use these research results to improve forecasts for rain and snow brought by the atmospheric rivers, some of which flow eastward from near Hawaii. Atmospheric rivers also affect Europe and Africa.

Deep Ocean Currents

Water from the Gulf Stream and other currents on the ocean surface are parts of a global conveyor belt that includes underwater currents and transports carbon dioxide and nutrients.

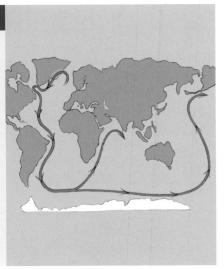

KEY FACTS

Oceanographers estimate that water takes 1,000 years to travel the complete circuit of global currents on and below the surface.

+ fact: Deep ocean water lies more than 6,000 ft (1,800 m) below the surface, where little light penetrates.

+ fact: Deep ocean water is very cold, usually from 32°F to 37°F (0°C to 3°C).

The Gulf Stream is a great river in the ocean that travels northward up the east coast of Mexico, eastward between Florida and Cuba, and northward along the U.S. and Canadian east coast. Water carried north cools and grows denser, with evaporation, leaving behind salt. East of Greenland, this water sinks to form Atlantic deep water, part of the global system of underwater currents. As these currents traverse the deep oceans, they carry organic matter including animal waste and parts of dead plants and animals. The currents also carry carbon dioxide that was absorbed by the water when it was cold. Eventually the water with its nutrients and carbon dioxide upwells along the west coasts of North and South America and Africa, and along parts of the Equator, creating rich areas for sea life.

Surface Ocean Currents

Global winds, such as tropical trade winds, drive oceanic surface currents. The currents carry heat from the tropics to the middle and polar latitudes with important effects on climate.

KEY FACTS

The warmth of the Gulf Stream can strengthen tropical and extratropical cyclones that cross it.

+ fact: Off the U.S. Atlantic coastline, the Gulf Stream moves as fast as 5.6 mph (9 kph).

+ fact: The California current, which moves south along the U.S. West Coast, helps keep coastal waters cool.

Earth's ocean currents form oceanwide gyres—clockwise in the Northern Hemisphere, counterclockwise in the Southern. Through the 20th century, scientists had thought that these currents did most of the work of transporting heat toward the poles. Now, however, there is strong evidence that the atmosphere in the Northern Hemisphere carries 78 percent of the heat moved toward the north, but that ocean currents do most of the work in the Southern Hemisphere, carrying 92 percent of the heat moving toward Antarctica. Researchers are also finding strong evidence that the Gulf Stream doesn't do as much to keep Europe warm as previously thought. Southwest winds also help warm Europe during the winter.

El Niño

El Niño, which happens every few years, occurs when unusually warm tropical surface water flows into parts of the Pacific Ocean. The shifting of warm water has global effects.

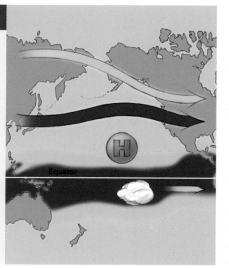

KEY FACTS

El Niño is part of an irregular global climate pattern called the southern oscillation.

+ fact: El Niño usually brings warmer-than-average fall and winter temperatures to the northern U.S. and Canada.

+ fact: El Niño produces high-altitude winds over the Caribbean Sea that can rip hurricanes apart.

Air flowing out of the tops of Pacific thunderstorms feeds global winds. El Niño pushes these thunderstorms farther east and disrupts jet streams downstream across the Americas. These disruptions, in turn, shift normal patterns of rain, dryness, and storminess as far away as Africa. In North America, the effect is increased rain in normally drier areas and noticeably arid weather in areas that usually get rain. An El Niño occurred in 1957–58, the International Geophysical Year, and scientists began to understand the connection between events that were long assumed to be unconnected. Today, measurements taken in the Pacific Ocean help to predict coming El Niño events, which can have serious economic repercussions.

La Niña

La Niña, the counterpart to El Niño, is a set of global atmospheric events set in motion when the eastern, tropical Pacific cools and the ocean's warmest weather moves to the west.

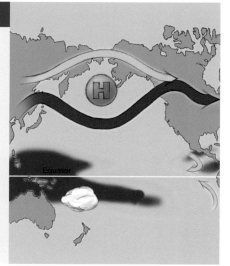

KEY FACTS

Paths of the subtropical and polar jet streams are more variable during La Niña years.

+ fact: La Niña increases the odds of a hurricane hitting the U.S. Atlantic or Gulf coast.

+ fact: La Niña years tend to have warmer-than-normal winters in the southeastern U.S. and cooler-than-normal winters in the northwestern U.S. and Canada.

Also part of the southern oscillation, La Niña is a Pacific Ocean phenomenon with global implications. El Niño begins with the shift of warm waters in the tropical Pacific to the east, and La Niña begins with enhanced upwelling of deep ocean waters along the South American coast, thus cooling this part of the ocean. Stronger trade winds push the warmest water to the west along with the thunderstorms above it. As with El Niño, winds flowing out of the tops of these storms affect jet streams, but in different patterns. Where El Niño would bring downpours, La Niña generally brings drought. For example, parts of Australia and Indonesia affected by El Niño droughts are wet during a La Niña. In 2011 some scientists linked that year's tornado outbreaks to La Niña, but other scientists disputed this. The question is far from settled.

Arctic Oscillation

The Arctic Oscillation (AO) is an irregular swing between opposite air pressure and wind patterns centered on the Arctic. It strongly affects winter weather in eastern North America.

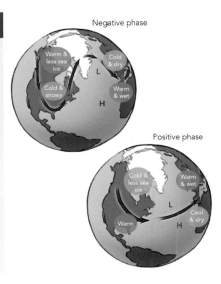

Negative phase

Positive phase

KEY FACTS

Better observations and computers enabled scientists to pin down details of the AO in the 1990s.

+ fact: The most negative AO on record was in February 2010, when three fierce snowstorms hit the United States and Canada.

+ fact: The National Weather Service mentions the AO in winter weather discussions.

The Arctic Oscillation's positive phase features lower air pressures over the Arctic and strong upper air winds around latitude 55° N, which blocks cold outbreaks from hitting the northeastern United States and Canada. The AO's negative phase includes higher air pressure over the Arctic and weaker upper air winds around 55° N, which allows more cold outbreaks to hit the Northeast. The AO can switch between phases in days, but sometimes one phase dominates for long periods. From the early 1960s until the mid-1990s, the AO was positive more often than negative. Since then, the AO has switched phases more often, with extreme negative phases dominating the winters of 2009–10 and 2010–11, which were cold and snowy in the Northeast.

Atlantic Multidecadal Oscillation

The Atlantic Multidecadal Oscillation (AMO) refers to swings in the surface temperature of the Atlantic Ocean between the Equator and Greenland that influence weather widely.

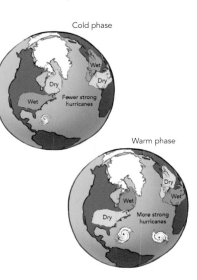

Cold phase

Warm phase

KEY FACTS

During a cold AMO phase, North America and the Caribbean experience more hurricanes.

+ fact: A cold phase from 1971 to 1994 averaged only 1.125 major hurricanes a year; a warm phase from 1995 to 2012 averaged 4 major hurricanes a year.

+ fact: The Dust Bowl drought of the 1930s occurred during a warm phase.

The Atlantic Ocean seems to swing between warm and cool phases lasting 20 to 40 years. Its average temperature during a warm phase is approximately 1°F (0.55°C) above that of a cool phase. Spread out over the ocean, this is a lot of heat, and it can energize hurricanes and affect patterns of high and low atmospheric pressure. Even pressure patterns far from the ocean appear linked to this cycle: African droughts in cold phases, North American droughts in warm phases. Paleoclimatic proxies, such as tree rings and ice cores, show that the AMO has been occurring for at least 1,000 years. It is not an effect of current climate change. Subtle changes in the speed of the Gulf Stream drive the AMO. When it slows, the Atlantic Ocean cools slightly; when it speeds up the Atlantic warms up.

Stationary Fronts

Like all fronts, stationary fronts separate large air masses with different densities—usually caused by temperature differences. Neither air mass is advancing along a stationary front.

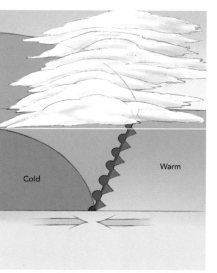

KEY FACTS

Map symbol: Red and blue lines with alternating red semicircles facing colder air and blue triangles facing warmer air

+ fact: Any kind of front can become stationary if upper air winds begin blowing parallel to the front.

+ fact: In the spring and fall, stationary fronts can lie across the Southeast for days at a time.

Cold

Warm

A stationary front forms when either a cold or a warm front stops moving. Warm, humid air can ride over the front to supply humidity for clouds and precipitation on the cold side of the front. Upper air disturbances can travel along the front, creating clouds and precipitation for days at a time. If an upper air pattern that encourages air to rise moves overhead, a low-pressure area will form on the front. Its counterclockwise winds around the low—in the Northern Hemisphere—begin pushing the warm air toward the north or northwest and the cold air toward the south or southeast to begin organizing an extra-tropical cyclone.

Cold Fronts

A cold front is the leading edge at the surface of a mass of cold air that is replacing warmer air. Showers and thunderstorms and wind shifts accompany most cold fronts.

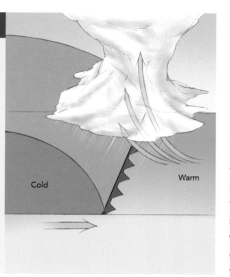

KEY FACTS

Map symbol: A blue line with blue triangles pointing in the direction of movement

+ fact: A "backdoor" cold front moves into the northeastern U.S. from the northeast instead of the northwest like most cold fronts.

+ fact: Frontogenesis is a front's formation; frontolysis is a front's dissipation or weakening.

Cold

Warm

As a cold front advances, the colder and denser air behind it wedges under the less dense warmer air, lifting it. If the warm air is moist and the atmosphere is unstable—the usual case in North America—this lifting forms showers and thunderstorms. These thunderstorms can be very strong, even severe, especially in the spring when the atmosphere is often unstable. During the winter, when a cold front reinforces dry, cold air already in place, little snow or rain might fall. A reliable sign that a cold front has passed is a wind shift from southwesterly to northwesterly. The coldest air is often a few miles behind the front.

Warm Fronts

A warm front is the boundary where warm air is replacing colder air. The clouds associated with a warm front can be more than 700 miles (1,100 km) ahead of the front.

KEY FACTS

Map symbol: Red line of half circles pointing in direction the front is moving

+ fact: A Northern Hemisphere warm front usually causes the wind to shift from blowing from the southeast to from the southwest.

+ fact: Warm fronts advance at an average speed of roughly 10 mph (16 kph), half that of cold fronts.

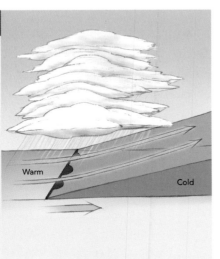

An advancing warm front doesn't arrive with the drama of a strong cold front, but it affects a much larger area. Because warm air is lighter than cold air, a warm front's air rises over the cold air. The warm air can be 6,000 feet (1,800 m) above the ground and 150–200 miles (249–320 km) ahead of the front. As a warm front approaches, you will first see high cirrus clouds, which become cirrostratus or cirrocumulus. These thicken and descend to become altocumulus and altostratus clouds. Snow or rain could then begin, as when you see nimbostratus clouds. After the surface front passes, the sky will begin clearing and temperatures will warm up.

Occluded Fronts

Unlike stationary, cold, and warm fronts that divide two air masses with contrasting densities, occluded fronts are more complex. They separate three air masses: cold, cool, and warm.

KEY FACTS

Map symbol: Alternating purple cold-front triangles and warm-front half circles

+ fact: Occluded fronts are the most common kind of fronts moving into western North America from the Pacific Ocean.

+ fact: More detailed observations and computer models are helping meteorologists better understand occlusions.

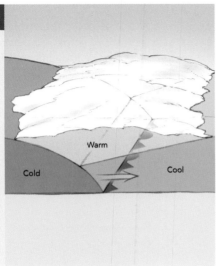

In a "cold" occlusion, the surface boundary separates very cold and cold air with the warm air appearing to have been shoved up by the very cold air to intersect the very cold air aloft. In a "warm" occlusion, cool air is riding over very cold air with the warm air above the cold air. Many textbooks say "warm air catching up with cold air" forms occlusions. Some meteorologists today, using more complete observations and computer models, say a better description involves the warm front and cold front wrapping around the cyclone's low-pressure center after the warm front separates from the low-pressure center.

Dry Line

A dry line, like a front, separates air masses of different densities. These differences are in humidity, not temperature, as with most fronts. Dry lines occur on the Southwestern Plains.

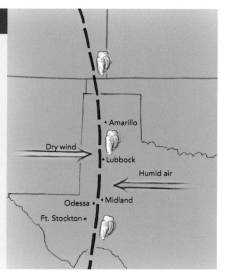

KEY FACTS

Weather map symbol: A gold line with adjacent half circles along the line

+ fact: Dry lines are most often located in western Texas but have been observed as far north as the Dakotas and as far east as the Texas–Louisiana border.

+ fact: Dry lines usually move east in the afternoon and back to the west at night.

The air masses that are in conflict along a dry line are very dry, warm, or hot air moving east from the Southwest and humid hot air moving west from the Gulf of Mexico. Because humid air is less dense than dry air of the same temperature, the dry air pushes under the humid air much as cold air shoves under warm air. This can trigger showers and thunderstorms, sometimes severe thunderstorms with tornadoes, much as advancing cold fronts cause storms. Dry air is denser than humid air because added water molecules are lighter than the nitrogen and oxygen molecules they replace as humidity increases. Thunderstorms tend to form where the dry line bulges and pushes air up. These storms usually are more isolated and severe than those that form elsewhere because they aren't competing with other storms.

Upper Air Troughs

Upper air troughs are elongated areas of low atmospheric pressures relative to adjacent air pressures at particular altitudes. They influence the locations, strengths, and paths of storms.

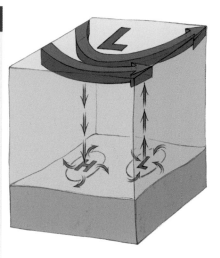

KEY FACTS

Air rises on the eastern side of a Northern Hemisphere trough aloft, which helps surface storms to intensify.

+ fact: Air sinks on the west side of a Northern Hemisphere trough, creating a dry, high-pressure area below.

+ fact: A trough aloft forms when the air below is colder than air on either side of the trough.

Weather forecasters pay particular attention to troughs aloft because they have a major influence on the weather below by helping storms to form or intensify. When you hear broadcast meteorologists talk about possible bad effects on local weather from "upper air energy" or an "upper air disturbance," they are probably talking about a trough or a "cutoff low." The low formed when the southern end of a trough was pinched off to become an upper-air low pressure area that is disconnected from the upper air wind flow. Cutoffs can hang around for days in the same place or even move to the west, causing cloudy skies and precipitation before they dissipate. A trough and an upper air ridge—where winds aloft turn to the north and back to the south—make one of the three to seven meandering waves that circle Earth.

The Aleutian Low

The Aleutian low is a semipermanent area of low atmospheric pressure that strengthens each fall and fades during spring. From fall through spring, it steers storms into the Pacific Northwest.

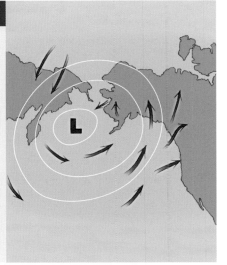

KEY FACTS

In winter, the Aleutian low regularly sends storms into the Pacific Northwest with only brief breaks.

+ **fact:** During an El Niño in the tropics, the Aleutian low tends to be deeper and spins off stronger storms.

+ **fact:** A weak Aleutian low can occur any year, but it is especially likely during a strong La Niña.

The Aleutian low forms each winter as Alaska begins turning frigid, leaving the Pacific Ocean around the Aleutian Islands the warmest surface in the region. Because the water is relatively warm, air over it begins rising, forming a low pressure center that continues until spring when the land warms up. Storms that form or strengthen here often have winds faster than 50 mph (80 kph), which create huge waves that surfers in Hawaii love—some as tall as 65 feet (20 m) high when they head south. Waves heading north are the ones that make the Discovery Channel's *Deadliest Catch* show exciting. The Aleutian low's Atlantic Ocean counterpart is the Icelandic low. In summer when the Aleutian low is weak, the North Pacific High moves north so it is west of California, strengthens, and keeps the West Coast mostly dry.

The Bermuda High

While the Bermuda high is centered far out over the Atlantic Ocean, it has several effects on North American weather, including steering Atlantic hurricanes toward or away from the U.S.

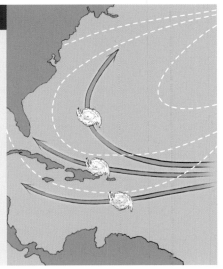

KEY FACTS

The high doesn't always protect Bermuda from hurricanes. One hurricane, on average, hits Bermuda every seven years or so.

+ **fact:** In 2004 the high was farther west than usual and steered a record four hurricanes to Florida.

+ **fact:** In winter and spring, the high is centered over the Azores as the Azores high.

Bermuda is in the global belt of high atmospheric pressure and mostly calm winds around latitude 30° N, where air rising in tropical thunderstorms descends to maintain high atmospheric pressure. Winds flowing clockwise out of the high contribute to the easterly tropical trade winds and often feed warm humid air into the eastern U.S. in the summer, especially the Southeast. The clockwise winds around the high steer hurricanes. Slight changes in the strength and position of the high help to determine whether a hurricane heads northward between Bermuda and the U.S., hits the Northeast, the Southeast, or heads into the Gulf of Mexico. Predicting how the Bermuda High and winds around it will change is a major aspect of forecasting the likely paths of hurricanes in the Atlantic Ocean.

Ground/Radiation Fog

Radiation fog forms when heat radiates from the Earth overnight, cooling the air enough for its water vapor to begin condensing into tiny water drops. The fog forms next to the ground.

KEY FACTS

Ground fog is usually thickest before sunrise when evaporation starts.

+ fact: Fog cuts visibility to 0.6 mi (1 km); mist cuts visibility to less than 6 mi (9 km).

+ fact: The NWS reports "fog" when it is less than 20 ft (6 m) deep, and "shallow fog" when it is less than 6 ft (2 m) deep.

Radiation fog, also called ground fog, is most likely to form when the sky has been mostly clear all night, which allows the most heat to radiate away from Earth. Winds should also be calm or nearly calm because stronger winds will mix the coldest air next to the ground with slightly warmer air a few feet higher. Chances of fog are much better when rain has soaked the ground the day before. Water from soaked ground evaporates into the air. The added moisture allows water vapor to began condensing at a higher temperature than it would in drier air. The fog usually begins evaporating shortly after sunrise. Afternoon rain that leaves water on the ground to evaporate, a sky that clears overnight, and 5 to 10 mph (8 to 16 kph) winds combine to make morning fog likely.

Steam Fog

The wisps of "steam" you see rising from ponds or lakes in the fall when the year's first cold air arrives are "steam" fog, which needs a combination of warm water and cold air to form.

KEY FACTS

Steam fog forms when warm water evaporates into cold air, making the air humid enough to form fog.

+ fact: All year, steam fog forms above thermal ponds in Yellowstone National Park.

+ fact: Steam fog begins a few inches above the water because the rising air needs to cool enough to form fog.

Arctic air that begins moving south over North America in the fall is too dry to form fog without some help. This help comes when it flows over ponds, lakes, or rivers, and some of the relatively warm water evaporates into the cold air, giving it enough humidity for condensation to begin at the air's current temperature. Steam fog that forms over an ocean is called "sea smoke." When frigid air moves over much warmer water, the rising fog can create steam devils up to 1,600 feet (500 m) high. In a case studied, the water was 39°F (22°C) warmer than the air. Most fog forms in light winds, but steam devils illustrate the turbulence associated with steam fog. It forms under extreme temperature differences between warm water and frigid air. The Great Lakes are a prime location for steam devils.

Advection Fog

Advection fog forms when winds push humid air over ground or water cold enough to chill the air to a temperature that causes its humidity to condense into tiny fog drops.

KEY FACTS

Coastal advection fog supplies 30 to 40 percent of moisture to California's redwoods.

+ fact: Warm Gulf Stream air advected over the nearby Labrador Current makes Newfoundland's Grand Banks one of Earth's foggiest places.

+ fact: Summer breezes across the cool Great Lakes form persistent advection fog.

Meteorologists use the word "advection" for the horizontal transport of meteorological property such as temperature or moisture. Some of North America's most troublesome fog is the widespread, dense and long-lasting advection fog across the Midwest created when warm, humid air moves north from the Gulf of Mexico. Southern California's famous "May Gray" and "June Gloom" marine layer is advection fog that is formed when tropical Pacific air flows over the cold California Current near the coast. This type of fog normally rolls in early in the morning and dissipates during the day. Radiation fog does not form over oceans as it does over land, which quickly radiates heat away on clear nights.

Ice Fog

Ice fog is made of tiny ice crystals that float in the air, just as water drops do in ordinary fog. It forms only at temperatures below -30°F (-35°C)

KEY FACTS

Ice fog particles are small enough for ten to fit side by side on the edge of a piece of paper.

+ fact: When the air is -40°F (-40°C), water from a car's tailpipe drops from 250°F (121°C) to the air temperature in less than 10 seconds.

+ fact: Ice fog is sometimes called pogonip, from a Shoshone word for "cloud."

Because ice fog needs such low temperatures, it occurs only in the northern provinces of Canada, in inland and northern Alaska, and in some high elevations in the Pacific Northwest. Ice fog can also be called frozen fog, but it should be distinguished from freezing fog, which is made of water drops that instantly freeze when they touch anything. Ice fog forms only in extremely dry air, but in built-up places such as Fairbanks, Alaska, exhaust from vehicles adds water vapor to the air, and that can instantly become tiny ice fog particles. In Fairbanks, the resulting thick ice fog—with other pollutants added—can drop visibility to near zero and push air quality to unhealthy levels. A few miles from the city's traffic, the air can be clear.

Natural Haze

Haze is a collection of very fine, widely dispersed, solid or liquid particles suspended in the air. They turn the sky milky white and subdue colors. It can be natural or from human activities.

KEY FACTS

The chemical properties of haze particles can change the effects of the haze on the view.

+ fact: Volatile organic compounds from trees help to form a bluish haze: That's what makes the Blue Ridge Mountains blue.

+ fact: Salt particles in the air affect beach scenes, creating a low, white haze in the daytime and red sunsets.

Haze has a direct effect on how far we can see, because its particles are just the right size in relation to wavelengths of light to scatter or absorb some colors, which reduces visibility. We often think of any such obstruction to visibility as pollution, but it's not always the case. Haze is sometimes air pollution—particles or gases in the air added by human activity with the potential for harming life or property. But haze can be natural, coming from a volcanic eruption, or smoke from wildfires started by lightning, or dust. Water vapor can condense on dry haze particles and make them larger, which further reduces visibility by scattering or absorbing more light. This condition most likely occurs in the morning or evening, when relative humidity is higher.

Pollution Haze or Smog

Photochemical smog, a brownish haze, is a mixture of hundreds of hazardous chemicals. It is most often found over and downwind of cities. It is unpleasant and unsightly, and can be deadly.

KEY FACTS

Henry Antoine des Voeux, a London physician, coined "smog" for smoke and fog in 1905.

+ fact: Reactions involving sunlight, nitrogen oxides, and volatile organic compounds produce photochemical smog.

+ fact: Ozone near the Earth's surface is a pollutant, but ozone in the stratosphere blocks dangerous ultraviolet light.

An inversion—air aloft that's warmer than ground-level air—sets the stage for smog: It blocks warm air from rising, so that it cannot be replaced by clean air descending from above. At times, an inversion can trap polluted air over a city for days as more smog brews. In addition to the visible smog, invisible pollutants are at work, including carbon monoxide and extremely small particles so tiny that they travel deep into the lungs, causing damage that can be fatal. Rain and snow wash pollutants out of the air, but civilization often produces so much pollution that this natural cleaning process cannot keep up. Volatile organic compounds (with noticeable odors) from industrial and natural sources, such as trees, are one (but far from the only) component of smog.

Rainbows

When sunlight, a mix of colors, enters and reflects out of a water drop, it bends, with each color bending at a different angle. The light reflects off the drop's back and is visible as separate colors.

Rainbows are visible phenomena, but are not actual objects, and your eyes must be in the right relationship between the sun and raindrops to see them. If the sun isn't directly in back of your head, you aren't looking at a rainbow. (You could be seeing iridescence or a circumzenithal arc.) Light hits the raindrops, which act like tiny prisms, dividing it into its component colors as it shines back out. Double rainbows are made by the same raindrops that make the primary rainbow. You often hear that rainbows have seven colors, but a rainbow has the whole range of colors from red at the top to violet at the bottom of a primary bow. Usually we see fewer than seven colors.

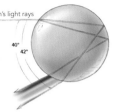

Fogbows and Moonbows

A fogbow is created the same way as a rainbow but by water drops in fog, not rain. Rain or a waterfall can disperse the light of a full moon to create a moonbow. Both are rare.

Fogbows are much fainter than rainbows because light is traveling through much smaller water drops, and their colors aren't as sharp as those of rainbows. They usually appear white, sometimes with faint red and blue. Moonbows will occur only when the moon is full and brightest, but not directly overhead: For a moonbow to occur, the moon cannot be more than 42°, or a little less than halfway up from the horizon to the zenith. As with a sunlit rainbow, the moon must be behind your head as you look into the rain or waterfall, where drops of water are bending the incoming light and dividing it like a prism into its component colors. Full moons set around sunrise, and the hours before sunrise are the best time to look for a moonbow.

Sun Dogs

"Sun dog" is an informal name for a parhelion, a splotch of light seen on one or both sides of the sun. They are the most common visual display caused by ice crystals floating in the air.

KEY FACTS

Sun dogs are brightest when the sun is low.

+ fact: Some Arctic people are said to call them "the sun's dogs."

+ fact: In 1461, King Henry VI reportedly inspired his troops to win a battle by calling the sun and two sun dogs they saw the Holy Trinity: God was on their side.

Parhelia—the plural of parhelion—are colorful, glowing spots formed by light bending as it enters and leaves plate-shaped, hexagonal ice crystals floating facedown in cirrus clouds. In most parts of the world, sun dogs appear at a 22° angle higher than the sun and on one or both sides of the sun. They can appear as often as a couple of times a week, most visibly when the sun is close to the horizon. Once you start looking for sun dogs when cirrus clouds are in the sky, you might be amazed how often you see them (and you can amaze others by pointing them out, and explaining how ice causes them).

Atmospheric Halos

Arcs or circles shining around the sun or moon, atmospheric halos are the visible effects of light passing through ice crystals in the atmosphere.

KEY FACTS

Scientists call all atmospheric displays caused by ice crystals "halos."

+ fact: A 22° halo is caused by randomly oriented, hexagonal columns—crystals shaped like tiny pencils, scattered through the atmosphere.

+ fact: Some atmospheric halos glow with colors, ranging from red on the inside to blue on the outside.

Ice crystals hanging in cirrus clouds high in the upper troposphere, 3 to 6 miles (5 to 10 km) above the ground, can create visible arcs, circles, and spots by reflecting and refracting light. As in rainbows, the white light is sometimes split into its component colors. Atmospheric halos have been observed and interpreted for millennia, sometimes as signals of weather to come and sometimes as spiritual messages. Today, the optics of the phenomenon are fully understood, but the marvel of halos remains. On January 11, 1999, for example, those at the South Pole saw 22 different kinds of halos.

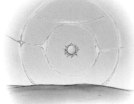

Circumzenithal Arc

A circumzenithal arc is a ring directly overhead or almost directly overhead. It is easier to miss than most other halos because you're not likely to look straight up.

KEY FACTS

If you see a sun dog with the sun 15° to 25° above the horizon, look straight up. You might see a circumzenithal arc.

+ fact: When looking at halos, you need to shield your eyes from the sun.

+ fact: Regular skywatchers might see a circumzenithal arc 25 times a year.

Unlike most halos that appear white, maybe with a tinge of color, a circumzenithal arc is colorful. It has been described as an "upside-down rainbow." The colors normally range from blue on the inside to red on the outside—the side toward the sun. Bending of light rays as they enter and leave ice crystals causes all halos. Such refraction separates sunlight into its colors. Hexagonal columns shaped like a pencil cause circumzenithal arcs when light enters the top face and leaves through one of the sides. The arc forms only when the sun is lower than 32.2° above the horizon. The arc can be wider than shown here. Some call them fire rainbows because they sometimes have a fiery look, but they are not rainbows.

Moon Dogs

A moon dog is the lunar version of a sun dog: a glowing spot or pair of spots visible alongside the bright full moon, created by moonlight passing through ice crystals in the atmosphere.

KEY FACTS

The scientific name for a moon dog is paraselene (plural paraselenae), meaning "beside the moon."

+ fact: You see little color in moon dogs because they aren't bright enough to activate your eye's cones, which perceive color.

+ fact: A few nights before and after the full moon, it is bright enough to make halos.

Because the moon is a source of light in the sky, it produces halos, including moon dogs, just as the sun does. If ice crystals are present in the atmosphere, moonlight coming through may reflect and refract that light, making it visible in these attractive patterns. Moon dogs and halos are much dimmer than those generated by the sun, because all of the moon's light is reflected. Moon halos are more common than moon dogs. Folklore says that a ring around the moon means that rain is on the way. Maybe. Cirrus clouds that create halos are sometimes, but not always, a sign that a warm front is arriving with rain.

Coronas

You see a corona when the sun or moon shines through thin clouds. It appears as a bright center (the sun or moon) surrounded by one or more reddish or brownish rings.

KEY FACTS

A safe way to view a corona around the sun is to look at its reflection in water.

+ fact: Coronas are often seen in altocumulus or altostratus clouds, which are often part of a storm that could bring rain or snow.

+ fact: Coronas have colors when the cloud drops or ice crystals in the cloud are the same size.

Rainbows and halos are visible when water or ice causes the refraction, or bending, of light rays. Coronas (and other sky phenomena such as iridescence) are visible when water drops in the atmosphere cause the diffraction, or spreading and rejoining, of light rays. Coronas are created by smaller water particles than halos, and they are more commonly seen. Sometimes the physics of the light and water droplets is such that colors are visible. When that happens, red is always the outermost color and sometimes the only color. This use of the word should not be confused with its astronomical meaning, referring to the sun's outer atmosphere. A corona's size depends on the diameters of the cloud drops. Small drops produce large coronas. A corona is clearest when the drops are mostly the same size.

Crepuscular Rays

Close to sunset or sunrise, beams of light will often appear as if they are radiating from the sun, coming through breaks in the clouds and fanning out as they come down to Earth.

KEY FACTS

"Crepuscular" is derived from the Latin word *crepusculum*, which means "twilight."

+ fact: People sometimes say "the sun is drawing water," harking back to the ancient Greek belief that sunbeams draw water into the sky.

+ fact: Crepuscular rays are often red or yellow because air molecules selectively scatter blue light.

You see the "beams" of light—called crepuscular rays—because haze, dust, or tiny water drops floating in the air scatter the sun's light in all directions, including essentially parallel lines toward you. From your vantage point, the beams appear to converge in the sky because of the principles of perspective—the same reason the edges of a straight road appear to converge at the horizon. While you're looking at crepuscular rays, turn around, put the sun at your back, and if the sky is clear, you may see "anticrepuscular" rays converging on the "antisolar" point opposite the sun. If the air were perfectly clean, you would not see crepuscular rays. Light from the sun is traveling a straight path from the sun to Earth, not toward you. Tiny particles such as dust scatter light in all directions, including toward your eyes.

Sun and Light Pillars

Near sunrise or sunset, bright columns of light appear to shoot above and below the sun: sun pillars. On a frigid night, you may see columns shooting into the sky from streetlights.

KEY FACTS

The second and third brightest objects in the sky, the moon and Venus, also form light pillars.

+ fact: Crystals for light pillars form only at temperatures below freezing.

+ fact: Gusty winds can rearrange a light pillar's crystals, making it shimmer like an aurora overhead.

Halos become visible when the sun's light rays bend as they pass through ice crystals, but both sun pillars and light pillars emerge when light reflects off ice crystals instead. This is why they are the color of the source of the light—red beaming above the red setting sun near the horizon—instead of being mostly white. Sun pillars usually extend at an angle of only 5° to 10° above the sun. Light pillars form on clear nights that are cold enough to form ice-crystal fog, composed of tiny crystals. This form of fog is also called diamond dust.

Auroras

Auroras, shimmering curtains of green and brown-red lights seen high in the skies of Earth's far north and far south, are visible evidence of our planet's direct connections to the sun.

KEY FACTS

Auroras follow the 11-year solar cycle and tend to be more frequent in the late autumn and early spring.

+ fact: Around the Arctic Circle in northern Norway and Alaska, auroras are visible almost every night.

+ fact: Auroras are seen only once or twice a century over the southern United States.

Solar winds send energetic charged particles from the sun toward the Earth, attracted especially into the atmosphere over the Arctic and Antarctic by the electromagnetic fields of our planet. Here they smash into the atoms of nitrogen and oxygen, the primary constituents of our atmosphere. These collisions send off photons, which create visible light: the eerie, shimmering, sky-filled curtains of the aurora borealis (northern lights) over the Arctic and aurora australis (southern lights) over Antarctica. The auroras are the only visible aspect of space weather: the many effects on Earth of high-energy particles from the sun, which can disrupt satellites.

Iridescence

Iridescent clouds have washed-out, mostly pastel colors visible on some or all of the cloud. Most often, iridescence appears as a border of red and green along the edge of a cloud formation.

KEY FACTS

Every cloud does not have a silver lining: Some have colored iridescent linings; most have no lining at all.

+ fact: Iridescence can be seen in cirrus, altocumulus, cirrocumulus, and lenticular clouds.

+ fact: Iridescence is more visible if the sun is shaded, either by a denser cloud or deliberately by the viewer stepping behind a tree or building.

Sometimes a cloud will seem to show colors similar to a rainbow contoured to the shape of the cloud. Iridescence—the appearance of colors within or along the borders of a cloud—is caused by tiny water drops or small ice crystals, each individually scattering and diffracting light. The optics that create an iridescent cloud are similar to those that create colors on the surface of an oily puddle. Iridescence develops primarily in thin clouds positioned relatively close to the sun. The colors can be very subtle or rather bright, but they will shift and change quickly. The colors you see in an iridescent cloud can be considered fragments of a corona. Unlike coronas that form in clouds or parts of clouds with drops of relatively uniform size, iridescence shows that the cloud is made of drops with different sizes.

Glories

From your window seat on an airplane just above the clouds, you're casually looking out when you see the shadow of your airplane with a ring of light around it. You're looking at a glory.

KEY FACTS

Before flying became common, the best way to see a glory was from a mountain above cloud tops.

+ fact: You sometimes see a glory around the shadow of your head on the cloud.

+ fact: The shadow glory is called a "Brocken spectre," for the highest peak in Germany's Harz Mountains, known for its glories.

A glory, a colorful halo encircling a shadow in the clouds below the viewer's eye level, is not as simple as light scattered back from the cloud to create colored rings. A phenomenon that appears at the antisolar point—directly opposite the sun in relation to the viewer— a glory is visible only to a person positioned between the sun and the top layer of clouds. It must be explained by a more complex physical process than simple reflection or diffraction. Physicists are still unsure of the process, and determining the atmospheric optics of glories is a challenge to this day. Their findings could apply to climate science because of the possibility that clouds reflect more sunlight than previously thought.

Green Flash

As you watch the sun set over an open-water horizon, you may see a green flash: a momentary change in the color of the sun to green before it disappears.

In locations where a large lake, bay, or the ocean stretches out to the west, far as the eye can see, many people gather in early evening to try to catch a glimpse of the elusive green flash. It may look like a green spot for a second or two or like a green ray shooting up from the sun. There is a physical explanation for this. The atmosphere refracts or bends light's different wavelengths in a standard order of colors. As the sun slips below the horizon, the red disappears first, then a second or so later the yellow, green, and finally blue and violet disappear. Because air scatters blue and violet the most, these colors don't reach your eye. But green might reach you, enhanced as a mirage caused by a layer of warm air over a relatively cool ocean.

Twilight

Twilight is the transition between night and day: from the time light first appears in the morning until sunrise, and from the time the sun reaches the horizon until light fades from the sky.

Twilight—the time between last or first light and the brightness of day—has been defined in many ways. Civil twilight is the period when the sun's center is between the horizon and 6° below the horizon, and the ambient light is usually still bright enough for outdoor activities. Nautical twilight is the period when the sun's center is between 6° and 12° below the horizon, and shapes are visible but not distinct. Astronomical twilight is the period when the sun's center is between 12° and 18° below the horizon. After that, twilight ends, and it is dark. The term "twilight" can also be applied to the same periods of time as the sun is rising.

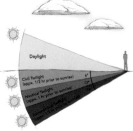

Thermal Inversions

A thermal inversion (usually called just an inversion) is a layer of warm air above colder air. Temperatures normally decrease with altitude, which makes a thermal inversion unusual.

KEY FACTS

Cold air flowing into valleys pushes up warm air, creating inversions.

+ fact: An inversion caused the 1948 pollution episode in Donora, Pennsylvania, that killed 20 people and sickened hundreds.

+ fact: August inversions at the South Pole can make surface temperatures 70°F (39°C) colder than those 1,000 ft (305 m) above.

Surface inversions form in winter when heat radiates away from Earth faster than solar warmth can replace it. The strongest inversions form in polar regions when the sun doesn't rise for weeks or months, and these are sources of cold waves. In warmer places, inversions can develop overnight. Inversions also form where air from aloft is sinking to create surface high pressure. Inversions sometimes form caps that keep warm air from rising, preventing thunderstorm development. When rising air finally breaks such an inversion, the hot air that has been bottled up can quickly rise to form severe thunderstorms.

Blue Sky

Anyone who isn't color-blind and who looks at the sky knows it's blue. But why? The sky's blue has baffled some of the world's best minds, from the ancient Greeks to 19th-century physicists.

KEY FACTS

Air scatters violet more efficiently than blue, but sunlight contains less violet than blue.

+ fact: Scattered light moves in all directions. Reflected light rebounds according to the angle of its arrival in the air.

+ fact: Outer space is black because it contains nothing to scatter light.

From Aristotle to Leonardo da Vinci and Isaac Newton, the greatest minds of science throughout world history have tried to explain why the sky is blue. Many ideas included reflection of one kind or another. Finally the British physicist John William Strutt, usually referred to as Lord Rayleigh, solved the puzzle in the 1870s. Since Newton, it had been known that sunlight is composed of all colors, which can be split apart by a prism. Lord Rayleigh suggested that some colors in light scatter more readily than others. Blue scatters the most; hence, the sky looks blue, except around sunrise and sunset.

Mirages

In meteorological parlance, a mirage is not a hallucination—it is an optical distortion caused by atmospheric conditions. In fact, you can photograph a mirage, as many people have.

KEY FACTS

Outside of polar regions, superior mirages are much less common but are more stable than inferior mirages.

+ fact: In May 1909, a superior mirage in Greenland showed explorer Donald B. MacMillan land that was 200 mi (320 km) away.

+ fact: Because hot air rises, inferior mirages are unstable and may distort images.

Mirages are caused when layers of air at different temperatures bend light in various ways, distorting an apparent object into an image that does not correspond with physical reality. The greater the temperature differences, the more striking the mirage. Air near the ground that is warmer than air higher can create an "inferior mirage," with objects appearing lower than they really are. For instance, the sky might appear as a pool of water in the desert. A thermal inversion—air near the ground that is colder than air higher—can create a "superior mirage," with objects appearing higher than they are. Striking superior mirages occur especially in the polar regions, including some that show objects that are below the horizon to the viewer's eye.

Fata Morgana

A fata morgana is a complex mirage with elements both compressed and stretched, both inverted and right side up, combining into visions that change quickly.

KEY FACTS

Fata morganas are named for Morgan le Fay, the shape-shifting sorceress of the King Arthur legends.

+ fact: In 1906 explorer Robert Peary named land he saw "Crocker Land." It was a fata morgana.

+ fact: A famous fata morgana is the *Flying Dutchman*, a ghost frigate doomed to sail forever, according to ancient lore.

Steep thermal inversions with warm air over cold water, cold air, or polar ice help create the atmospheric ducts needed to evoke a fata morgana mirage. An atmospheric duct is a layer in the lower atmosphere that guides light along Earth's curvature. The components of a fata morgana can shift between being inferior and superior mirages. Some fata morganas create multiple images, alternately expanded and compressed vertically. At times they appear to be buildings of a city or hills of an island where none exists. Over the years, fata morganas have led polar explorers to map islands or other lands that don't exist. Numerous reports of ghost ships, sometimes in the air, are likely the results of fata morganas.

Thunderstorm Safety

An estimated 100,000 thunderstorms hit the United States yearly. Ten percent of thunderstorms are severe—with winds of 58 mph (93 kph) or faster, large hailstones, or a tornado.

KEY FACTS

When thunderstorms are likely, make sure to stay near shelter in case a storm develops.

+ fact: A severe thunderstorm watch means severe thunderstorms are possible; you should be ready to take shelter.

+ fact: A severe thunderstorm warning means a severe thunderstorm has been spotted; you should take shelter.

Lightning is the big danger in all thunderstorms. Severe thunderstorms add the danger of winds that can topple trees and send debris flying. A sturdy building is the best shelter against both. Your best defense is not being caught far from safe shelter—hiking up a mountain, two hours or more away from safety, for instance—when a thunderstorm hits. Although forecasters can give advance warning about hurricanes and winter storms, they can rarely predict when and where a thunderstorm will hit more than about a half hour ahead. They can, however, pin down general areas where thunderstorms are likely to occur hours ahead of time. Use these general alerts to plan time outdoors to ensure that you won't be caught in the open.

Lightning Safety

Lightning is one of the top three weather killers most years in the United States. If you are anywhere outdoors when you hear thunder or see a lighting flash, lightning could hit you.

KEY FACTS

If you see lightning or hear thunder, no matter how far away, immediately take shelter.

+ fact: Enclosed buildings with electrical wiring and plumbing or enclosed metal vehicles are the best shelters.

+ fact: If lightning catches you outdoors, don't squat or lie on the ground and don't hide under a tree. Run for shelter as fast as you can.

The lightning to be concerned about is a powerful but very brief electrical current seeking a path to the ground with the least resistance. To avoid death or injury, you need to avoid being that easiest path to the ground that it is seeking. When you are inside an enclosed building with electrical service and plumbing, lightning will find the building's wires or pipes to be a better path to the ground. But stay away from water and from anything plugged in, including a telephone with a wire, when you are inside. (A cell phone is safe.) An enclosed vehicle is safe because electricity goes through the metal and to the ground, but a strong lightning strike could blow out the tires.

Heat Exhaustion and Heatstroke

High temperatures and humidity upset the body's systems for regulating temperature. Perspiration is the body's natural cooler, but humidity hinders evaporation, making you hotter.

KEY FACTS

Heatstroke, also called sunstroke, is a medical emergency worth a 911 call.

+ fact: The National Weather Service Heat Index combines temperature and humidity for an "apparent temperature."

+ fact: Many cities have cooling centers where the elderly and others endangered by heat can relax in air-conditioned places.

The dangers presented by extreme heat and humidity are different for healthy men and women than for the elderly and people in ill health who don't have air-conditioning. If you know someone in this latter category, checking on them through a heat spell could be a lifesaver. For those in good health, the dangers arise when a person is so focused on a workout or sport that signs of trouble, such as cramps or thirst, are ignored. Heat exhaustion with heavy sweating, weakness, and pale, clammy skin is serious. Victims should get out of the sun—into air-conditioning if possible—and sip water. Heatstroke with a high body temperature could be next, leading even to unconsciousness. This would be a medical emergency.

Tornado Safety

Tornadoes are the strongest storms on Earth, but they are small in reach and relatively rare, with the strongest ones extremely rare. It's still a good idea to know what to do if one threatens you.

KEY FACTS

Flying debris is a tornado's greatest danger; avoiding it should be your main safety goal.

+ fact: Don't waste time opening windows as a tornado approaches; flying debris will do it for you.

+ fact: Don't even think of sheltering under a highway overpass; wind squeezing through makes it a debris-laden wind tunnel.

Because a tornado's path cannot be predicted until it is on the ground, you and your family need to have a tornado plan. You could even conduct a tornado drill, as schools do. Because your goal is to avoid flying debris, you should take shelter in a low room with no windows. A basement under a strong workbench or table is best. In a high-rise building, the interior stairway is the best shelter. Forget the elevator—tornadoes bring power failures. Many schools and shopping centers on the Great Plains, where tornadoes are more prevalent, have marked tornado shelters. Walk-in coolers have saved many people when tornados hit restaurants or convenience stores. If a tornado threatens, you should seek shelter.

Hurricane Safety

Two good rules for responding when a hurricane threatens your home or where you are staying are: Run from the water and hide from the wind. To do so, you need to know the flood danger.

KEY FACTS

Water is the biggest hurricane killer: both the water that a storm pushes ashore and the flash floods it can create inland.

+ fact: Wind can destroy a well-built house if the windows are not protected against flying debris.

+ fact: You need to check flood maps to see whether a hurricane could flood your home.

Over the centuries and around the world, hurricanes and tropical cyclones have killed more people with water than with wind. Usually, the highest death tolls are from flash floods from downpours dumped by dying storms far inland. If inundation maps show that your home is well clear of any possible surge, you can think about hiding from the wind at home. But, as Hurricane Andrew in 1992 proved in South Florida—the region of the U.S. where hurricanes are most likely to hit—expensive homes were not necessarily built to withstand storms. If you expect a need to evacuate, plan where you'll go and what you'll take before hurricane season begins. You should have storm centers or plywood window covers ready before hurricane season begins.

Cold Weather Safety

Cold weather brings two unique health threats: frostbite (freezing of tissue such as your fingers) and hypothermia (a lowering of your body's core temperature).

KEY FACTS

The U.S. Antarctic survival schools stress staying hydrated to help avoid hypothermia and frostbite.

+ fact: The U.S. National Weather Service and Environment Canada used wind-tunnel tests with volunteers to develop windchill charts.

+ fact: Hypothermia can occur at temperatures above 40°F (4.4°C) if a victim is wet.

Frostbite is not life threatening unless it is untreated and leads to gangrene. Hypothermia is always life threatening if not stopped in time. Windchill figures can alert you to the danger of both. Windchill does not change the temperature. If the temperature is above freezing with the windchill below freezing, your fingers and toes cannot freeze. Nevertheless, a frigid windchill means wind is carrying warmth away from your body faster than in calm air. This can speed hypothermia, when your body's core temperature falls below 95°F (35°C). First aid for hypothermia includes removing wet clothing and carefully warming a victim without burning him or her. Mental confusion can be a symptom of hypothermia.

Winter Storm Safety

If a winter storm traps you at home or in your car, be prepared to stay warm to avoid hypothermia. Carbon monoxide poisoning is a major winter storm danger.

KEY FACTS

Roughly 70 percent of deaths related to ice and snow occur in vehicles.

+ fact: A winter storm watch means you should prepare for snow or ice that closes roads and interrupts electric power.

+ fact: About 50 percent of those who die from hypothermia are over age 60.

A winter storm watch should signal you to stock up on food, water, prescription medications, and other things you will need if you find yourself trapped at home with roads closed by ice or snow and possibly having no electrical power for a few days. These threats are worse in the South, where snow and ice are uncommon and people are less well prepared. Anything that burns fuel—a generator, a charcoal grill, or backpacking stove—produces carbon monoxide, so be sure to use these in ventilated places. If you must go anywhere in a car, you should be dressed to survive for a few hours in the conditions outside, so that if you slide off the road or find yourself otherwise immobilized, you can survive the cold.

Flood Safety

The main flood danger, especially in flash floods, is drowning when trying to walk or drive into floodwaters. The danger continues after waters recede; bacterial contamination remains.

KEY FACTS

You shouldn't wade in floodwater if the water is moving faster than you can walk, or if you cannot see the bottom.

+ fact: A mere 18 in (46 cm) of water is enough to lift and carry a car or SUV downstream with a good chance of it rolling over.

+ fact: More than half of those killed in floods are inside vehicles.

An essential element of flood safety is keeping healthy after a flood. When you clean up, you may not want to think about what was in the water that soaked everything, but you must think about it enough to protect yourself from diseases and hazardous chemicals. If possible, find out what could have been in the floodwater. Your hepatitis A and tetanus shots should be current. The Federal Emergency Management Agency recommends wearing boots for flood cleanup and having bleach and water to decontaminate them before getting into your car or going into your home or garage. Also, remember that floods can seriously weaken buildings, so be aware of structural dangers as well. Flooded buildings can harbor venomous snakes.

Surface Weather Observations

Weather data such as temperature, atmospheric pressure, wind speed, and wind direction are raw materials for weather forecasts and the ground truth used to check forecast accuracy.

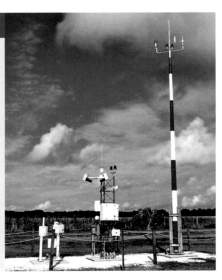

KEY FACTS

All of the world's weather stations use Coordinated Universal Time (formerly Greenwich [mean] time) for observation times.

+ fact: Weather averages for locations are based on 30 years of observations, updated every 10 years.

+ fact: Almost all U.S. weather observations are collected and transmitted automatically.

As the computers used to produce weather forecasts become more powerful, they can handle more and more data. Data are being collected and fed into the National Weather Service (NWS) network not only from traditional weather stations but also from automated stations at numerous small airports and locations operated by state, local, and private highway operators. Electronic devices that sense data such as atmospheric pressure make automated stations possible. In addition to traditional observations, the NWS receives data including water levels in streams, dryness of woodland areas, and soil temperatures and moisture levels.

Aviation Weather

Aviation is extremely weather dependent, hence the effort that the Canadian Weather Office and the U.S. National Weather Service put into collecting data for aviation forecasts.

KEY FACTS

Ceiling is the height above ground of the lowest level of clouds covering more than six-tenths of the sky.

+ fact: Sky condition refers to the heights of layers of clouds and how much sky each layer covers.

+ fact: Pilots use local altimeter readings to adjust their altimeters for the airport's air pressure.

Automated stations at airports provide all sorts of information essential to pilots. The height of the cloud ceiling and current visibility are essential elements in flying and air traffic management. Pilots normally point into the wind for takeoff and landing, so wind speed and direction are necessary knowledge. If the wind is not blowing directly down a runway, pilots prefer to use the runway that is nearest to pointing into the wind. Pilots flying into small airports can use a special radio frequency to hear the airport's automated weather report. The U.S. National Weather Service produces most U.S. aviation forecasts. The Aviation Weather Center in Kansas City supplies nationwide predictions, and local NWS offices make forecasts for airports. Some airlines employ forecasters.

Measuring Snow

Scientists are working on high-tech instruments to measure the amount of snow that falls, but for now, plywood left on the ground and rulers are still the best way to track snowfall.

KEY FACTS

To be considered measurable, at least 0.1 in (0.25 cm) of snow has to fall; less is reported as a "trace."

+ fact: With powder snow, approximately 20 in (51 cm) amounts to an inch of water.

+ fact: In a heavy, wet snow, only 5 in (13 cm) of snow may amount to an inch of water.

National Weather Service volunteer observers use a "snowboard"—a 16-by-16-inch (41 cm) piece of plywood on the ground in a place without drifts or bare spots as snow falls. Each hour, the volunteer uses a ruler marked in tenths of an inch to measure the snow and then brushes the board clean. Another board that is not brushed off measures the snow on the ground after natural settling and melting. In the western mountains, the U.S. Natural Resources Conservation Service uses automated instruments that measure the weight of snow and then convert it into water content. That information helps forecasters to predict water flows in western rivers when the snow melts. Melting snow is also an important source of soil moisture for farmers.

Upper Air Weather Observations

Data about winds, temperatures, humidity, and other atmospheric properties measured as high as 100,000 feet (30,000 m) above Earth are vital for making today's forecasts.

KEY FACTS

Radiosondes rise until the balloon bursts, above 100,000 ft (30,000 m); tiny parachutes lower them to the ground.

+ fact: If you find a radiosonde, directions on the side tell how to return it.

+ fact: From 1898 to 1933, the Weather Bureau used kites to collect data up to 10,000 ft (3,000 m).

Since the 1940s, basic upper air data have come from weather balloons carrying radiosondes, small boxes that collect and radio back the data twice a day. Approximately 800 global weather stations launch balloons once or twice a day. Extra balloons are launched when more data are needed on a storm. The main supplement to balloon data is automated reports that many airliners send every few minutes measuring temperatures, wind speeds, and wind directions. When all U.S. airline flights were grounded September 11–13, 2001, the accuracy of forecasts dropped significantly. Forecasts for three hours ahead were only as accurate as 12-day predictions using constant airline data. The forecasts most affected were those especially intended for aviation, but others were also less accurate.

High-Altitude Turbulence

Although turbulence isn't likely to bring down one of today's airliners, it regularly causes injuries to passengers en route. Researchers are seeking ways to warn pilots of turbulence.

KEY FACTS

Most lower-altitude severe turbulence is in or near clouds; pilots know which clouds to avoid.

+ fact: FAA statistics show that passengers not wearing seat belts account for 98 percent of turbulence injuries.

+ fact: High-altitude turbulence encounters are mostly in clear air with no warning.

Most high-altitude turbulence is in or near jet-stream winds with few clouds. Because radar tracks by sensing cloud water drops or ice crystals, turbulence outside of clouds is invisible to an airliner's radar. The best forecasters can do now is predict where turbulence is likely to occur. An airplane's turbulence encounter is a good detector for other airplanes, and pilots and controllers exchange real-time turbulence information via radio. Researchers are developing a LIDAR (a radar that uses light instead of microwaves) for airplanes, which shows promise of spotting turbulence. Others are testing ways to use data from ground-based Doppler radars, which collect more data than airliners' radars, to spot high-altitude turbulence.

Weather Research Airplanes

Unlike many other scientists, meteorologists can get inside the phenomena they study. Since weather extends to the edge of space, scientists need aircraft to examine the weather in full.

KEY FACTS

NOAA's two WP-3D hurricane hunters began flying into hurricanes in 1977 and are still at it.

+ fact: In 1987, NASA's ER-2, based in Chile, collected Antarctic ozone hole data.

+ fact: In 2012, NASA's Global Hawk, a large unmanned airplane, made the first of many planned flights observing hurricanes from above.

Weather balloons collect needed data, but to study the weather scientists must be able to say: "What's that? Let's take a look." Research flights into hurricanes grew out of the reconnaissance flights by the U.S. military during World War II. Meteorologists on some flights learned new things about hurricanes and saw the value of studying storms from airplanes. This has led to today's research fleet, which includes NASA's ER-2, a civilian version of the U-2 spy plane; two Gulfstream business jets equipped as flying research labs; and NOAA's two WP-3Ds, which are best known for hurricane flights but have helped scientists investigate weather phenomena globally, including El Niño, winter storms, ocean winds, Great Plains thunderstorms, and low-level jet streams.

Tornado Chasing

Since the 1970s, scientists have realized that the only way to improve tornado forecasts is to collect extensive data on what happens inside the supercells that spawn the strongest twisters.

KEY FACTS

VORTEX-1 produced the first full documentation of the life cycle of a tornado.

+ fact: The worst tornado in U.S. history happened in 1925, killing 695 people in Missouri, Illinois, and Indiana.

+ fact: A person is more likely to fall off a cliff or contract leprosy than be killed by a tornado.

Tornado researchers began using portable Doppler radar devices in 1995 to collect unprecedented close-up views of tornadic supercells. VORTEX (Verification of the Origins of Rotation in Tornadoes Experiment) is the largest tornado research project ever, designed to study how, when, and why tornadoes form. In 1994–95, 18 vehicles collected data as part of VORTEX-1; in 2009–10, roughly 100 men and women in 40 vehicles collected data from 11 supercells as part of VORTEX-2. The goal of these storm chases was to find changes inside supercells that could be detected, measured, and used in the future as indicators that a strong tornado is likely to form. VORTEX-2 scientists will be studying data and presenting analyses for another decade.

Weather Radar

Radar transmitters emit microwaves that objects such as raindrops scatter, some back to the radar. Computers convert the signals into information such as precipitation location.

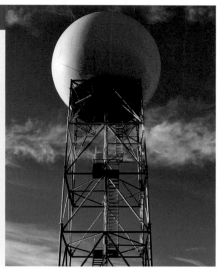

KEY FACTS

Radar is an acronym for RAdio Detection And Ranging.

+ fact: Today's Doppler radars detect precipitation intensity and locations plus wind speeds and directions.

+ fact: There are nearly 200 weather radars in the United States and Canada.

The U.S. and the United Kingdom developed radar during World War II to detect and track enemy ships and aircraft. When researchers realized that they were encountering interference caused by precipitation—rain or snow—they began applying radar to track weather phenomena as well. The U.S. National Weather Service (NWS) completed its first nationwide radar network in 1967, and radar quickly became important for both forecasters and researchers. Today, radar meteorology is a separate branch of the science. In 2012, NWS finished updating NEXRAD to dual polarization, which supplies more data about the nature of precipitation.

Weather Forecasting

Computerized predictions are the core of all of today's professional weather forecasts. As computers have gained power and speed since the 1950s, predictions have improved.

Forecasts begin with weather data from around the world flowing into the National Weather Service (NWS) National Centers for Environmental Prediction in College Park, Maryland. Supercomputers use the data to run several "models." These are computer programs that use equations of fluid dynamics and thermodynamics to predict weather around the world, with more detailed forecasts for the U.S. These are sent as maps and data that local NWS offices and other meteorologists use as starting points for their predictions. The results of different models enable forecasters to get what amount to second, or even third, opinions on what's likely to happen. These products are available on the Internet for anyone to use.

Weather Satellites

Weather satellites are such a part of our lives today that we're no longer amazed by their stunning images, such as of hurricanes. They collect other data as well as taking pictures.

Almost all the satellite images you see come from geostationary satellites, which orbit 22,238 miles (35,788 km) above the Equator. At this altitude their orbital speed matches Earth's rotation, keeping them above the same spot. They scan an entire disk of Earth. The U.S. has two of these, which with similar satellites of other nations give global coverage. Polar orbiters circle Earth from north to south and back 540 miles (869 km) high. Their closer view allows them to collect more detailed atmospheric data. For example, one polar orbiter's cloud-top temperature sensor detects changes 540 miles below—equivalent to detecting whether a lightbulb 22.6 miles (36 km) away is 100 watts or 101.6 watts.

Weather Radio

Some weather warnings, such as those for tornadoes and flash floods, require immediate action. The consequences of missing a warning could be dire.

KEY FACTS

The NWS began Weather Radio in 1967. From then until the 1990s, local meteorologists recorded on tape.

+ fact: In the late 1990s a text-to-computer system using "Paul" was introduced, but many didn't like his voice.

+ fact: The text-to-voice system introduced in 2000 has "Donna," "Tom," and a Spanish voice, "Javier."

All National Weather Service (NWS) offices broadcast warnings on NOAA Weather Radio, which requires special receivers to hear. The necessary radios, which pick up the seven VHF frequencies used for weather radio, are available at most stores that sell electronics. Many of these radios can be used also as ordinary radios. Most have a feature that will automatically turn the radio on and sound an alarm when a warning is broadcast. Newer weather radios have Specific Area Message Encoding (SAME), which allows those interested to program the radio to turn on and sound the alarm only for warnings for an area that is specified, not other parts of the region the local NWS office covers.

Ocean Observations

Weather doesn't begin at the ocean's edge. Forecasters need data from along the coast and far out at sea to make forecasts. Automated buoys and coastal stations meet this need.

KEY FACTS

The U.S. maintains Pacific Ocean buoys collecting data to track El Niño.

+ fact: Hurricane Sandy in 2012 hit a New York Harbor data buoy with a record 32.5-ft (10 m) wave.

+ fact: A 141 mph (198 kph) wind at the Fowey Rocks C-MAN station was the highest recorded when Hurricane Andrew hit Dade County, Florida, in 1982.

The short towers and moored buoys marked as belonging to NOAA are supplying weather data NWS forecasters said they required, as in the 1980s. The U.S. Coast Guard was automating lighthouses, whose keepers were also weather observers. The NWS set up its National Data Buoy Center to maintain the Coastal-Marine Automated Network (C-MAN) stations along the U.S. coast as well as buoys moored in the water to collect data. When a hurricane or extratropical cyclone is battering the shore, forecasters and the public turn to reports from these stations to see how bad the storm is. Even more important, the stations also supply vital information on average winds and waves as well as extreme events in the same way weather stations on land supply the climate data needed for planning.

Short-term Forecasts

A general rule in meteorology is that the further into the future a weather forecast projects, the higher are the odds that the forecast will not be accurate.

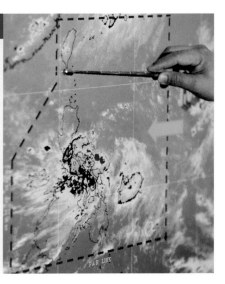

</br>

KEY FACTS

The NWS also produces maps showing 6- to 10-day outlooks for the entire period instead of day-by-day forecasts.

+ fact: Maps show odds for normal, above, or below-normal precipitation and temperatures.

+ fact: Forecasters use map color shades to show their confidence in forecasts.

The U.S. National Weather Service (NWS) and the Canadian Weather Office, along with many private forecasters, regularly produce day-to-day forecasts ten days ahead. In general, it is wisest not to make any important decisions based on a forecast for more than two or three days into the future. The atmosphere is an extremely complex system in which minor differences in the present can have major consequences in the near future. Furthermore, the atmosphere is chaotic, with random changes that no model or computer can predict accurately. When a forecast contains a percentage—"a 40 percent chance of rain," for instance—the meteorologist has calculated both physical likelihood and confidence in the prediction.

Long-term Forecasts

The NWS Climate Prediction Center produces outlooks for short and long periods: from the next 6 to 14 days to the 3-month period ending 12 months ahead.

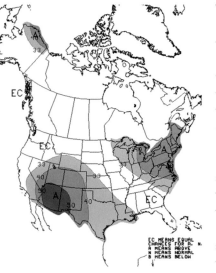

KEY FACTS

The National Weather Service says outlooks for temperatures are most accurate in late winter and late summer.

+ fact: All precipitation outlooks are less accurate than those for temperatures.

+ fact: A strong El Niño makes the precipitation outlook as likely to be correct as some temperature forecasts.

Weather outlooks are intended for rather sophisticated users who are comfortable with probabilities. All outlooks are for averages over monthlong or three-month periods. They will never be specific enough to select a fall weekend for a college homecoming, but they will have general information one could use to estimate fuel oil needs through the period. Still, it's a gamble: If the long-term forecast calls for a cold winter, you can contract to buy oil for the season at a low but fixed price. But if the winter turns out to be warmer, the cost could be below the total price you paid. Nevertheless, by using long-term forecasts, you would be like a professional poker player who loses occasionally but wins more often.

Weather Watches

The terms "watch" and "warning" for potentially dangerous weather may sound somewhat the same, yet they have distinctly different meanings.

KEY FACTS

Severe storm, tornado, and flash-flood watches should make you extra alert for a potential event.

+ fact: Hurricane and winter storm watches give you a small amount of time to get last-minute supplies.

+ fact: You and your family should spend time deciding how to respond to different kinds of weather watches.

The U.S. National Weather Service (NWS) and Canada's Weather Office issue a weather watch when forecasters see a chance that a dangerous event will occur but do not yet consider it a sure thing. A given locale will be put under storm watch, for instance, when the storm is still far enough away not to pose an immediate threat. What you should do when you hear a storm watch depends on the kind of event. A tornado could happen in an hour or so after NWS issues a watch. A hurricane could arrive 48 hours after the watch begins. Depending on the kind of event, it can mean that you should stay alert and be ready to act, or it can mean that you should prepare for an event that could greatly impact your daily life, such as a hurricane or winter storm.

Weather Warnings

A weather warning indicates imminent danger. When it is issued, you should already have used the weather watch period to prepare to keep you and your family safe from the hazard.

KEY FACTS

Each weather warning includes specific advice for the wisest steps to take for safety.

+ fact: A flash-flood warning means leave quickly if you're in a location subject to flash floods.

+ fact: A red flag warning means conditions are right for wildfires to start and spread rapidly, so great care must be taken and no open fires should be lit.

A weather warning means dangerous conditions are occurring or will occur soon. NWS and Canada's Weather Office consider a warning to indicate a threat to life or property. Ideally, you heard the watch and already planned what to do in response. Tornado warnings often advise people to take shelter in a place safe from flying debris. When a winter storm warning is in place, you should stay off the road, remaining safe in your home or elsewhere with the supplies you'll need for a day or two. Hurricane warnings are issued approximately 36 hours before the brunt of the storm is expected, because responses to an oncoming hurricane are more complex and take more time. You should evacuate if you're in a possible flood zone or a home that won't withstand high winds.

The Milky Way and other celestial attractions illuminate
the night sky above Death Valley, California.

5 | Night Sky

Nature's Nocturnal Entertainment

We tend to overlook the night sky, because nighttime itself is something we work around; often nothing but a light switch stands between our day and night activities. But the night sky is worth knowing: It is our window on the past, present, and future. Knowledge of night sky phenomena—moon, stars, planets, comets, and galaxies—draws us in toward the origins of the universe, helps us understand the flow of the seasons, and makes us ponder our planet's destiny. Above all (no pun intended), observing the night sky is the ultimate free pastime, available to all without a reservation.

The Basics

The naked eye remains the most important piece of "gear" for sky watching, combined with a vantage point as free as possible from ambient light. Getting your night vision in shape is the first step. Allow your eyes to adapt to the dark (which takes a minimum of 15 to 20 minutes) and then keep them that way. A glance at a car headlight can undo the process. Because you'll need a flashlight to read this chapter or sky charts, tape a piece of red cellophane over a regular flashlight lens or buy a red-lensed light, preferably LED. Dim red light interferes less with night vision.

Choose the darkest location you can find. At home, turn off your indoor and outdoor lights. A hill, field, or park is a better choice, as far from city light as possible. In the country, find an observation point that keeps city lights to your back. A black cloth or jacket over your head can help block out further light.

Look for a clear night; the sky is especially clear of humidity and haze following a cold front or late-afternoon storm. Developing high-pressure systems also create clear skies. Be sure to prepare for the chill and dampness that can creep in as the night wears on. A beach-type lounge chair offers a relaxing viewing angle, especially when settling in for a meteor show.

Getting Oriented

Star maps show the night sky as if all the celestial objects revolved around the Earth. It helps to picture the Earth as encased in a sky globe with the stars attached to it. This globe has an equator and poles, just as the Earth does. Stars are assigned celestial coordinates similar to latitude and longitude, and these appear on many sky charts. Declination, like latitude, describes how far the star appears above or below the celestial equator. Right ascension is similar to longitude, divided like

Earth's day into 24 hours, with each hour equal to 15° of circumference. From our point of view, the stars pass above from east to west. It is important to know your approximate latitude and the reference latitude of any star guide you're using. The charts in this book, for example, are based on a view from latitudes close to 40° N.

To use the large seasonal sky charts on pages 394 through 401, choose the appropriate one for the season and try to observe at the time indicated for each month. Turn the map so the name of the direction you are facing appears right side up. Then use this orientation of the chart to help locate constellations and other objects.

The seasonal charts and the smaller maps that accompany constellation descriptions in this book show other objects such as star clusters, nebulae, and galaxies. The meanings of these symbols appear on the seasonal charts. Most objects shown are numbered. Many numbers have an M in front; those were listed and catalogued by 18th-century French astronomer Charles Messier and range from 1 to 110. An overlapping and much larger list is the New General Catalogue; objects from this list carry the letters NGC.

■ What's Out There

So what am I looking at? Are those little lights the stars, planets, planes, meteors, or what? Where are those gods, goddesses, and animals with their connect-the-dot outlines? There are no outlines, just stars. But with time and practice, you can discern the shapes of many constellations and learn to tell apart stars and satellites, comets and spacecraft, planets and airplanes.

Stars move from east to west through the night. They twinkle because their pinpoints of light are perturbed by fluctuations in the atmosphere. Planets don't usually twinkle because they are much closer to us. Unlike starlight, planets' disk shapes reflect light from the sun, and this relatively broader width of light is less affected by the atmosphere. Five planets are visible to the naked eye: Mercury, Venus, Mars, Jupiter, and

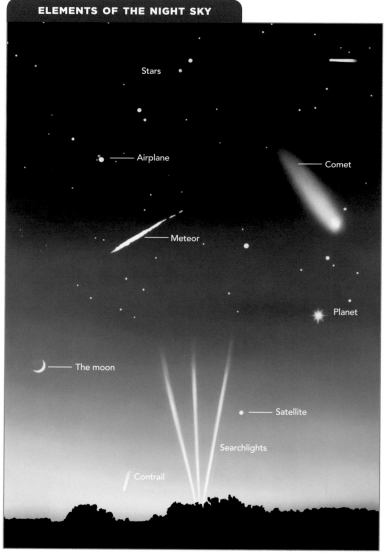

ELEMENTS OF THE NIGHT SKY

Stars

Airplane

Comet

Meteor

Planet

The moon

Satellite

Searchlights

Contrail

Time-lapse star trails etch the darkness at Arches National Park in Utah.

Saturn. Mercury is a bit elusive, but can be viewed well at certain intervals during the year. Venus is the second brightest object in the night sky after the moon. It shines a brilliant white. Mars glows orangish red, while Jupiter is a steady white, and Saturn is pale yellow. Planets take the same path that the sun and moon take across the sky.

The blurry band of light running across the sky, high in winter and summer, is the Milky Way. Yes, we also reside in the Milky Way, but we are able to view other parts of it. Small blurry blobs in the sky are star clusters, nebulae, or perhaps even a galaxy, such as the Andromeda. It appears as a little blur in the constellation Andromeda and is the most distant object visible to the naked eye.

Fast-moving objects include meteors ("shooting stars"), which are bits of solar system debris that flame out in Earth's atmosphere. Slower and steadily moving objects with steady, blinking lights are airplanes. Satellites are visible just after sunset or

But when I follow at my pleasure the serried multitude of the stars in their circular course, my feet no longer touch the earth.
—PTOLEMY

before sunrise, or during the night depending on latitude; they appear as small, starlike lights that move smoothly before disappearing into Earth's shadow.

■ Equipment and Resources

As with any hobby, there is a range of gear that can take sky watching to the nth degree, if you want to. But beginners need to acquire no more than a basic pair of binoculars to enhance their viewing.

The night sky's wonders are accessible with or without equipment.

your best choice. This size is large enough to let in enough light and small enough to wear around your neck and hold steady.

A telescope is a much larger investment and should be carefully chosen. Avoid the low-end ones sold in toy stores and "big box" stores. The poor quality and results of these mass-produced instruments will only disappoint. Instead, go to a specialized store where experts can help.

Much more of what you would like to know and learn about the night sky is just a few taps away on your computer. There are many wonderful instructional and even interactive websites that offer all kinds of practical information about every aspect of sky watching. An example is the website for *Sky & Telescope* (www.skyandtelescope.com), a print publication that has been guiding amateur astronomers for many decades. Among other features, it allows you to create a custom-made naked-eye map of the sky for any location. And of course there are apps by the dozens for your mobile phone that will show you the stars and planets overhead and help you locate them. Many of these are available free or for only a small fee. But, before you turn to your gadgets, you might want to learn about the night sky in the low-tech way that humans have done for millennia. Your well-trained eyes on their own can take you a long distance into its wonders.

Binoculars can reveal a fantastic array of objects and details such as the larger craters on the moon, the distant planets Uranus and Neptune, and even beautiful star fields along the Milky Way. A pair of 7x50 binoculars (7 being the power of magnification and 50 the diameter of the outer lens) is probably

HOW TO MEASURE DISTANCE

Distance across the night sky is typically measured in degrees. You can roughly measure degrees and distance using only your hands. Outstretch your arms and hold out your hands against the sky. At arm's length, a thumb covers approximately 2° of sky, a fist covers about 10°, and a fully outstretched hand about 20°. These approximations can be used to estimate the width and height of constellations. Each constellation entry in this book gives a width measurement using this system. Of course, hands come in all sizes, so these measurements are just approximations for average adult hands.

NIGHT SKY

||||||||||||||||||||||||||||||||||||||||||||||||||||||||||||||||||||||||||||||||||||||

The Atmosphere

Any sky watching we do is affected by the atmosphere—miles-high layers of gases, liquids, and solids that envelop the Earth. The first few layers contain most of the atmosphere's mass.

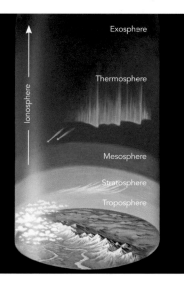

KEY FACTS

The troposphere (to 6–10 mi) is where clouds form and weather occurs.

+ fact: The stratosphere (10–30 mi/16–48 km) is the radiation-absorbing ozone.

+ fact: The mesosphere and thermosphere (30–210 mi/48–338 km) are where temperatures drop, then rise steeply.

+ fact: The exosphere (above 210 mi) contains hydrogen and helium.

Light from the sun, stars, and other celestial objects travels great distances to reach Earth. The portion of light that reaches Earth's atmosphere must navigate the properties and activities that each layer of atmosphere contains. These cause further reductions in the light that will eventually reach the surface. The most challenging transit is through the troposphere, the layer between the surface and six to ten miles above it. This layer holds the largest concentration of atmospheric mass, including more than 97 percent of water vapor. Clouds, dust, chemical pollution, and light pollution combine there to hinder our observations of the night sky.

Nightfall

Sunset initiates nightfall, but the often spectacular colors that accompany the event reveal only part of the interactions among the Earth, the sun, the moon, and the atmosphere.

KEY FACTS

The higher the latitude, the longer twilight lasts because of the angle at which sunlight reaches Earth.

+ fact: In the far north during and near the summer solstice, twilight lasts all night long.

+ fact: The opposite phenomenon occurs in the far north's winter, when little or no sunlight arrives during the day.

As the sun sets in the west, the atmosphere disperses the sun's light like a prism. The long wavelengths of red, orange, and yellow in the spectrum reach us even after the sun has set. Meanwhile, the atmosphere scatters short-wavelength blue and violet at the opposite end of the spectrum, and almost none of these colors reach us. Look to the east and you may see a deep blue band across the horizon—the shadow of Earth, cast on the sky. The moon and stars replace the sun. The amount of moonlight we see at night depends on the reflected light of the sun; changing amounts cause the moon's phases. Atmospheric conditions also affect the moon's brightness.

Noctilucent Clouds

In northern locations, twilight sometimes illuminates high-flying noctilucent clouds that originate far above the region where most clouds occur.

KEY FACTS

Noctilucent clouds were first observed in 1885, two years after the Krakatau Volcano erupted, and initially linked to the eruption.

+ fact: When Krakatau's ashes disappeared, noctilucent clouds remained.

+ fact: Noctilucent clouds (NLCs) are also called polar mesospheric clouds (PMCs).

+ fact: Atmospheric methane increases formation of NLCs.

Clouds typically occur in the troposphere, which ranges from 6 to 10 miles (10 to 16 km) above the Earth. Noctilucent (night-shining) clouds form at much higher altitudes in the mesosphere, which extends to 50 miles (80 km). These thin, wispy clouds appear pale white or an eerie electric blue, and they often occur above the typical orange-red layer of sunset. The arid, frigid mesosphere offers poor conditions for cloud building, but noctilucent clouds assemble from tiny ice crystals and possibly meteor smoke. They are visible in summer when the sun has dipped about 10° below the horizon and usually at altitudes above 40° N. Astronauts on the International Space Station have viewed the clouds from their unique vantage point and have taken scientifically valuable photographs of the phenomenon.

Magnetosphere

A magnetic shield sustained far beyond Earth's atmosphere by the planet's inherent magnetism, the magnetosphere deflects solar and cosmic radiation.

KEY FACTS

On the side of Earth facing the sun, solar wind compresses the magnetosphere to about 40,000 miles.

+ fact: Geologic and other studies show that the magnetic field changes over time.

+ fact: The location of the north magnetic pole has wandered over arctic Canada.

+ fact: In the future, the north magnetic pole will move into arctic Russia.

Earth's molten metal core generates a magnetic field on a cosmic scale. Magnetic current flows outward from the two poles and loops back toward the center, forming a magnetosphere that extends tens of thousands of miles into space. It deflects the steady stream of atomic particles known as the solar wind, which can knock out communications and electronics during intense geomagnetic storms. Buffeting by the solar wind causes the magnetosphere to change shape during the day. It becomes compressed on the side facing the sun, and it elongates into a long tail extending past the orbit of the moon on the opposite side. The sun and the other planets also have magnetospheres, but Earth's is the strongest among rocky planets of the inner solar system.

Van Allen Belts

Consisting of two concentric zones of intense radioactivity that gird the Earth, Van Allen belts are named for James Van Allen, the physicist who discovered them in 1958.

KEY FACTS

Van Allen designed special Geiger counters that traveled on the first U.S. satellite.

+ fact: The satellite transmitted evidence of trapped cosmic and solar radiation.

+ fact: Discovery of the belts was one of the first major breakthroughs of the space age.

+ fact: Spacecraft traveling through the belts require protective shielding.

The Van Allen belts harness atomic particles that have penetrated the magnetosphere, including protons and electrons from outside the solar system and helium ions from the sun. The trapped, highly charged particles mingle with the atmosphere and travel a spiral path as they bounce back and forth between Earth's magnetic poles. The belts can be visualized as concentric doughnuts, thickest at the Equator and weaker at the poles. The inner belt begins approximately 600 miles (966 km) above Earth's surface and extends to about 3,000 miles (4,800 km). The outer belt extends from about 10,000 to 25,000 miles (14,000 to 24,000 km). In 2013, NASA announced the detection of a third belt by twin satellites named the Van Allen Probes.

Auroras

A by-product of solar storms, auroras are visible mainly in polar areas. The light show is called aurora borealis in the Northern Hemisphere and aurora australis in the Southern Hemisphere.

KEY FACTS

Auroras shimmer because solar storms stretch the magnetosphere, and it bounces back into place.

+ fact: Northern peoples developed myths to explain them.

+ fact: Southern Hemisphere auroras appear mostly over uninhabited land.

+ fact: At the peak of a solar-storm cycle in 1989, auroras were seen as far south as the Caribbean.

During violent solar events, such as mass coronal ejections and solar flares, the discharge of highly charged particles from the sun increases astronomically, shooting out billions of tons of matter at speeds as fast as 2 million miles (3 million km) an hour. This barrage breaches the magnetosphere and Van Allen belts, reaching perhaps as close as 50 miles (80 km) to the Earth's surface. The abundant solar material energizes molecules of oxygen and nitrogen in the atmosphere, causing dazzling curtains of green and red and pink light that intensify in proportion to the intensity of the solar storms in their 11-year cycle.

Stars

From their earliest formation about 13 billion years ago, stars have been a universal sky feature. A few thousand can be seen with the naked eye on a clear, moonless night.

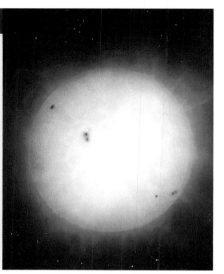

KEY FACTS

The sun is a third-generation star with elements from two previous star explosions.

+ fact: The sunlight currently shining on Earth was generated in the star's core 50 million years ago.

+ fact: Sunlight takes 8.3 seconds to reach Earth.

+ fact: A neutron star fragment the size of a sugar cube would weigh about 100 million tons on Earth.

Elementally simple—composed almost entirely of hydrogen and helium gases—stars have not yet revealed the precise details of their formation. In basic outline, gravity pulls inward dense patches of gases and dust, causing pressure and heat to build at the center of the ball of gas and triggering nuclear fusion. The energy created, which we see as starlight, pushes outward and is balanced by the pull of gravity, establishing equilibrium. Stars are born in nebulae, large gas clouds found throughout the Milky Way. They also come into being from the supernova explosions of older stars that create raw material for new stars. The nuclear reactions within stars form the heavy elements carbon, oxygen, and nitrogen—the essential building blocks of life as we currently understand it.

Types of Stars

Astronomers classify stars by the characteristics of size, temperature, and color. They use hydrogen spectral analysis of stars' hydrogen emission lines to make their determinations.

KEY FACTS

Stars are classified by a letter system that runs O, B, A, F, G, K, M (largest/hottest to smallest/coolest).

+ fact: A mnemonic to remember the order is "Oh boy, a flying giraffe kicked me!"

+ fact: A Hertzsprung-Russell diagram shows the relationship between color, temperature, and luminosity of stars.

+ fact: The sun is a modest G star.

A star's size is determined by its mass and not by a linear measurement, such as diameter. Mass regulates the other properties of a star, including temperature, color, and longevity. Smaller, cooler stars burn fuel at a slower pace and live longer than larger, fuel-guzzling hot stars. Red dwarfs are the most common smaller stars and have potential life spans of tens of billions of years. Massive blue giants are the hottest and may burn out in a million years or so. The supergiant is the largest star; Betelgeuse in Orion is a red supergiant, about 1,000 times bigger than the sun. Despite its large size, a red supergiant is a relatively cool star. It burns approximately ten times cooler than hotter and smaller blue O stars.

Star Brightness

We talk about star brightness in terms of magnitude, a number that often represents how bright the star is from our vantage point on Earth.

KEY FACTS

Sirius, a blue-white binary star in the constellation Canis Major, is the brightest in the Northern Hemisphere.

+ fact: Second brightest is Arcturus, an orange giant in Boötes.

+ fact: Fifth is Vega in Lyra. (Third and fourth are visible only in the southern sky.)

+ fact: Sixth is Capella in Auriga.

In the second century B.C., the Greek astronomer Hipparchus created a catalog that ranked the brightest stars as of the first magnitude, less bright stars as the second magnitude, and so on. Adopted by Ptolemy and refined in later centuries, the system became a logarithmic scale. As observation methods improved and fainter stars came into view, the scale was extended at the higher (dimmer) end. The need also arose to accommodate differences among first-magnitude stars. With nowhere to go but down, negative numbers were added. Magnitude is now calculated in many ways, including absolute magnitude, which is brightness if all stars were placed at the same distance from Earth. Under average night sky conditions, the naked eye can see stars to magnitude +6.

Star Death

The state of equilibrium that keeps a star together is a temporary one. As a star's core consumes its fuel, it sets in motion the scenario of its death.

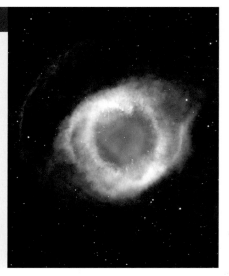

KEY FACTS

A spent white dwarf often hooks up with a vibrant star and travels with it.

+ fact: Brightest star Sirius travels with the Pup, not a young star, but an old white dwarf.

+ fact: Neutron stars, also known as pulsars, emit beacon-like pulses of radiation.

+ fact: A supernova explosion creates large fields of new material for star formation.

As a star burns up its fuel, its nuclear core begins to shut down. Gravity can no longer be held at bay, and the star contracts. This raises its temperature and triggers renewed fusion, allowing the star to expand. Smaller stars will become pulsing red giants. Eventually, they shed outer layers and become dead stars known as white dwarfs. Larger stars morph into supergiants, with contracting cores that may reach 1,000,000,000°F. More intense gravity in the core leads to an implosion followed by an explosion—a supernova—that blows apart the outer layers. Next follows either a neutron star or a total collapse into a black hole.

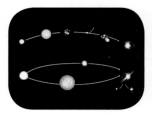

Sun

Our sun came into being the way any star does—in a cauldron of condensed gas and fused atoms. This life-sustaining star sits at the center of our solar system.

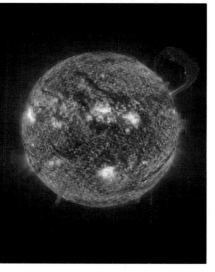

KEY FACTS

For a star, the sun is of middling size and heat—10,000°F (5500°C) at surface.

+ **fact:** Sun's core temperature: about 27,000,000°F (14,999,820°C)

+ **fact:** Composition of sun's outer layer: 74 percent hydrogen, 25 percent helium, 1 percent heavy elements

+ **fact:** The sun provides 1,400 watts of energy to each square yard of Earth.

Despite its seemingly extreme statistics, the sun is the perfect star for our planet. One of about 200 billion stars in the Milky Way, it resides 93 million miles from Earth. Its dimensions, dynamics, and distance provide just what we need in terms of energy and light to support life without overwhelming the planet with violent radiation. The sun is classified as a yellow dwarf star and lies in the middle of the classification scheme for color and temperature. Earth's life-supporting relationship with the sun is jeopardized by depletion of the upper ozone layer in the atmosphere, which allows more ultraviolet light to reach the planet. As with any star, the sun has an expiration date, estimated at about five billion years from now.

Sun Anatomy

The sun is a roiling, complicated body of violent extremes, with its own weather systems and a disturbing hint of unpredictability.

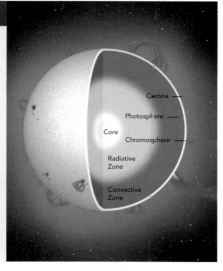

KEY FACTS

The sun's equatorial region rotates in 25 days, but its polar regions take 34 days to complete rotation.

+ **fact:** In the core, about half the sun's mass is packed into 7 percent of its volume.

+ **fact:** Nuclear fusion in the core generates 400 trillion watts of energy a second.

+ **fact:** Holes in the corona allow the flow of a steady stream of solar wind.

The sun has a dense, high-energy core, where fusion takes place and hydrogen is converted to helium; emitted photons begin a journey to the surface that can last millions of years. Three-fourths of the way they reach the convection zone that continues to draw gas and heat outward. They then encounter the 300-mile-thick (483 km) photosphere, which gives the sun the appearance of a solid with gas bubbling at the surface. The chromosphere is a pinkish layer that harbors solar flares. Beyond that lies the corona, a billowing megahot halo with temperatures reaching 2,000,000°F (1,111,093°C) that extends millions of miles. Holes in the corona, caused by the sun's magnetic field, allow a steady stream of particles known as solar wind to break free.

Sunspots

Sunspots peak and ebb in an 11-year cycle that coincides with other solar activity. Chinese astronomers made note of sunspots in 28 B.C.

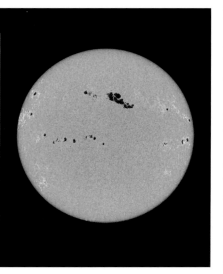

KEY FACTS

A 70-year absence of sunspots in the 17th and 18th centuries coincided with a period called the "Little Ice Age" in the Northern Hemisphere.

+ fact: Sunspots measure up to several times Earth's diameter.

+ fact: The period of low activity during the cycle is known as the solar minimum.

+ fact: Sunspots appear in magnetically linked pairs.

Disrupted currents in the sun's magnetic field can slow the flow of solar material to the sun's surface, causing it to cool and darken in color. This produces sunspots, dark, visible splotches on the sun's photosphere. The center of a sunspot, depressed slightly below the surrounding gases, can measure some 3600°F (1980°C) cooler than the sun's typical surface temperature of 10,000°F (5500°C). Within a cycle, sunspot activity usually starts in the regions of latitudes 30° N and 30° S and then moves near the sun's equator. The 11-year cycle is part of a 22-year period in which the magnetic fields in the sun's upper and lower hemispheres switch polarities. NASA's Solar Dynamics Observatory is designed to study the variations in solar activity that affect Earth and near-Earth space.

Solar Flares

One thing leads to another: Solar flares, immense eruptions of high-energy radiation from the sun's photosphere, are associated with sunspots in ways not completely understood.

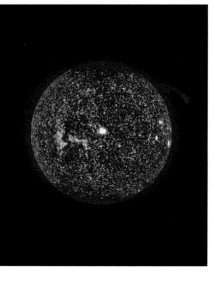

KEY FACTS

One solar flare can unleash more energy than all the atomic bombs ever detonated.

+ fact: The first solar flare was recorded in astronomical literature in 1859.

+ fact: In 1989, intense solar activity knocked out much of Quebec's electrical grid.

+ fact: NASA and other science organizations predict solar storms.

When magnetic energy that is pent up as a result of sunspot activity breaks free, it releases intense bursts of radiation across the entire electromagnetic spectrum, from radio waves to gamma waves. Solar flares extend into the sun's corona and take the form of arcs of highly charged superheated material known as coronal loops. The solar storms produced by flares sabotage electronics, communications, and electrical grids as they unleash their disruptive radiation. The most troublesome kinds of solar flares are known as X-class flares, which produce prolonged radiation storms in the upper atmosphere.

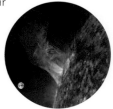

Sun's Path

For centuries, we have known that the Earth orbits the sun, but to conveniently describe all kinds of solar and celestial phenomena, we talk of the sun's apparent path across the sky.

KEY FACTS

Earth rotates from west to east, making the sun's path appear to be from east to west.

+ fact: Perihelion is the point at which a planet or other sky object in orbit is closest to the sun.

+ fact: Aphelion is the point at which a planet or other sky object is farthest from the sun.

We on Earth cannot track in a useful way our planet's daily rotation or yearly orbit without reference to the sun. The sun's apparent daily journey across the sky—and the changes to its route over the course of a year—have long formed the means by which we tell time, mark the seasons, and plan annual celebrations. The path of the sun across the sky is known as the ecliptic, which is also the term given to the Earth's route around the sun. Most sky charts indicate the sun's ecliptic on the night sky as a useful reference for locating constellations and planets and describing their whereabouts. The constellations that form the zodiac follow the path of the sun's ecliptic over the course of a year, the basis of the 12 astrological signs.

Sun & Seasons

Earth and sun are oriented to each other and interact in a way that produces predictable periods of difference in sunlight and temperature throughout the year.

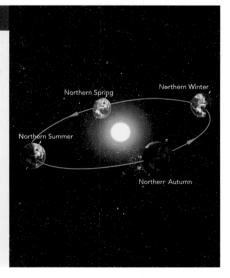

Northern Spring

Northern Winter

Northern Summer

Northern Autumn

KEY FACTS

During the equinoxes, the division of day and night is about the same around the globe.

+ fact: Summer at the North Pole means nearly 24 hours of daylight.

+ fact: The Earth's axis moves in position a little more than a half degree a century.

+ fact: The change of direction in the Earth's axis is known as precession.

As Earth orbits the sun, its rotational axis is tilted 23.5 degrees off the perpendicular. Because of this, the sun's direct light will hit different parts of the Earth at different times of the year, causing the seasons. Astronomical convention marks the transitions in two ways: as equinoxes in the spring and autumn, when the sun at noon shines directly on the Equator; and as solstices in winter and summer, when the sun shines on Earth as it is tilted away from or toward the sun. The Northern and Southern Hemispheres experience seasons at opposite times of the year, and equatorial areas experience little seasonal variation at all. Seasonal differences in sunlight and temperature affect many of the life processes on Earth.

Viewing the Sun

Many of us quickly reach for our sunglasses when we hit the bright sun—a good instinct. Viewing the sun directly with the naked eye can cause permanent damage.

KEY FACTS

Telescopes can be used for safe viewing.

+ fact: A filter on the objective lens allows safety; never use one only on the eyepiece.

+ fact: Use a scope as a projector. Attach cardboard to the objective lens, leaving a 2–4 in (5–10 cm) opening.

+ fact: Stand white paper behind the eyepiece, and view the sun on the paper.

Here is a mantra to repeat over and over for sky watchers young and old: Never, ever look directly at or point your binoculars or telescope toward the sun—not even for a split second. That said, there are safe ways to observe the sun. Welders' goggles rated 14 or higher are one option, as are approved solar filters and "eclipse glasses." These items can be found online. It is also easy to make a pinhole projector by piercing a card with a small hole and placing a whole card below it. Position the top card toward the sun, without looking. A binocular lens can be used the same way; leave the caps on the unused lens. Using exposed film, medical x-rays, or smoked glass as filters is dangerous. Eye damage may be painless, as it occurs because it is caused by invisible infrared rays.

Solar System

We tend to talk of *the* solar system—ours—but there are many of them. Our sun's gravity draws planets and other objects into orbit around it.

KEY FACTS

More than 99 percent of the solar system's mass is contained within the sun.

+ fact: Primitive meteorites provide a measure of the age of the solar system: about 4.5 billion years old.

+ fact: The solar system orbits the center of the Milky Way every 220 million years.

+ fact: The word "planet" means "wanderer."

The sun holds sway over a diverse array of objects ranging from planets and dwarf planets to moons and asteroids—and even to dust. Its influence spreads from its own surface to the distant Oort cloud that encircles the solar system. The planets under the sun's control—including Earth, of course—spread across a distance ranging from Mercury, on average about 36 million miles (58 million km) from the sun, to Neptune, with an orbit averaging about 2.8 billion miles (4.5 billion km) from the sun. Pluto and other dwarf planets range beyond that. Of course, our solar system is only one of many that have formed or at this moment are forming around distant stars, as ongoing discoveries attest.

Moon

The moon is a stargazer's nearly constant companion, visible to the naked eye even in competition with extreme light pollution. Binoculars bring out details of its features.

KEY FACTS

Made of lighter elements than Earth, the moon has ¼ the diameter of the Earth, contains about ¹⁄₈₀ the mass, and has about ⅙ the gravity.

+ fact: Diameter is 2,159 miles (3,465 km).

+ fact: Distance from Earth is 238,855 miles (384,400 km).

+ fact: Orbital period is 27.3 days; cycle of moon phases is 29.5 days.

Moons are natural satellites that orbit celestial bodies such as planets and asteroids, and exercise their own gravitational pull. Most planets in our solar system have moons, usually multiples. Earth's solitary moon likely formed from debris ejected when an object about the size of Mars struck the young Earth, about 4.5 billion years ago. Evidence from moon missions shows a similar composition of their respective rocks. Characteristics of their orbits also support this theory. Over time, the moon acquired its craters and seas from bombardments by space debris and the eruption of molten rock from the interior. The "man in the moon" is an impression that the unaided eye sees of contrasting lunar surface features. Some people see a woman, a hare, or a toad.

Moon in Motion

The motion of the moon shares complex interactions with that of the Earth. The planet and its satellite regulate and influence each other in various ways.

KEY FACTS

Friction from Earth-moon interaction causes the Earth to slow slightly.

+ fact: The slowdown is offset by a slight pulling away of the moon—about 1.5 in (3.8 cm) a year.

+ fact: The moon's orbital period is 27.3 days.

+ fact: The Earth's rotation accounts for much of the moon's visible motion.

The moon spins on its axis in the same manner as the Earth. But the moon's rotation takes the same amount of time as its orbit around the Earth—27.3 Earth days. This is because the Earth's gravity produces a bulge, or "land tide," in the moon that slows it down. As a result of this synchronization of rotation and orbit, the moon always presents the same side to the Earth. The moon still exerts its own influence: For example, it creates a gravitational tug responsible for the tides on Earth. The moon's gravity requires spacecraft engineers to take it into consideration when calculating trajectories for lunar landings. World cultures have attached other influences to the moon, such as influencing the course of love or initiating the legendary transformation from human to werewolf.

Moon Phases

The moon "borrows" light from the sun to create its shine. The phases, or appearance, of the moon at different times of its orbit, reflect changes in available sunlight.

KEY FACTS

The moon takes 29.5 days to complete the cycle of its phases.

+ fact: Moonrise occurs 12 degrees farther to the east each night.

+ fact: A full moon rises with the setting sun and sets at sunrise the next day.

+ fact: View crescent moons best in the evening in spring and just before sunrise in autumn.

Observers on Earth cannot see the moon during its phase known as new moon. It occurs when the moon lies between the Earth and the sun and is fully backlit. About two weeks later, the Earth lies between the sun and moon, and we see the latter's maximum appearance—a full moon. In the weeks leading to the full moon, the moon comes into view in gradual increments: a waxing crescent, crescent, first quarter, and then waxing gibbous. (Gibbous is from the Latin for "hump.") In the weeks after the full moon, its light starts to diminish, and it becomes a waning gibbous, last quarter, crescent, and then waning crescent. Worldwide, many agricultural tasks as well as religious observations are linked to the phases of the moon.

Moon's Near Side

Collisions with meteors and comets, combined with volcanic activity from the moon's interior, produced the landscape of ridges and craters we see on the near side of the moon.

KEY FACTS

Each part of the moon's surface is visible for about two weeks in a phase cycle.

+ fact: Galileo observed the moon's surface through his telescope in the 1600s.

+ fact: First and last quarter phases show greater detail on the moon's surface.

+ fact: The last major crater to form was the Tycho crater about 109 million years ago.

Earth's moon began in violence that continued for the first half billion or so years. Debris bombarded the surface, creating large, wide, and shallow areas that later filled with dark lava pouring from the moon's core. These areas are called "maria" (MAH-ree-uh), the Latin word for "seas." The most famous is the Mare Tranquillitatis, the Sea of Tranquillity, site of the first moonwalk by Neil Armstrong in 1969. Smaller debris gouged smaller, deeper holes known as craters. These appear much lighter on the moon's surface than the dark maria. Over time, the bombardment decreased dramatically. NASA's Astronomy Picture of the Day (APOD), available at *apod.nasa.gov/apod/astropix.html,* often features amazing images of the moon, including some interactive ones.

Moon's Far Side

Although it is not a visual presence in the night sky, understanding the moon's far side rounds out our appreciation of Earth's satellite.

KEY FACTS

Many far-side features have names that reflect Russian space exploration, such as the Moscow Sea.

+ fact: The far side experienced fierce asteroid bombardment, causing craters.

+ fact: Bombardment on the far side caused volcanic activity on the near side.

+ fact: The far side has a thicker crust with fewer seas than the near side.

The moon's far side is often referred to as its "dark side," but it is by no means dark on the far side; we just cannot see the light because, due to its synchronous orbit with Earth, the moon presents only its near side to our view. The far side experiences phases just as the near side does; there is a time each month when it is in full sunlight and a time when it is in full shadow. The far side of the moon was first mapped from space-based observations by the Russians in 1959, with pictures the spacecraft Luna 3 took. After that, three decades of probes and space missions created detailed maps. The quest continues to provide even more detailed and scientifically advanced topographical images of the moon's far side with the Lunar Reconnaissance Orbiter, launched by NASA in 2009.

Moon Exploration

On July 20, 1969, men on the moon—the astronauts of Apollo 11—met the man in the moon, so to speak. Remote reconnaissance and robotic landings preceded the historic event.

KEY FACTS

Scientists believe that most of the 6 U.S. flags planted on the moon still stand.

+ fact: In the 1960s, orbiters photographed 99 percent of the moon's surface.

+ fact: The last Apollo mission, in December 1972, collected more than 250 pounds of moon rocks.

+ fact: India, China, and Japan are developing lunar exploration projects.

The manned Apollo missions with their moon landings and space buggy rides were preceded by remote data gathering. The Russians sent the first probes to check out the moon in 1959: Luna 1 flew by it, Luna 2 crash-landed, Luna 3 took photographs of the far side; and in 1966, Luna 9 made a soft landing. NASA followed with several series of lunar probes, including orbiters and surveyors that paved the way for six manned Apollo missions and their 12 astronauts.

NASA maintains a Lunar Reconnaissance Orbiter that acquires sophisticated data, and the space agency is committed to returning humans to the moon.

Solar Eclipse

The sun and moon collaborate to produce three kinds of solar eclipses, obscurations of the sun by the moon. These are called annular, partial, and total.

KEY FACTS

Viewing solar eclipses requires eye protection.

+ fact: Many cultures feared solar eclipses, believing they portended bad fortune.

+ fact: The slender trail of a total solar eclipse varies year to year in an 18-year cycle.

+ fact: The next path of a total solar eclipse to cross the U.S. will occur in 2017.

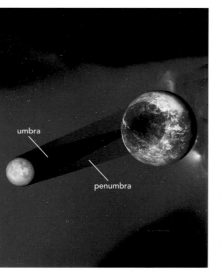

An annular eclipse occurs when the moon is too far from the sun to cover it completely. Instead, a ring or annulus, of the sun's bright edge is visible. A partial eclipse happens when only part of the sun is obscured if viewed from within the moon's penumbra, the outer region of its shadow. Annular and partial eclipses only dim the sun. During totality of a total solar eclipse, the moon's disk completely covers the sun for those on Earth within the moon's umbra, or central shadow. Daylight fades to dark twilight, and stars are visible during totality. Sky watchers can see the sun's corona feathering out around the edges of the moon's shadow.

Lunar Eclipse

Like the sun, the moon experiences three kinds of eclipses: penumbral, partial, and total. Unlike solar eclipses, they can be viewed with the naked eye.

KEY FACTS

Columbus used his knowledge of an impending lunar eclipse to gain the cooperation of Arawak Indians, who were amazed by his ability to predict it.

+ fact: Ancient sky watchers tracked lunar eclipses and kept tables of their occurrences.

+ fact: Use binoculars or a telescope with a low-power eyepiece to see the whole moon.

If the moon passes through the outer region of the Earth's shadow, the penumbra, it causes a penumbral lunar eclipse. This will dim the moon a bit. If some of the moon passes through the Earth's dark central shadow, the umbra, a partial eclipse will occur. When the moon moves completely into the Earth's umbra, a total eclipse takes place. Even then, the moon does not become completely dark. Sunlight bends around Earth's surface and shines dimly on the moon. The refracted light casts a reddish or orange glow. If Earth's atmosphere is unusually dusty, the moon might be barely visible for a while.

Satellites

The U.S.S.R. launched the first artificial satellite, Sputnik 1, into orbit on October 4, 1957. Since then, thousands more have been launched for scientific and commercial purposes.

KEY FACTS

Sputnik 2 carried a dog named Laika. Her flight proved that an animal could survive launch and weightlessness, although she perished shortly after.

+ fact: In 1959, U.S. satellite Explorer 6 photographed Earth for the first time.

+ fact: In 1962, Telstar, the first communications satellite, was launched.

A satellite is any object that orbits a planet; Earth's moon is a natural satellite. Artificial satellites were created and sent into space for purposes of data gathering and signal transmission. The launch of Sputnik 1 in 1957 ushered in the space age. The United States followed on January 31, 1958, with the launch of American Explorer 1—and the space race between the U.S.S.R. and the U.S. was under way. In the meantime, satellite launchings have become commonplace. There are roughly 6,000 satellites in orbit now, not all of them functional, with specialized missions such as communications, weather, navigation, and astronomy.

Cassini-Huygens Spacecraft

A joint venture of NASA, the European Space Agency, and the Italian Space Agency, the Cassini-Huygens spacecraft launched in October 1997 to begin the long journey to Saturn.

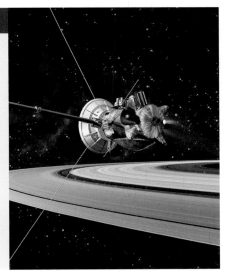

KEY FACTS

The Huygens probe was the first craft to land on a body in the outer solar system.

+ fact: Seventeen countries collaborated in the Cassini-Huygens mission.

+ fact: It takes more than an hour for radio signals from Cassini to reach Earth.

+ fact: The Cassini mission has been extended to 2017.

Saturn, with its rings suggestive of the sun's encircling gases and its many moons representing diverse aspects of planet formation, was an ideal target for the Cassini-Huygens mission. The hyphenated name reflects the craft's two components: the plutonium-powered Cassini orbiter and the wok-shaped Huygens probe, named for Italian-French and Dutch astronomers, respectively. Cassini-Huygens took nearly seven years to reach Saturn. Huygens parachuted to a soft landing on Saturn's moon Titan in January 2005. The duo has made pathbreaking discoveries, including strange weather patterns, new moons, methane lakes on Titan, and ice geysers on the moon Enceladus. Cassini is on a second mission extension until 2017, when the Saturn summer solstice can be observed.

Chandra X-ray Observatory

The Chandra X-ray Observatory, named for Indian American astrophysicist Subrahmanyan Chandrasekhar, provides a view of the universe not possible from Earth's surface.

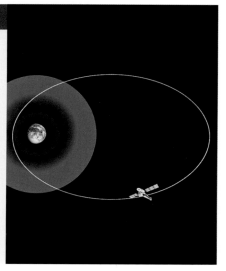

KEY FACTS

Chandra can see the x-rays from particles at the edge of black holes.

+ fact: Its orbit takes it 200 times higher than the Hubble Space Telescope.

+ fact: Its resolving power compares to reading a stop sign from 12 miles away.

+ fact: Chandra needs only the wattage of a hair dryer to run the spacecraft.

Launched in July 1999 by the space shuttle Columbia, NASA's Chandra X-ray Observatory is a telescope that focuses on the high-energy objects in space that radiate energy in x-ray wavelengths, such as supernovae, binary stars, and black holes. The world's most powerful x-ray telescope, it can detect x-ray sources only one-twentieth as bright as earlier telescopes could. To do so, it orbits above Earth's atmosphere, up to an altitude of 86,500 miles (139,000 km). Chandra, a NASA project that is hosted by the Harvard-Smithsonian Center for Astrophysics, has thrilled astronomers with its data, which it shares with scientists around the world.

Hubble Space Telescope

Launched in 1990, the Hubble Space Telescope is the astronomical equivalent of the greatest thing since sliced bread. It looks at the universe in mind-boggling detail.

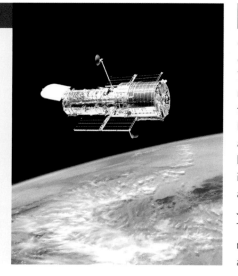

KEY FACTS

The Hubble travels at 17,000 mph (28,000 kph) and orbits 353 miles (569 km) above the Earth.

+ fact: Every year, the Hubble makes more than 20,000 observations.

+ fact: Space shuttle missions have upgraded and repaired equipment, including cameras on the Hubble.

Its ability to view far-flung reaches of the universe without atmospheric interference is the hallmark of the Hubble Space Telescope. Roughly the size of a large bus, the Hubble carries a primary mirror 7.9 feet (2.4 m) in diameter and an array of instruments, including cameras, spectrographs, and fine-guidance sensors used to aim the telescope. In the more than two decades since its launch, a relentless stream of images from the Hubble has helped determine the age of the universe, informed us about quasars, and enlightened us about dark energy. Its successor, the infrared James Webb Space Telescope, is in development. The Webb, planned to launch in 2018, will have a mirror nearly three times the size of the Hubble's and a sun shield the size of a tennis court.

Voyagers 1 & 2

Launched more than 35 years ago, NASA's Voyagers 1 and 2 are now headed out of the solar system, prepared for encounters with intelligent life.

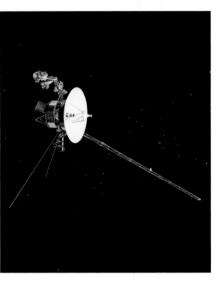

KEY FACTS

The probes' mission originally was to study Jupiter and Saturn and their moons.

+ fact: Early success led to an expansion and continuation of the Voyager mission, with flybys of all outer planets.

+ fact: Voyager 1 is the farthest artificial object in existence—119 times farther from the sun than the Earth is.

The deep-space probes Voyager 1 and 2 were launched in 1977 to study the outer solar system and, eventually, interstellar space. These nuclear-powered craft are headed out of the solar system on two separate trajectories, with Voyager 1 well ahead. Should either encounter intelligent life, each carries a gold-plated 12-inch (30 cm) phonograph record—very "old school," but the go-to durable format at the time. Drawings on the cover demonstrate how to play the record, and they show a map of our solar system relative to 14 beaconlike pulsars. Music included ranges from Bach to Javanese gamelan to Navajo chant to Chuck Berry's "Johnny B. Goode." The record also contains spoken greetings in 55 world languages and natural sounds, such as thunder, birds, and whales.

Space Junk

What goes up into space may come down, but it often stays up and accumulates with other outmoded hardware to become orbiting space junk.

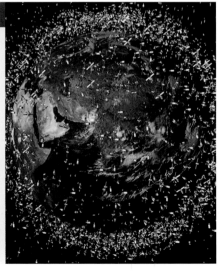

KEY FACTS

The now nonfunctional Vanguard 1 satellite, launched in 1958, is expected to orbit another 150 years.

+ fact: Space junk endangers weather, communications, and GPS satellites.

+ fact: When space junk falls out of orbit, it can cause potential damage on Earth.

+ fact: In 1978, a Soviet satellite spewed radioactive debris into the Arctic.

With the exception of manned spacecraft, most satellites and other equipment sent into space do not make the return journey. Even when no longer functional, these objects remain in orbit to become part of a growing array of space junk. Some 600,000 items ranging from rocket boosters to explosion debris to actual space station garbage float out there, creating hazards to functioning spacecraft and satellites from time to time. At most peril is the International Space Station, or ISS, and its inhabitants. In 2001, the ISS had to be nudged into a slightly higher orbit by the space shuttle *Endeavour* to avoid a collision with the upper stage of a spent Soviet rocket.

Space Station

Sky watchers are often treated to the sight of the International Space Station, or ISS, in the night sky. NASA's website and its SkyWatch 2.0 applet provide information on its whereabouts.

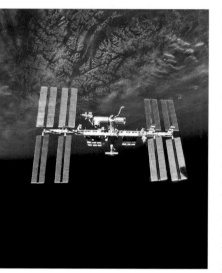

KEY FACTS

Skylab fell from orbit prematurely in 1979, breaking up over the Indian Ocean and Australia.

+ fact: The ISS is the largest space station ever constructed.

+ fact: Human habitation on the ISS has gone on uninterrupted since it began in 2000.

+ fact: Plants are grown experimentally in the ISS lab and service module.

Prolonged sojourns in orbit require a space station. Both the Soviet Union (and later Russia) and the United States committed to this endeavor, beginning with Salyut 1 in 1971. The Soviet Salyut series was succeeded by the Mir space station, which operated from 1986 through 2006. The U.S. launched Skylab in 1973, and it was manned through 1974. Today, the International Space Station is the sole station in orbit, and it serves the exploration and research needs of the U.S., Russia, Japan, Canada, and 11 European countries. The ISS received its first crew in November 2000, and is expected to remain in operation until at least 2020. A yearlong mission is planned for 2015 that would further examine human physiological responses to prolonged space travel.

UFOs

Few topics are as controversial as unidentified flying objects—UFOs. Despite many unexplained sightings, UFOs often turn out to be ordinary phenomena or objects.

KEY FACTS

The term "flying saucer" was coined by a pilot in 1947.

+ fact: Over the decades, NASA astronauts have reported a number of unexplained phenomena.

+ fact: In 2012, Mars rover Curiosity appeared to capture a UFO image that turned out to be dust kicked up by the impact of its sky crane with the surface.

Beginning sky watchers see a lot of objects they cannot identify, but that doesn't make those objects UFOs. The watchers soon learn to sort out planes, larger satellites, and sometimes even the International Space Station from relatively stationary objects in the sky. Otherwise, UFOs in general can usually be accounted for as meteors, clouds, birds, weather balloons, atmospheric phenomena such as sun dogs, or military aircraft. The often secret nature of military aircraft leaves many sightings unexplained. A large number of unresolved UFO sightings may well be unfamiliar atmospheric phenomena with unusual physical properties. The SETI Institute searches for signs of extraterrestrial life using radio and optical telescopes. Many prominent scientific institutions, as well as NASA, have collaborated on the project.

Comets

Dirty balls of frozen gases and dust, comets are leftovers from the formation of the solar system. Nudged out of their original orbits, they streak through the inner solar system.

KEY FACTS

Comets typically are large enough to survive hundreds of trips through the inner solar system.

+ fact: British astronomer Edmond Halley never saw the comet named for him.

+ fact: The most recurrent comet is Encke, cycling past Earth every 3.3 years.

+ fact: A stream of dust shed by a comet's nucleus forms a second tail.

When an icy ball of gases and dust enters a new orbit and approaches the sun, the warming object forms a glowing head—the coma—and a flow of charged ions is shaped by solar wind into a glowing tail. Long-period comets come from the Oort cloud, a belt of icy debris that encircles the solar system. Short-period comets, with orbits easy to predict, originate in the Kuiper belt beyond Neptune or in a far-flung area known as the scattered disk. Comets can be viewed with the naked eye and, of course, through optical equipment. To tell a comet from a star, check a star chart to see if it's there, look for a tail, and look for movement over time. The NASA website offers an Asteroid and Comet Watch that provides up-to-date information.

Meteors

Called "shooting stars," meteors are bits of material, often left behind by a comet, that ignite and burn up as they pass into Earth's atmosphere.

KEY FACTS

More than 60 meteorites found are pieces from the moon, and more than 30 are fragments of Mars.

+ fact: Comets under the pull of Earth's gravity reach speeds of 20,000–160,000 mph (32,000–257,000 kph).

+ fact: Atmospheric friction heats meteors to 2000°F (1093°C).

+ fact: Most meteors vaporize before they get within 50 mi (80 km) of Earth.

Meteors originate as meteoroids, bits of interstellar dust and debris. The particles may be very small or quite large, approaching asteroid size, although the difference between the two is somewhat arbitrary. Larger meteoroids may be broken-off pieces of asteroids, planets, or our moon. Many smaller meteoroids represent burned-off ashes from comets, which are left in massive trails throughout the solar system. As Earth travels in its orbit, it encounters perhaps a half billion meteoroids each year. Objects entering the atmosphere are known as meteors and those that reach the Earth are called meteorites. On a given night, there may be several visible shooting stars each hour. Up to 10,000 tons of meteoric material, much of it dust size, fall on Earth daily.

Meteor Showers

About 30 times a year, Earth passes through a dense dust trail, typically left by a comet, giving rise to a meteor shower or a rare, spectacular meteor storm.

KEY FACTS

The International Meteor Organization depends on information from amateurs to expand its database.

+ fact: Meteor viewing is affected by atmospheric conditions and light pollution.

+ fact: Halley's comet left a dust trail that produces two meteor showers each year.

Meteor showers, unlike the sporadic meteors, occur in steady streams at predictable times from specific locations in the sky known as radiants. Radiants typically are identified with constellations. The Leonid meteor shower, for example, originates in Leo. Meteor showers may produce several dozen meteors an hour; meteor storms can produce that many a second. Meteor showers are well publicized, with tips for best viewing times. No equipment is needed to view a meteor shower, but binoculars help focus on fainter shooting stars. A lounge chair provides comfort for all that looking up. Scan the whole sky; meteors may emerge anywhere. The Great Meteor Storm of November 1833 astonished and awed Western Hemisphere observers with an estimated 240,000 meteors emanating from Leo.

Asteroids

Rocky bodies left from the formation of the solar system about 4.5 million years ago, asteroids mainly occur in a belt between Mars and Jupiter.

KEY FACTS

A NASA-sponsored program tracks larger asteroids near Earth to gain important information.

+ fact: Asteroids are rich in metals and one day could be mined.

+ fact: Asteroids reveal details of solar system formation and planet migration.

+ fact: Asteroid names include Jabberwock and Dioretsa, a backward-orbiting asteroid.

Hundreds of asteroids are theoretically within range of even a 3-inch telescope. But asteroids are notoriously difficult to spot; their light is faint and they are indistinguishable from stars. Comparing the field of view through a telescope with a sky chart may confirm a starlike object not on the map as an asteroid. Or, you can map what you see through the telescope and compare it with the same view several hours later to determine which objects have moved. Those objects likely are asteroids. The millions of fragments in the asteroid belt collide and sometimes break free from the belt, streaking toward Earth as meteors. In February 2013, an asteroid half the size of a football field came harmlessly within 17,200 miles (27,700 km) of Earth, inside the orbits of some satellites.

Kuiper Belt

A major comet breeding ground, the Kuiper belt is a region of ice and rocky debris that begins just past Neptune and extends outward through the solar system.

KEY FACTS

The Kuiper belt has an outer extension of icy and rocky debris—the scattered disk.

+ fact: The Kuiper belt is about 20 times wider than the asteroid belt between Mars and Jupiter.

+ fact: Kuiper belt objects (KBOs) are difficult to measure because they are so distant.

+ fact: KBOs are also called trans-Neptunian objects, or TNOs.

The Kuiper (kind of rhymes with "wiper") is named for 20th-century Dutch astronomer Gerard Kuiper. It is a fertile field of debris, loaded with objects large and small that divulge information about the history of our solar system. The belt also contains a number of almost planet-size objects. Pluto is the biggest object in the Kuiper belt, at about 1,400 miles (2,300 km) wide. The demoted planet shares the belt with other dwarf planets Makemake and Haumea (Eris resides in the scattered disk). A number of Kuiper belt dwarf planet candidates are under investigation. Some of these objects seem to be in unusual orbits.

Oort Cloud

The Oort cloud is an enormous spherical cloud that houses a vast population of icy objects and encircles the solar system.

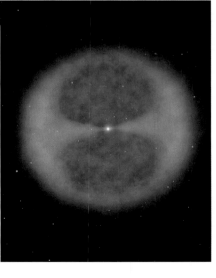

KEY FACTS

The trans-Neptunian object Sedna may be the first observed planetlike object to originate in the Oort cloud.

+ fact: The Oort cloud is so distant that telescopes have never seen it.

+ fact: There are up to 2 trillion icy rocks in the Oort cloud.

+ fact: Dislodged Oort cloud comets may get flung from the solar system.

In the mid-20th century, astronomers Ernst Öpik and Jan Oort theorized the existence beyond Pluto of an even more distant cloud of icy debris left over from the solar system's formation. Now called the Oort cloud, or sometimes the Öpik–Oort cloud, it is one of three solar system nurseries for comets, along with the Kuiper belt and its nearby scattered disk. Comets dislodged from the Oort cloud are long-period comets that can take thousands and even millions of years to orbit the sun. Short-period comets, with orbital periods less than 200 years, originate from the Kuiper belt. The Oort cloud's nearest edge lies some 465 billion miles (750 billion km) away. Its outer edge may extend 20 times farther than that, where the sun has less gravitational influence than nearby stars.

Planets

In 2006, the International Astronomical Union changed the rules for what defines a planet, demoting Pluto and ending more than seven decades of belief in a nine-planet solar system.

KEY FACTS

Current planets are Mercury, Venus, Earth, Mars, Jupiter, Saturn, Uranus, and Neptune; dwarf planets are Pluto, Ceres, Eris, Makemake, and Haumea.

+ fact: In the 1850s, the official list of planets numbered 41.

+ fact: If Earth's moon had an orbit independent of the Earth, it could be a planet.

Discoveries of new objects larger than Pluto, some with their own moons, made it necessary to rethink the definition of a planet. The International Astronomical Union established criteria an object needed to meet to be considered a planet. It must orbit a star, not be a satellite of another object, remain spherical by the pull of its own gravity, and be substantial enough to clear objects from its orbit. Pluto didn't meet the last benchmark. Currently, 13 sky objects qualify as planets or dwarf planets, 8 from the traditional list and 5 dwarf planets, and there is a waiting list of other objects under consideration. The four planets closest to the sun in the inner solar system have solid, rocky natures and are known as terrestrial planets; the four outer ones are known as gas giants.

Planet Viewing

Observing the planets can be done "old school," without gear or charts, or can involve more effort and a large array of equipment and resources.

KEY FACTS

Some planets can be observed with the naked eye; others require optics.

+ fact: Venus, brightest object in the sky after the sun and moon, is highly visible to the naked eye.

+ fact: Jupiter is also visible; its moons are starlike points through binoculars.

+ fact: Saturn can be viewed unaided; viewing its rings requires a telescope.

Both hardware and software greatly aid planet viewing. Good binoculars and a telescope with a lens aperture of 3 inches (7.5 cm) or more are basic tools to be combined with charts, almanacs, and other resources that help determine the best viewing times. The Internet is a tremendous source of detailed information; Space.com is a great place to start, with its monthly summaries and daily updates of planetary locations. Mobile apps provide another level of assistance, and many are available free or for only a small fee. Add a clear night and a location with unobstructed viewing and minimal light pollution, and you're all set. With the demand for commercial space travel, it is only a matter of time before non-astronauts can add Earth to their planet-viewing wish lists.

Mercury

Smallest and innermost of the planets, Mercury is close enough to be seen with the naked eye but is one of the more difficult planets to spot.

KEY FACTS

Mercury, slightly larger than Earth's moon, has a highly eccentric orbit at irregular distances from the sun.

+ fact: Mercury has a diameter of 3,031 mi (4,878 km).

+ fact: The planet's distance from the sun is 29–43 million mi (47–69 million km).

+ fact: Its orbital period is 88 Earth days, and its rotational period is 58.6 days.

The surface of Mercury is barren, rocky, and pocked with craters. A molten iron core makes up about 65 percent of the planet, which essentially has no atmosphere. Mercury circles the sun every 88 days, making it invisible much of the time due to competition with the sun's glare. The planet is best viewed in the west in evenings in March and April and in the east in mornings in September and October. Extremely clear viewing conditions are usually needed to see it, and it can be a challenge even with binoculars or a telescope. Consult resources such as almanacs and astronomy sites on how best to hunt for it. Mercury is named after the fleet Roman messenger god.

Venus

Neither similarities in size, mass, and orbital year with Earth nor having the goddess of love for a namesake can mask the treacherous nature of Venus.

KEY FACTS

+ fact: Earth's closest neighbor, Venus has phases somewhat similar to those of Earth's moon, but has no moon of its own.

+ fact: Venus has a diameter of 7,521 mi (12,103 km).

+ fact: The planet has a distance from the sun of 67.2 million mi (108.2 million km).

+ fact: Its orbital period is 225 Earth days, and its rotational period is 243 days.

Although farther from the sun than Mercury, Venus has a greenhouse-type atmosphere that traps the sun's heat and sends temperatures soaring to 860°F (460°C). The atmosphere is thick with carbon dioxide and is blanketed with a 40-mile (64 km) layer of sulfuric acid, which makes Venus highly reflective and bright, but hides its surface. Known as both the "evening star" and the "morning star," Venus is visible for some months each year, low to the horizon in the western sky at nightfall or rising in the east before sunrise. Telescope users generally get a better view at twilight when there is not so much contrast with the dark sky because the planet's brightness can be overpowering.

Transit of Venus

This rare phenomenon, the passing of Venus across the face of the sun, last took place in June 2012. The next transit does not occur until December 2117.

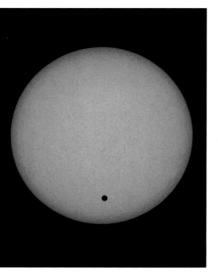

KEY FACTS

The ancient Babylonians may have recorded the transit as early as 3000 B.C.

+ fact: Captain James Cook set up an observation post in Tahiti to take transit measurements in 1769.

+ fact: In 1882–83, John Philip Sousa wrote the "Transit of Venus March."

+ fact: The June 2012 transit was celebrated with observing parties, webcasts, and music.

The alignment that places Venus directly between the Earth and sun affords a chance to see Venus as a dot that slowly moves across the face of the sun. The rare transits occur in pairs eight years apart, with more than a hundred years between pairs. In recent centuries, each transit was observed and recorded with increasingly sophisticated technology. The anticipated transit of 1769 inspired England's Royal Society to send a team, with Captain Cook at the helm of the H.M.S. *Endeavour,* to observe the transit from southern latitudes. Transits provide an opportunity to learn about the size of the solar system. NASA's webcast of the June 2012 transit of Venus as viewed from Mauna Kea, Hawaii, is archived on its website.

Mars

The stuff of alien fantasies, Mars in reality is just as fantastic—and knowable, thanks to concentrated efforts to study the red planet.

KEY FACTS

Mars has two moons called Phobos and Deimos—"fear" and "panic"—and the solar system's largest volcano.

+ fact: Mars has a diameter of 4,222 mi (6,794 km).

+ fact: Mars is 141.6 million mi (227.9 million km) from the sun.

+ fact: Its orbital period is 687 Earth days, and its rotational period is 24.6 Earth hours.

Mars and Earth share a number of similarities. A day on Mars is just a tad longer than an Earth day. Mars has seasons, an atmosphere, clouds, and polar caps, but it is much smaller than Earth. As numerous remote and on-site investigations attest, the surface of Mars is barren and a dusty red, the result of a high iron oxide, silicon, and sulfur content. Its equator is marked by immense volcanoes and dramatic canyons. Mars is visible to the naked eye, but its apparent size and brightness vary significantly during its 687-day orbit. A modest telescope reveals details such as dark lava, boulder fields, and frozen polar caps. Binoculars will show the planet's reddish color.

Mars Rovers

Mars has had more than its fair share of scientific inspection in the form of flyby probes, orbiters, landers, and rovers, including the latest—the Curiosity rover.

KEY FACTS

The Mars Pathfinder spacecraft delivered rover Sojourner to Mars in 1997.

+ fact: Rover Opportunity, originally expected to travel 2,000 ft (0.6 km), has racked up more than 22 mi (35 km).

+ fact: Rover Opportunity's twin, Spirit, went off-line in March 2010.

+ fact: Rover Curiosity is about the size of a Volkswagen Beetle.

Robotic vehicles festooned with scientific equipment provide invaluable information about the surface of Mars. The rover Opportunity was deployed in 2004, and was expected by some to survive there only 90 days; it completed its ninth year on Mars in January 2013, providing an ongoing bonus of data. The Curiosity rover, a car-size vehicle carrying a payload of ten of the most sophisticated science instruments ever designed, landed there in August 2012. It is looking at the planet's geology, past planetary processes, surface radiation, and biological potential. In early 2013, it sent back the first nighttime photos of Mars. Yet much of the excitement of Curiosity's mission involves the evidence it has detected of water-bearing minerals in the planet's rocks.

Asteroid Belt

A gap of 340 million miles (547 million km) between Mars and Jupiter, the asteroid belt marks the division between the solar system's terrestrial planets and its gas giants.

KEY FACTS

The dwarf planet Ceres comprises one-third of the mass in the asteroid belt.

+ fact: Jupiter's orbit contains a group of asteroids known as the Trojan asteroids.

+ fact: Earth has its own small Trojan asteroid belt located at about 60° east or west of the sun.

+ fact: Each of the Beatles has an asteroid named for him.

Jupiter's massive gravitational field prevents the rocky debris in the asteroid belt—leftovers from the formation of the solar system—from coalescing into planets. The asteroids in the belt are like a warehouse of raw materials that under different circumstances could have made the transition to full-fledged planets. The debris won't achieve that destiny, but larger objects in the belt, such as Ceres, have the chance to be considered dwarf planets. Astronomers have cataloged more than 120,000 bodies in the asteroid belt, and more than 13,000 are named, sometimes very personally for girlfriends and favorite writers. According to the guidelines set out by the International Astronomical Union, near-Earth objects are supposed to have names taken from the mythology of any world culture.

Jupiter

A planet of superlatives, Jupiter is the largest by far and has the largest number of satellites with 65 moons, making it a kind of solar system unto itself.

KEY FACTS

Jupiter has a ring, but it is small and thin compared to other planets' rings.

+ **fact:** Jupiter has a diameter of 88,846 mi (142,984 km).

+ **fact:** The planet's distance from the sun is 483.7 million mi (778.4 million km).

+ **fact:** Its orbital period is 11.9 Earth years, and its rotational period is 9.9 Earth hours.

Jupiter has a massive influence in the solar system with its strong magnetic field and gravitational pull, intense emissions of radio waves, and bursts of radiation. It has a metallic core but otherwise is a swirling mass of gases: hydrogen, helium, methane, and ammonia. Its complete rotation lasts just short of ten hours; at this clip, the atmosphere appears to us as faint pastel bands. Persistent storms, including the Great Red Spot, a perpetual high-pressure zone, add to the spectacle. At the other extreme, Jupiter takes nearly 12 years to orbit the sun, spending nearly a year in each constellation of the zodiac and becoming an easy target for sky watchers.

Jupiter's Moons

In 1610, Galileo spotted the four largest moons of Jupiter with his telescope. Io, Callisto, Ganymede, and Europa became known as the Galilean moons.

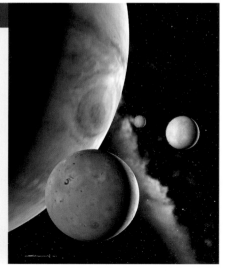

KEY FACTS

Jupiter has a chart-topping 65 moons; presently, only 51 of them have names.

+ **fact:** Callisto is the most heavily cratered object in the solar system and lacks a molten core.

+ **fact:** Moon Io is the most volcanically active object in the solar system.

+ **fact:** Many of Io's more than 400 volcanoes erupt almost continuously.

Jupiter's intense gravitational pull, second only to the sun's, has lured many objects into its orbit. It governs these moons like the masterful planet that it is, and its four largest moons do the planet proud. Jupiter's Galilean moons are large enough that if they orbited the sun, they would qualify as planets. Among them, Ganymede is the largest moon in the solar system, larger than Mercury, Ceres, Pluto, and Eris combined. It generates its own magnetic field. Europa may harbor a saltwater ocean under its icy surface, and it may have twice as much water as Earth does. The Galilean moons can be viewed easily from Earth with binoculars. In orbit, all the moons keep the same face toward Jupiter, just as Earth's moon does.

Saturn

Gas giant Saturn, the second largest planet, is best known for its rings, but it has also acquired a multitude of moons. Despite its massive dimensions, Saturn is less dense than water.

KEY FACTS

Saturn with its rings is the most recognizable planet.

+ fact: Saturn has a diameter of 74,898 mi (120,536 km).

+ fact: Its distance from the sun is 885.9 million mi (1.4 billion km).

+ fact: Saturn's orbital period is 29.4 Earth years, and its rotation period is 10.7 Earth hours.

Known to the ancient Mesopotamians, Saturn was dubbed "the old sheep" for taking more than 29 years to orbit the sun. The plodding orbit is countered by speedy rotation; a day on Saturn lasts just under 11 hours. And although the planet's volume could contain 763 Earths, its low density means that it would float in water if a big enough basin could be found. It also means that the planet appears considerably squashed in profile; the diameter through the poles is significantly shorter than its equatorial diameter. Saturn is visible to the naked eye, but binoculars or a telescope are needed to view details. The Cassini orbiter has captured some amazing images.

Saturn's Rings

Saturn's iconic rings are composed of rubble, dust, and ice. The rings may also contain the pulverized remains of moons, comets, and asteroids that strayed too close.

KEY FACTS

Galileo viewed Saturn through his telescope in 1610, but he thought the side bulges he saw were separate bodies.

+ fact: Saturn's rings vary in width and thickness; they extend 170,000 mi (274,000 km) from side to side.

+ fact: Saturn's rings were named alphabetically in the order of their discovery.

+ fact: Saturn's rings may disappear.

A ring around Saturn was first proposed by Dutch astronomer Christiaan Huygens in 1655. We now know that Saturn has multiple rings fashioned from billions of icy particles—some as small as grains of sand and others as big as buildings. The rings are complex and not solid, containing ringlets and gaps between them. The appearance of the rings, visible with a small telescope, changes over time. When the tilt of Saturn is toward Earth, the rings are visible from a broad top-down view. Roughly every 14 years, the rings are tilted edgewise toward Earth, making them all but disappear from view—an event that will next occur in 2023. The Cassini-Huygens spacecraft is named in part for the Dutch astronomer.

Saturn's Moons

Saturn's impressive assemblage of moons includes many that exhibit unusual features. Only 53 of the 62 known moons currently have names.

KEY FACTS

Saturn's largest moon is the massive Titan, larger than the planet Mercury.

+ fact: Titan's atmosphere contains more than 90 percent nitrogen.

+ fact: Moons Titan and Enceladus have the potential for life.

+ fact: Some of Saturn's moons share orbits, traveling the same paths.

Among Saturn's fascinating moons, Titan has surface liquids in the form of two large methane lakes. The poles of Enceladus shoot geysers of water vapor and ice particles that suggest liquid water beneath its frigid surface. Moon Mimas sports a large impact crater of unknown origin. Several moons appear to wrangle the billions of ice particles that comprise Saturn's rings. These "shepherd moons"—Pan, Atlas, Pandora, and Prometheus—straddle two of the rings and keep them intact. Much of what we know about Saturn, its moons, and its rings comes from the data being collected by the Cassini orbiter since its arrival at Saturn in 2004. Cassini transmits many raw images that are made available to the public before they are calibrated and analyzed by NASA scientists.

Uranus

Uranus is a tilted planet, spinning on its side at an angle of 98° to its orbital plane, possibly the result of a massive collision.

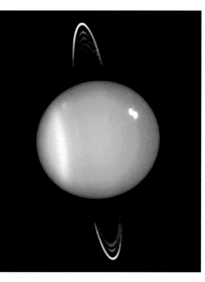

KEY FACTS

Uranus has a retrograde rotation, revolving in the opposite direction of most other planets.

+ fact: Uranus has a diameter of 31,764 mi (51,119 km).

+ fact: The planet's distance from the sun is 1.8 billion mi (2.9 billion km).

+ fact: Its orbital period is 84.02 Earth years, and its rotation period is 17.24 Earth hours.

Distant Uranus, barely visible to the naked eye, was incorrectly identified as a star by ancient astronomers before its discovery in 1781 by astronomer William Herschel. Spotted more realistically through binoculars, Uranus appears as a small blue-green disk, the color due to methane gas in its upper atmosphere. Uranus is considered an ice giant, with temperatures reaching down to –357°F (–216°C). Uranus's tilted rotation means that the planet's poles point toward the sun: first one pole and then, 42 years later, the other. In addition to 5 large moons and more than 20 smaller ones, 13 narrow rings surround Uranus. The ring system was imaged by the Hubble Space Telescope.

Neptune

Named for the Roman god of the sea, Neptune displays a vivid blue color, a product of the planet's surrounding clouds of icy methane.

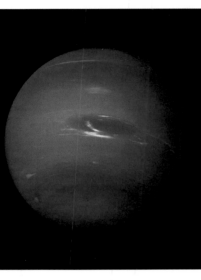

KEY FACTS

Neptune's moon, Triton, has a retrograde orbit, orbiting opposite the direction of Neptune's rotation.

+ fact: Neptune has a diameter of 30,776 mi (49,529 km).

+ fact: The planet's distance from the sun is 2.8 billion mi (4.5 billion km).

+ fact: Its orbital period is 164.79 Earth years, and its rotational period is 16.11 Earth hours.

The outermost major planet in the solar system, Neptune is an ice giant similar in size to Uranus, although denser. A deep atmosphere containing hydrogen, helium, and a small amount of methane surrounds it. The planet seems to have an internal heat source and a puzzling tendency toward very turbulent weather, despite its great distance from the sun. Winds on Neptune can reach more than 1,300 miles (2,100 km) an hour. Neptune is surrounded by a system of rings that are faint and lumpy, and appear to be younger than the planet itself. The rings could be the remnants of a former moon pulled apart by Neptune's gravity.

Dwarf Planets & Plutoids

Dwarf planets are almost planets, which tend to be small and roundish and travel with debris in their orbits. Plutoids are dwarf planets in the trans-Neptunian Kuiper belt.

KEY FACTS

Haumea is named for a mythical Hawaiian sorceress; its moons are named for her daughters.

+ fact: Haumea is sometimes called the "cosmic football" for its shape and motion.

+ fact: Makemake is named for the chief god of Easter Island in the South Pacific.

+ fact: Eris was originally called Xena after the TV warrior princess.

When Pluto was demoted in 2006, it was put into a category of dwarf planet. Besides Pluto, several other objects in the Kuiper belt qualify. Generally, dwarf planets are larger than Mercury, roundish, and are not satellites of other objects. Haumea, a rocky, egg-shaped object, tumbles end over end as it spins every four hours. Makemake seems to have an atmosphere containing nitrogen and methane. It shines red and has no moon. The orbit of Eris and its moon, Dysnomia (Greek for "lawlessness") takes it 10 billion miles (16 billion km) beyond the Kuiper belt. Dwarf planet Ceres is not a plutoid; it appears in the interior asteroid belt beyond Mars. Astronomers are investigating hundreds of objects to determine whether the objects qualify for dwarf planet status.

Pluto

Pluto enjoyed 76 years of planetary status until, one day in 2006, it didn't. Demoted to dwarf planet, the celestial body named for the god of the underworld retains legions of fans.

KEY FACTS

Some astronomers consider Pluto and its moon Charon to be a double planet.

+ fact: Pluto has a diameter of 1,430 mi (2,301 km).

+ fact: The planet's distance from the sun is 3.7 billion mi (5.9 billion km).

+ fact: Its orbital period is 247.92 Earth years, and its rotational period is 6.38 Earth days.

In 1930, an assistant at the Lowell Observatory found a long-anticipated planet beyond Neptune. Excitement about the newbie generated a naming contest, which an 11-year-old British girl won. She proposed the name of the Greek god of the underworld that coincidentally was the name of Mickey Mouse's popular pup. By 2006, the International Astronomical Union, or IAU, had its doubts about Pluto and demoted it to the status of dwarf planet. Pluto did not meet one of the prerequisites the IAU had drawn up: It didn't clear its orbit of debris. Pluto now lends its name to dwarf planets that inhabit the Kuiper belt—the plutoids. Many of Pluto's devoted fans mourn its demotion.

Ceres

Ceres, a dwarf planet, rose out of the asteroid belt between Mars and Jupiter. Since its discovery in 1801, Ceres has worn several solar system hats: planet, asteroid, and now dwarf planet.

KEY FACTS

Ceres has a rocky inner core and a mantle of water ice, and is one of two dwarf planets without a moon.

+ fact: Ceres's diameter is 590 mi (950 km).

+ fact: The dwarf planet's distance from the sun is 258 million mi (415 million km).

+ fact: Its rotational period is 9.075 Earth hours.

In 2006, Ceres was promoted to the new group of dwarf planets. About one-fourth the diameter of Earth's moon, Ceres is the largest object in the asteroid belt, an area holding thousands of objects that are remnants left from the formation of the solar system. Dwarf Ceres is the fifth planet from the sun and is classified among the terrestrial planets because of its rocky nature. It is suspected of having water ice under its crust because it is less dense than Earth and its surface shows evidence of water-bearing minerals. Named for the Roman goddess of agriculture, Ceres takes 4.6 Earth years to complete one orbit around the sun. Ceres was discovered by Italian astronomer Giuseppe Piazza.

The Universe

The word "universe" evokes all the superlatives one can imagine, only to take on more with each breakthrough in cosmic understanding enabled by sophisticated technology.

KEY FACTS

Many physicists and astronomers believe that our universe is just one of an unknown number of universes.

+ fact: The age of our universe is about 13.7 billion years old.

+ fact: The first stars were formed about 100 million years after the big bang.

+ fact: The first galaxies formed less than 600 million years after the big bang.

The unfolding story of the universe involves greater stretches of the imagination than most science fiction tales could devise. Consider the plot: An infinitely small point of infinite density and infinite gravity erupts to create both space and time, unleashing radiation that eventually forms matter that eventually forms stars, galaxies—and us. The initial radiation of the universe's chaotic beginnings was left by hot plasma (ionized gas) that emitted a microwave signal we can detect by space probes such as the Wilkinson Microwave Anisotropy Probe. The probe accomplished a full-sky survey of background microwave radiation between 2001 and 2010, which scientists scrutinize for information about this phenomenon.

Big Bang

More of an expansion than an explosion, the big bang set in motion the formation of the universe about 13.7 billion years ago.

KEY FACTS

Events immediately after the big bang occurred at an incomprehensibly fast speed.

+ fact: In 10^{-35} seconds after the big bang, its energy turned into matter.

+ fact: In 10^{-5} seconds after the big bang, the universe's natural forces took shape.

+ fact: In 3 seconds after the big bang, the nuclei of simple elements formed.

The term "big bang" was coined in the 1950s by British astronomer Fred Hoyle. Ironically, he used it derisively, but it stuck. Current thinking holds that the universe began as a singularity, a point of infinite gravity, where space, time, and all subsequent matter and energy were contracted into an object without size. At the point in which contents of the singularity escaped, a dramatic expansion began that has played out over billions of years. Radiation from the initial event persists as detectable cosmic background radiation. The origin of the singularity remains scientifically elusive, but predicted scenarios of the universe's demise include a big crunch, a big chill, or a big rip. Astronomers have a few tens of billions of years to look for answers.

Galaxy

Gravity holds together the vast assemblage of stars, interstellar dust and gas, and dark matter (little understood matter that emits no light) that form a galaxy.

KEY FACTS

There may be one hundred billion galaxies in the universe.

+ fact: Galaxies occur in clusters throughout the universe.

+ fact: Galaxy clusters form larger associations, giving the universe a "clumpy" nature.

+ fact: The most distant galaxy yet identified is more than 13 billion light-years away.

Shape determines galaxy classification. Spiral galaxies, by far the most numerous, rotate around a bright nucleus. Arms spiral out from that point. Elliptical galaxies may be nearly spherical or stretched to an oblong; stars in them are mostly older red giant stars. Irregular galaxies, which lack a coherent shape, are amorphous collections of stars that include star-forming nebulae. The most recently identified type is the starburst galaxy, so-called because the stars in it seem to burst out. Galaxies are so massive that the strength of one can rip apart another, and they can collide even in the relative emptiness of space. The Hubble Space Telescope's website (*www.hubblesite.org*) features an extensive gallery of amazing, downloadable photos that demonstrate galaxy diversity—and beauty.

Milky Way

Home sweet home, the Milky Way galaxy contains our solar system, just a tiny enclave in the vast spiral galaxy of several hundred billion stars.

KEY FACTS

The Milky Way is composed of some 200 to 500 billion stars.

+ fact: Our solar system is located about 25,000 light-years from the galaxy core.

+ fact: The Milky Way's core contains a supermassive black hole.

+ fact: In ancient times, the Milky Way was a more prominent feature in the sky and many myths developed around it.

Thought to be about 13 billion years old, the Milky Way is a bit younger than its oldest stars, which are estimated to be about 13.5 million years old. The galaxy's disk shape bulges at the center with orange and yellow stars. A corona of old stars and globular clusters extends above and below the disk. The younger, brighter stars and nebulae crowd the arms of the galaxy. The entire spiral structure spins, completing a rotation approximately every 200 million years. Overall, the Milky Way has a diameter of about 100,000 light-years; its thickness ranges from 1,000 to 13,000 light-years. Dark nights and clear skies give the best chance of viewing the Milky Way. In the Northern Hemisphere, the summer months usually offer the best views of the galaxy.

Andromeda Galaxy

A faint oval smudge in the gap between the constellations Cassiopeia and Andromeda, the Andromeda galaxy is the most distant object visible to the naked eye.

KEY FACTS

Astronomer Edwin Hubble's study of the Andromeda galaxy was the first confirmation of the existence of galaxies other than the Milky Way.

+ fact: The light we see from the Andromeda galaxy left the galaxy 2.5 to 3 million years ago.

+ fact: The galaxy has a magnitude of 4.4, making it one of the brighter Messier objects.

Possessing a spiral shape like the Milky Way, the Andromeda galaxy is the larger of the two, containing an estimated one trillion stars. Nevertheless, recent observations indicate that the Milky Way may contain greater mass, if dark matter is included. The Andromeda galaxy is accompanied by several smaller galaxies, including M32 and M110, visible through binoculars. Under good conditions, especially during autumn, the Andromeda galaxy is visible to the naked eye. Astronomers as far back as at least A.D. 964 observed and recorded the feature. Many observers comment that its beauty compares to its namesake, the mythical princess Andromeda. A distance "only" about 2.5 million light-years away turns the Andromeda galaxy into the girl next door.

Galaxy Collisions

Despite the general emptiness of space, galaxies have a tendency to move toward each other and collide, a mechanism that allows galaxies to increase in size.

KEY FACTS

A merger from a galaxy collision can take millions of years.

+ fact: When a large galaxy merges with a smaller one, the larger one usually keeps its shape.

+ fact: Galaxy collisions cause friction that can trigger shock waves that lead to star formation.

+ fact: Galaxy collisions provide vital information about galaxy evolution.

In the scheme of the universe, galaxy collisions occur with different frequencies. More often, larger galaxies collide with dwarf galaxies and incorporate the mass of the dwarf galaxies into their own. Collisions of larger galaxies happen less frequently. Collisions can produce a galaxy of a different shape; two spiral galaxies, for example, can form an elliptical galaxy. The Hubble Space Telescope has revealed details of the approaching collision—4 billion years from now—between the Milky Way and the Andromeda galaxy. The news seems good for our solar system: displacement, perhaps, but not destruction. This future merger of the two galaxies sometimes is referred to as Milkomeda or even the Andromeda Way.

Nova

A spectacular sudden surge in the brightness of a star, a nova is the equivalent of a supermassive atomic explosion, but one from which the star often can recover.

KEY FACTS

Novae are not often spotted, but when they are, you can find them with binoculars along the Milky Way.

+ fact: At times, a nova can be viewed with the naked eye.

+ fact: The density of a white dwarf allows it to draw in hydrogen from the larger star.

+ fact: A classical nova erupts only once.

Typically, a nova occurs when stars in a binary system reach different points in their stellar evolution, with one star collapsed into a dense white dwarf and the other in a red giant phase. The white dwarf siphons off hydrogen from the red giant, and the temperature and pressure rise until the gas erupts in a thermonuclear explosion. This causes the white dwarf to brighten for a period of hours or days by as many as ten magnitudes before dimming over a period of months as the remnants of the blast dissipate. Novae are not necessarily fatal to the stars involved and often occur in cycles in some binary pairs. The Keck Interferometer, a two-telescope system at the W. M. Keck Observatory in Mauna Kea, Hawaii, has captured invaluable data about a nova in the constellation Ophiuchus.

Supernova

Unlike the nova, a supernova is a one-time event. It signals the death of a star, an irrevocable transformation of its material.

KEY FACTS

A supernova can briefly release as much energy as all the stars in the Milky Way.

+ fact: A supernova creates raw material for the formation of new stars.

+ fact: The Crab Nebula (M1) in the constellation Taurus is the remnant of an event in 1054.

+ fact: The brightest supernova in recent history occurred in 1987.

Like the nova, a supernova can also occur in a white dwarf if its core takes in so much hydrogen that it implodes. A supernova also announces the end stage for a red supergiant when its core implodes from intense heat and pressure, unleashing a violent shock wave that blasts away the surrounding cloud of gas. The supernova was first recorded by the Chinese in A.D. 185. Supernovae again were recorded in 1006, 1054, and 1572, and in the modern era. The bright flash of a supernova can occasionally be viewed by the naked eye, but usually requires an 8-inch or larger telescope. More than 25 years after its explosion in the Large Magellanic Cloud, Supernova 1987A continues to glow as it transforms into a supernova remnant, which the Hubble Space Telescope has imaged.

Black Hole

A black hole provides a one-way ticket to visual oblivion for the contents of a supernova explosion. The resulting gravitational field collapses in on itself, allowing no light to escape.

KEY FACTS

The point of no return at the edge of a black hole is called the event horizon.

+ fact: The energy from stellar gases swirl into a black hole like a whirlpool.

+ fact: The black hole in a galaxy in the constellation Virgo seems to have the density of 3 billion suns.

+ fact: The Chandra X-ray Observatory provides critical data on black holes.

Born in one of the universe's most dramatic and awe-inspiring events, a black hole is one of the two options that follow a supernova explosion. The last explosive gasps of a red supergiant create a gravitational field so intense that the force of gravity causes matter to collapse inward. This forms the black hole, in which distance between subatomic particles reduces to zero, and density and gravity expand to their maximum expression. Black holes affect the stars and gas around them, which is one way they can be detected. Small black holes formed at the creation of the universe, and supermassive black holes formed when the galaxies they inhabit were formed. Supermassive black holes spin furiously as they swallow stars and merge with other black holes.

Quasar

The bright light of a quasar is fed by a black hole in the form of a brilliant stream of energy that a black hole releases as it consumes the contents of nearby stars.

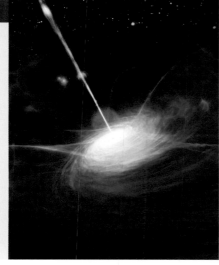

KEY FACTS

Coined in the 1960s, the word "quasar," comes from "quasi-stellar," as in quasi-stellar radio source.

+ fact: More than 120,000 quasars have been identified.

+ fact: Quasars are among the universe's very ancient objects.

+ fact: The Sloan Digital Sky Survey catalogs positions and absolute brightness of celestial objects such as quasars.

Bright, deep sky objects with a starlike appearance, quasars puzzled astronomers from the time they were first detected from their radio emissions. When spotted, quasars seemed too bright to be so far away. A boost from association with a black hole was proposed for the anomaly. When quasars were found to be abundant, they came to be considered a part of normal galactic evolution. Though distant, a quasar is within the reach of a skilled amateur astronomer. The quasar 3C 273 can be spotted some 3 billion light-years away through an 8-inch-aperture telescope just 5 degrees northwest of the figure's head in Virgo. Sometimes, though, even the Hubble Space Telescope can struggle to observe a quasar embedded in a galaxy obscured by a large amount of cosmic dust.

Binary & Multiple Stars

"Twinkle, twinkle, little stars"—plural—might be more appropriate, because about 80 percent of stars exist with a companion or companions.

KEY FACTS

The star pair Mizar and Alcor in Ursa Major are about 3 light-years apart and are not gravitationally linked.

+ fact: Mizar is actually a double binary, and Alcor is a binary, making a 6-star system.

+ fact: Binary stars, each star with its own orbiting planet, have been discovered.

+ fact: Binary stars, each with multiple planets in orbit, have also been discovered.

Our singleton sun aside, the massive gas clouds that nurture stars usually produce them in pairs or greater multiples that remain gravitationally attached to each other throughout their lives. Binary stars, triples, and larger groups are the norm, whether the companion is a twin of similar size and luminosity or a group of small siblings that stick close to a more dominant star. Double stars can be optical doubles, which only appear to be close, or true binary stars, with two stars orbiting a common center of gravity. Powerful binoculars or a telescope are often needed to pick out the individual members of a binary pair. Alpha Centauri is a binary star that seems to be bound to Proxima Centauri, which is far from Alpha Centauri but closer to us, as a so-called wide binary.

Nebula

Giant clouds of stellar gas, nebulae provide the raw materials for stars, planets, and galaxies. They are present at both star birth and star death.

KEY FACTS

Despite the name, a planetary nebula does not have an association with planets.

+ fact: Dark nebulae are so dense that they hide the light of stars within them, not only behind them.

+ fact: The Helix Nebula is the closet nebula to Earth.

+ fact: The North American Nebula is a large emission nebula in the constellation Cygnus.

Various types of nebulae represent those giant gas clouds in their different and dynamic roles and relationships with stars. Emission nebulae are star-forming clouds energized by the young stars within them, such as the middle star of Orion's belt. Planetary nebulae are formed of the gas departing from a dying red giant, as seen in the Ring Nebula in the constellation Lyra. Dark nebulae are collections of interstellar dust that obscure the light of the stars behind them. Reflection nebulae shine from the light of nearby stars. Many nebulae are visible with a 4- or 6-inch telescope, maximized with higher magnification and filters. The Orion Nebula, which is 1,500 light-years from Earth, is considered a textbook example of a planetary nebula, showing a number of different star-forming processes.

Eagle Nebula

The Eagle Nebula, located in the tail portion of the constellation Serpens about 7,000 light-years away, is an active region of star formation.

KEY FACTS

The dark towers are 56 trillion mi (90 trillion km) high.

+ fact: Images from the Chandra X-ray Observatory show that the pillars are low in x-ray content.

+ fact: The scant content perhaps signals the pillars' end.

+ fact: Further study suggests that a supernova destroyed the pillars 6,000 years ago, but evidence hasn't yet reached Earth.

The stellar nursery in the Milky Way known as the Eagle Nebula is actually a combination nebula and star cluster. Philippe Loys de Cheseaux discovered it between 1745 and 1746, as did Charles Messier independently in 1764. The nebula has dark pillars of dense material that rise at its center and can be seen with a 12-inch telescope. The Hubble Space Telescope captured these pillars in a now iconic image known as the Pillars of Creation. Newborn stars sculpt the pillars by burning away some of the gas within the nebula. The nebula in general can be viewed with low-powered telescopes and even binoculars. But to view the Eagle Nebula in all its dramatic glory, you will want to check out the online images available on HubbleSite (*www.hubblesite.org*).

Exoplanets

Exoplanets, short for extrasolar planets, are celestial bodies beyond our solar system that orbit stars and share other characteristics of known planets.

KEY FACTS

More than 2,300 exoplanets have been identified, with many more expected.

+ fact: Exoplanets include "hot Jupiters," gas giants that orbit closer to their stars than Mercury does to the sun.

+ fact: Hot Jupiters are also known as "roaster planets."

+ fact: Super-Earths refer to rocky exoplanets about twice the mass of Earth.

Radio astronomers discovered the first extrasolar planet in 1991. Since then, ground-based and space-based telescopes have confirmed the presence of hundreds of objects that orbit stars of other systems or that freelance as nomads—planets "kicked out" of their star systems. Various scientific organizations use different criteria to confirm an exoplanet's status, coming up with different totals, but all concede that the numbers are increasing. The ultimate find among exoplanets would be an Earthlike body in the habitable zone of its solar system that not only had a potential for life, but also signs of life itself. To keep up with the latest exoplanet news, visit the PlanetQuest website (*planetquest.jpl.nasa.gov*). It documents exoplanet research at the Jet Propulsion Laboratory.

Kepler Space Telescope

Launched in 2009, just south of Cape Canaveral, the Kepler Space Telescope is on a mission to monitor more than 150,000 stars, looking for Earthlike exoplanets.

KEY FACTS

The Kepler instrument is a photometer (light meter) telescope with a large field of view.

+ fact: The telescope observes a large field of stars over time, continuously monitoring the brightness of all.

+ fact: Kepler has found a 6-planet system orbiting an 8-billion-year-old star at 2,000 light-years away.

+ fact: The Kepler mission has been extended to 2016.

The Kepler Space Telescope scrutinizes our portion of the Milky Way galaxy for extrasolar planets and the stars they orbit. The big prize would be an Earthlike planet technically capable of supporting life (known as a "just right Goldilocks" planet), but all kinds of exoplanets are given attention: gas giants, hot super-Earths with short-period orbits, and ice giants. Kepler finds potential candidates through the concept of the transit—a planet crossing in front of its star from the observer's point of view. When this happens, the star's brightness dims by a factor related to the size of the object. A periodic occurrence at four equal intervals confirms an object as a planet candidate. Follow-up observations are made to eliminate other possibilities before the planet is verified.

Dark Energy

Einstein rightly realized that space was not just empty space. Though only inferred, dark energy may be the "dark force" involved in rapid expansion of the universe.

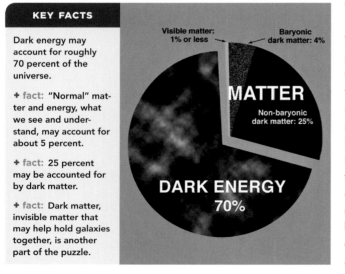

KEY FACTS

Dark energy may account for roughly 70 percent of the universe.

+ fact: "Normal" matter and energy, what we see and understand, may account for about 5 percent.

+ fact: 25 percent may be accounted for by dark matter.

+ fact: Dark matter, invisible matter that may help hold galaxies together, is another part of the puzzle.

Too much was going on in the universe for there not to be something in space besides space. When a study of supernovae showed an unexpected acceleration of the expansion of the universe, some force had to be held accountable. That force has been designated as dark energy, a concept that creates more questions than it answers at present. It can be seen as the reverse of gravity: Dark energy pushes objects away from each other rather than pulling them together and coalescing them. How it operates keeps astrophysicists in a quandary, because it calls into question the laws of gravity and the theory of relativity. Observations by the Hubble Space Telescope in 1998 set this discussion in motion, but a lot more data obviously are needed to arrive at some answers.

Constellation

Over the millennia, stargazers have teased out shapes of humans, animals, and objects from patterns of stars in the night sky. A recognized star pattern is called a constellation.

KEY FACTS

Ptolemy's list of constellations appeared in his astronomical treatise known as the *Almagest*.

+ **fact:** The name "Almagest" comes from a later Arabic title for the book meaning "the greatest."

+ **fact:** The constellation Orion, visible on the celestial equator, has inspired many elaborate mythologies.

The International Astronomical Union (IAU) recognizes 88 official constellations. These include the original 48 constellations listed by Ptolemy in the second century A.D., which have names and backstories rooted in Greco-Roman mythology. The Greek constellation Argo Navis has since been divided into three separate constellations. The list also includes a series of "modern" constellations that have been added since about 1600 to fill in gaps, especially in the southern sky. According to the IAU, a constellation refers not just to the star pattern itself, but also to an agreed-upon, bounded segment of the sky in its vicinity. The convention helps us locate objects in the night sky and allows astronomers to convey information about deep-space objects that are not part of a constellation.

Asterism

When is a constellation not a constellation? When it's an asterism, a small group of stars that makes a recognizable, attention-grabbing shape in the sky.

KEY FACTS

Many asterisms feature distinct geometric shapes.

+ **fact:** The Summer Triangle asterism is a geometrically interesting 30°-60°-90° triangle.

+ **fact:** The westernmost stars of the Great Square of Pegasus point southward to the star Fomalhaut.

+ **fact:** Late winter skies offer the Great Hexagon asterism.

A look at the official list of 88 constellations will show that some prominent celestial names, like the Big and Little Dippers, are missing. These well-known star formations are asterisms, small groups of stars that have a distinct shape and may form part of a constellation but are not constellations themselves. The iconic dippers, for example, are contained within Ursa Major and Ursa Minor, respectively. Asterisms are useful points of orientation in the sky and may include stars from multiple constellations. The Winter Triangle, for example, connects the alpha stars Betelgeuse of Orion with Procyon in Canis Minor and Sirius in Canis Major. Similarly, the Summer Triangle asterism connects the bright stars Vega in constellation Lyra, Deneb in Cygnus, and Altair in Aquila.

Cluster

Two kinds of star clusters appear in the deep sky. These dramatic formations are sometimes visible to the naked eye, but a telescope reveals their scale and complexity.

KEY FACTS

In an open cluster, the individual stars often can be resolved through a telescope.

+ fact: Individual stars often escape open clusters and travel on their own.

+ fact: Globular clusters are prominent in the central bulge of the Milky Way.

+ fact: Many of the globular clusters in our galaxy have highly eccentric orbits.

Star clusters are classified as either open or globular clusters. Open clusters are loose concentrations of a few to several thousand associated stars that are bound weakly by gravity. They also share a common star birth "cloud," or source of raw materials. A good example is the Pleiades in the constellation Taurus, a young open cluster that will eventually disperse over time. Another type is the globular cluster, a massive star association that may contain up to a million stars and may be up to 13 billion years old. About 150 of them have been identified, mostly in the outer reaches of the Milky Way. Star clusters in proximity to each other appear to merge, creating larger clusters. Astronomers hope that the James Webb Space Telescope will reveal details of such interactions.

The Pleiades

Any Subaru owner should recognize this asterism, which appears in stylized fashion as the logo for the Japanese auto manufacturer named for the star cluster.

KEY FACTS

The Pleiades cluster (M45) contains hundreds of stars; the 6 or 7 brightest stars make up the famous Seven Sisters asterism, most of them visible to the naked eye.

+ fact: months best viewed: January–February

+ fact: seasonal chart location: Winter, southwest quadrant

+ fact: noted sky-mark: Taurus

Also known as the Seven Sisters, the Pleiades cluster is one of the most easily identified objects in the night sky. A line traced from alpha stars Betelgeuse in Orion through Aldebaran in Taurus brings the asterism into view. The Pleiades' seven bright stars feature prominently in many world mythologies; they get their names from seven sisters in Greek mythology, daughters of the Titan Atlas and the Oceanid Pleione. Zeus set them in the sky to thwart Orion's advances. The brightest star is Alcyone, with a magnitude of 2.86. A Kiowa Indian tale tells of seven sisters chased by giant bears onto a tall rock and then whisked into the sky. The rock is Devil's Tower and its long cracks are the claw marks of the frustrated bears.

Big Dipper

The first asterism many people learn to locate, the Big Dipper is an abiding feature of the northern sky in all seasons, because of its proximity to the celestial North Pole.

KEY FACTS

This 7-star asterism, which forms a distinct ladle or dipper, is incorporated into the constellation Ursa Major. Its orientation changes according to time of the year.

+ fact: months best viewed: March–April

+ fact: seasonal chart location: Spring, center of chart

+ fact: noted skymark: Ursa Major

To locate the Big Dipper, face the north horizon. As it circles the celestial pole during the year, the asterism faces downward in spring and upward in autumn. The Big Dipper also points to its counterpart in Ursa Minor, the Little Dipper. Follow an imaginary line from the front edge of the Big Dipper's ladle to find Polaris on the end of the Little Dipper's handle. Visible year-round, the Big Dipper served as a constant guide for escaped slaves traveling along the Underground Railroad and is mentioned in many stories and songs, including "Follow the Drinking Gourd." It is known as the Plow in Britain and the Great Cart in Germany. In India it is the Seven Sages. Cultures in much of the world have given locally significant names to this familiar asterism.

Polaris

Not the brightest star in the sky—as many assume—Polaris is actually the 50th brightest star, but its year-round presence makes it an invaluable "skymark."

Polaris

KEY FACTS

Polaris is a yellow supergiant star that varies in brightness throughout the year. It is part of a triple star system, although its other two stars are not visible to the naked eye.

+ fact: months best viewed: Year-round

+ fact: seasonal chart location: All star charts

+ fact: noted skymark: Little Dipper

Located directly above Earth's northern axis, Polaris appears to be the star around which the rest of the sky revolves. The stars in the front edge of the Big Dipper asterism line up with it, making the North Star, or polestar, easy to find. Due to the wobbling of the Earth on its axis as an effect of gravitational pull from the sun, moon, and planets—a motion known as precession—the Earth's pole points to different stars over time. The pole will be closest to Polaris about the year 2100, and after that will move through a series of new polestars and reach Vega in Lyra about 12,000 years from now. In the southern sky, Beta Hydri in the southern constellation Hydrus (different from the northern sky constellation Hydra) is the closest bright star to the celestial south pole.

Northern Sky

As most of the world's population lives in the Northern Hemisphere, the northern sky has gotten the lion's share of attention, from the ancient astronomers onward.

KEY FACTS

Northern sky constellations are named mostly for figures in Greco-Roman mythology.

+ fact: Some constellations are visible in the Northern Hemisphere all year.

+ fact: Ursa Major is an example of a north circumpolar constellation.

+ fact: Others near the celestial equator are visible in northern and southern skies.

When Greek astronomer Ptolemy codified the constellations in the second century A.D., he followed traditions handed down from ancient Mesopotamian observers who had watched the skies at about latitude 35° N. They would have seen the northern sky up to the celestial north pole and, in theory, a good bit of the southern sky, although objects near the horizon are difficult to see. Constellations and features farther south would have been blocked by Earth. This partially accounts for the fact that ancient astronomers limited their constellation list to 48, although they saw some of the southern constellations some of the time. The "Tonight's Sky" feature at HubbleSite (*www.hubblesite.org*) offers a preview of constellations, deep sky objects, planets, and events visible each month in the skies of the Northern Hemisphere.

Southern Sky

Not fully appreciated until voyages of exploration sailed southern routes beginning in the 1500s, the southern sky harbors extraordinary sights.

KEY FACTS

Crux, or the Southern Cross, is the smallest in area of all the constellations.

+ fact: The Southern Cross appears on the flags of Australia and New Zealand.

+ fact: Many beautiful globular sky clusters are visible in the southern sky.

+ fact: The brightest part of the Milky Way passes over southern skies in winter.

The skies of the Southern Hemisphere contain a glorious array of stars, clusters, galaxies, and nebulae. The newer names for southern constellations often honor the tools of science and art. The iconic Southern Cross, or Crux, includes several notable stars as well as the "Jewel Box" open cluster and the Coalsack Nebula, a dense cloud of gas and dust silhouetted against the Milky Way. The southern sky also contains the huge constellation Centaurus—location of Alpha Centauri, neighbor to the sun and only 4.3 light-years away—as well as the Large and Small Magellanic Clouds, irregular dwarf galaxies that orbit the Milky Way. If you have an opportunity to visit the Southern Hemisphere, check out the wonders of its night sky. Many online resources can guide your tour.

Zodiac

A group of star patterns recognized from ancient times, the 12 zodiac constellations represent animals, people, creatures—and one inanimate object, Libra—that appear in an annual cycle.

KEY FACTS

The zodiac constellations are Aries, Taurus, Gemini, Cancer, Leo, Virgo, Libra, Scorpius, Sagittarius, Capricornus, Aquarius, and Pisces.

+ fact: Several constellations of the zodiac are always in view.

+ fact: These constellations are located along the ecliptic, the sun's apparent yearly path across the sky.

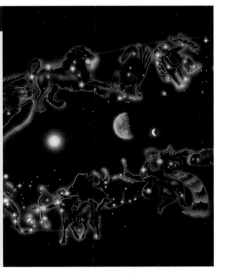

The word "zodiac" comes from the Greek for "circle of animals," and most of the zodiacal constellations do have animal associations. In the sky, these constellations form an annual procession along the ecliptic, the apparent path taken by the sun across the sky as the Earth orbits the sun. As each of the 12 zodiacal constellations reaches its westernmost phase, it is in alignment with the sun, creating the idea of "sun sign" that forms the basis of the pseudoscience of astrology. Constellation Ophiuchus also reaches into the ecliptic path, but it usually is excluded from the zodiac. In Arab astronomy, even Libra (The Scales) had an animal connection, as reflected in the names of its alpha and beta stars, which were considered the claws of the constellation Scorpius.

Astrology

In the mystical and mind-bending 1970s, "What's your sign?" was a frequent conversation starter, reflecting the influence of astrology—at least superficially—in those changing times.

KEY FACTS

For millennia, astrology was considered a scholarly pursuit, and it remains so in some cultures.

+ fact: The Babylonians developed the first organized system of astronomy.

+ fact: Astrology helped promote the advancement of astronomy.

+ fact: Arab astrologers made many notable contributions to astronomy.

Few people are unaware of their "sign," the constellation of the zodiac where the sun resided when they were born. The correlation between signs and personal characteristics and the alignment of the sun, moon, stars, and planets at a particular point with future events forms the basis of astrology. Astrology is a practice as old as stargazing itself. The ancient Egyptians, for example, looked to the rising of the star Sirius to predict the flooding of the Nile, events that did coincide. Astrologers deem such occurrences to be cause-and-effect situations, using reliable observations to support their claims. In some cultures, predictions based on astrological interpretations are still taken into consideration when planning life events such as marriages and making other important decisions.

Sky & Constellation Charts

The following section of this chapter presents the night sky in the Northern Hemisphere through a series of four seasonal sky charts or maps and following those, through focused charts that highlight discussions of 56 different constellations.

The seasonal sky charts—one each for winter, spring, summer, and autumn—show the visible constellations, the brightest stars, and the locations of many galaxies, nebulae, and other deep sky objects. These maps reflect the fact that the celestial sphere operates on a constantly changing continuum. The maps are prepared from the perspective of an observer at latitude 40° N, at the dates and times indicated on each chart.

Getting Oriented

These charts show the sky at a particular time but are useful any night. Like the sun, the stars and constellations appear to move from east to west, in some cases falling from view below the horizon for several months.

Observers at latitudes above 40° N will find the northern constellations higher in the sky, for longer periods of time, while losing sight of some below the southern horizon. The reverse is true for observers at positions toward the Equator.

The seasonal sky charts include a silhouetted landscape around the border indicating the horizon. Constellations appearing within about

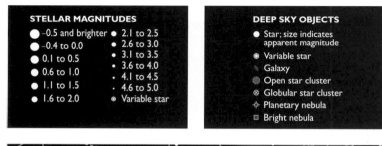

NOTES ABOUT SKY CHARTS

+ **Stars:** The brightest stars have common names, like Antares, which often relate to ancient mythology. But most bright stars that form constellations are typically known by Johann Bayer's 1603 method of calling the brightest star in a constellation "alpha," the second brightest "beta," and so on. For example, Deneb is the common name for the brightest star in the constellation Cygnus (the Swan), but the star is also known as Alpha Cygni, and is often labeled with both names.

+ **Deep sky objects:** Most of the galaxies and other "fuzzy" objects seen by backyard astronomers take their names from either the Messier catalog (M) or the New General Catalog (NGC), where they are assigned a number. Charles Messier

published his catalog of about 100 objects in 1771. The New General Catalog, with 7,840 objects. came in 1888. There are alternative names in these catalogs and others. For example, in addition to the common name Andromeda galaxy, the object is also known as M31 and NGC 224, as well as by other names.

+ **Mythological icons:** These small drawings are designed to help relate the shapes of the constellations to the figures they represent in mythology. They are not necessarily oriented in the direction they are typically seen in the night sky, and their arbitrary line patterns do not always match what is shown on the detailed star charts on the following pages.

Star maps often show line patterns to help illustrate the shape typically recognized as a constellation. These line patterns vary from source to source. Choosing which stars to show as part of a constellation drawing is arbitrary. The International Astronomical Union (IAU) defines constellations as areas of the night sky, not as strings of specific stars. The IAU's well-defined technical boundaries are not shown on these charts.

10 degrees of that edge will be difficult to spot. On the charts, stars with magnitudes down to 5 are connected by lines to form the constellations. Many stars of magnitude 3.5 or brighter (meaning a lower numerical value) are labeled for reference, and some have been tinted to indicate color. A star's size reflects its magnitude, indicated in a table on the margins of the chart, as are the symbols for various deep sky objects. The ecliptic— the sun's apparent path during the year and the line of travel for the zodiacal constellations—is the dotted line across each chart. The faint white field represents the Milky Way.

Constellation Charts

The pages that follow the seasonal sky charts describe 56 of the 88 recognized constellations. Many of the original 48 in Ptolemy's *Almagest*, an astronomical handbook, are included, as are newer constellations added to fill in the sky in the Northern Hemisphere. Those not included are mainly the "deep south" constellations that cannot be viewed from the mid-northern latitudes. You can learn about southern sky constellations online or in astronomy guides that often include such information.

Key Facts

Each constellation heading shows the "Size on the Sky," using a thumb length, closed fist, and outstretched hand or hands held at arm's length to indicate the constellation's approximate size in the sky. The Key Facts for the constellations include the number of brighter stars found in the constellation. "Best viewed" indicates the best months to hunt for the constellation. "Location" indicates the appropriate seasonal sky chart for reference. The constellation's alpha star also is noted.

Each constellation entry features a star map showing the stars that make up a constellation and a portion of the surrounding sky. Light lines connect the brighter member stars of each constellation; prominent asterisms are named. The brightest stars are labeled with letters of the Greek alphabet, beginning with alpha. Some of the brighter and well-known stars are labeled with their proper names and, where appropriate, tinted to indicate the approximate color in the sky. Background stars, neighboring constellations, and nearby deep sky objects also appear on the maps. Messier (M) or New General Catalog (NGC) numbers indicate deep sky objects. See "Notes About Sky Charts," above, for M and NGC histories.

Winter

Orion makes a good orientation point in the winter sky. Bright star Betelgeuse appears on Orion's right shoulder and Rigel forms the left foot. At the sword, you'll find the Orion Nebula (M42). Taurus lies northwest of Betelgeuse and so do two famed star clusters, the Pleiades and the Hyades.

Find Auriga, and draw a line north from Betelgeuse to bright star Capella. Northeast of Orion, Gemini is marked by the stars Castor and Pollux. To the southeast lies alpha star Sirius of Canis Major. Betelgeuse, Sirius, and Procyon in Canis Minor form the Winter Triangle asterism.

NORTH

DRACO

CYGNUS
Deneb
NW

CEPHEUS
M39

LACERTA

Polaris

CASSIOPEIA
M52

ANDROMEDA
M110
M31
M32

PEGASUS
β
μ

Great Square

Double
Cluster
M76

M103

PERSEUS
Algol
M34

TRIANGULUM
M33

Circlet

WEST

AURIGA

M45
Pleiades

ARIES
M74

Hyades

PISCES

Ecliptic

TAURUS
M77

Mira

CETUS

ERIDANUS

FORNAX

SW

WIL TIRION

CAELUM

HOROLOGIUM

DATE	TIME
12/21	11 p.m.
1/21	9 p.m.
2/1	8 p.m.

STELLAR MAGNITUDES

- −0.5 and brighter
- −0.4 to 0.0
- 0.1 to 0.5
- 0.6 to 1.0
- 1.1 to 1.5
- 1.6 to 2.0
- 2.1 to 2.5
- 2.6 to 3.0
- 3.1 to 3.5
- 3.6 to 4.0
- 4.1 to 4.5
- 4.6 to 5.0
- Variable star

DEEP SKY OBJECTS

- Star; size indicates apparent magnitude
- Variable star
- Galaxy
- Open star cluster
- Globular star cluster
- Planetary nebula
- Bright nebula

Spring

Spring is a good time to hunt for galaxies, as the sky offers a good view of elusive deep sky objects. Earth's position relative to the Milky Way has shifted so that the core of our galaxy lies near the horizon in the west. In spring, Ursa Major swings highest above Polaris, bringing the Big Dipper along with it. Virgo is prominent, together with its dense collection of galaxies, including the Virgo supercluster, the Sombrero galaxy (M104), and many other Messier items.

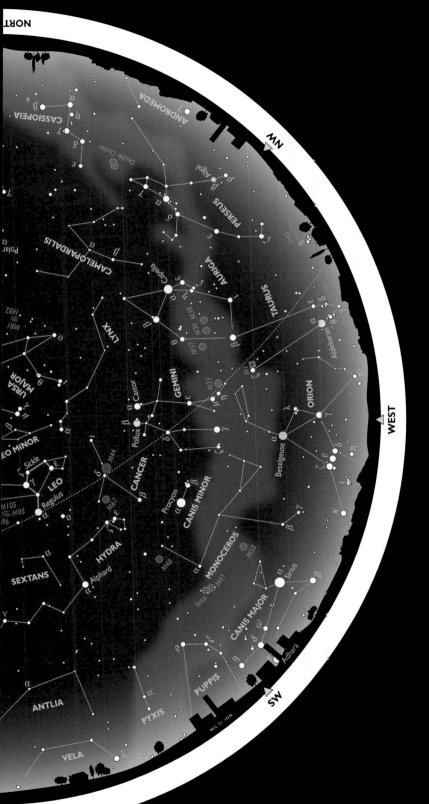

DATE	TIME
3/21	11 p.m.
4/1	10 p.m.
4/21	9 p.m.

STELLAR MAGNITUDES

● -0.5 and brighter	● 2.1 to 2.5
● -0.4 to 0.0	● 2.6 to 3.0
● 0.1 to 0.5	• 3.1 to 3.5
● 0.6 to 1.0	• 3.6 to 4.0
● 1.1 to 1.5	· 4.1 to 4.5
● 1.6 to 2.0	· 4.6 to 5.0
	⊙ Variable star

DEEP SKY OBJECTS

● Star; size indicates
    apparent magnitude

⊙ Variable star

＼ Galaxy

⊛ Open star cluster

⊗ Globular star cluster

✧ Planetary nebula

□ Bright nebula

Summer

Summer places the Summer Triangle overhead, highlighting three constellations. Vega in Lyra, Deneb in Cygnus, and Altair in Aquila form the seasonal asterism. Vega will be almost directly overhead, unmistakably bright at magnitude 0. Sagittarius is an area rich in star clusters and nebulae. Locate the Teapot asterism within it to spot notable objects surrounding the lid: the Lagoon (M8) and Trifid (M20) nebulae, and the great Sagittarius star cluster (M22).

NORTH

AURIGA

LYNX

LYNX

CAMELOPARDALIS

LEO MINOR

URSA MINOR

M82

M81

URSA MAJOR

Big Dipper

M108

M97

M109

M106

DRACO

M101

M51

M94

M63

CANES VENATICI

COMA BERENICES

LEO

Sickle

NW

WEST

Keystone

M13

M3

M64

M85

M88 M100

M53

M99

M91

M90 M89

M60 M58 M98

M59 M87 M86 M84

M49 M61

CORONA BOREALIS

Arcturus

BOÖTES

M5

M12

SERPENS (Caput)

VIRGO

Spica

CORVUS

Ecliptic

M107

M80

Antares

M4

LIBRA

LIBRA

HYDRA

R

LUPUS

LUPUS

CENTAURUS

WIL TIRION

SW

DATE	TIME
6/21	11 p.m.
7/1	10 p.m.
7/21	9 p.m.

STELLAR MAGNITUDES

- ⬤ -0.5 and brighter
- ⬤ -0.4 to 0.0
- ⬤ 0.1 to 0.5
- ⬤ 0.6 to 1.0
- ⬤ 1.1 to 1.5
- ⬤ 1.6 to 2.0
- ● 2.1 to 2.5
- ● 2.6 to 3.0
- · 3.1 to 3.5
- · 3.6 to 4.0
- · 4.1 to 4.5
- · 4.6 to 5.0
- ⊛ Variable star

DEEP SKY OBJECTS

- ● Star; size indicates apparent magnitude
- ⊛ Variable star
- ╲ Galaxy
- ⚞ Open star cluster
- ⊗ Globular star cluster
- ✧ Planetary nebula
- ▢ Bright nebula

Autumn

The Great Square of Pegasus, an asterism at the center of constellation Pegasus, is central to locating several other autumn constellations, including Andromeda. Alpha star Alpheratz is shared by Andromeda and the asterism. Autumn brings four major meteor showers: the Orionids in October, the Taurids and Leonids in November, and the Geminids in December. The season also offers rewarding galaxy viewing: The Andromeda galaxy (M31) in that constellation is visible to the naked eye.

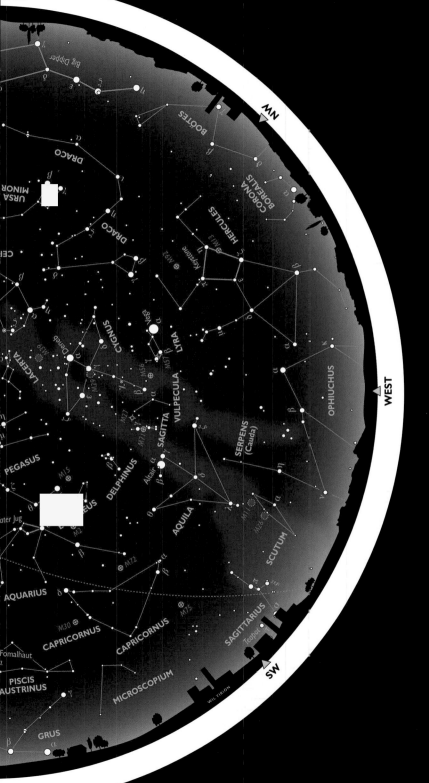

9/21	11 p.m.
10/21	10 p.m.
11/1	8 p.m.

STELLAR MAGNITUDES

- -0.5 and brighter
- -0.4 to 0.0
- 0.1 to 0.5
- 0.6 to 1.0
- 1.1 to 1.5
- 1.6 to 2.0

- 2.1 to 2.5
- 2.6 to 3.0
- 3.1 to 3.5
- 3.6 to 4.0
- 4.1 to 4.5
- 4.6 to 5.0
- Variable star

DEEP SKY OBJECTS

- Star; size indicates apparent magnitude
- Variable star
- Galaxy
- Open star cluster
- Globular star cluster
- Planetary nebula
- Bright nebula

Andromeda

The Chained Maiden   Size on the sky: ✋

This constellation notably contains the Andromeda galaxy (M31), a spiral-shaped deep sky galaxy similar to our own Milky Way. Andromeda galaxy is visible to the naked eye.

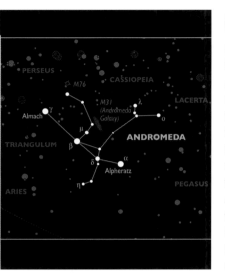

KEY FACTS

The figure of Princess Andromeda, most often seen upside down, contains 7 main stars. Locate by tracing a line northeast from the northeast corner of the Great Square of Pegasus.

+ months best viewed: October–November

+ seasonal chart location: Autumn, center of chart

+ alpha star: Alpheratz

Cassiopeia and Cepheus, the mythical rulers of Ethiopia, angered the Greek gods when Cassiopeia boasted that her daughter Andromeda's beauty surpassed that of the daughters of Nereus, god of the sea and Poseidon's father-in-law. In retaliation, Poseidon sent the sea monster Cetus to destroy the kingdom. Cassiopeia and Cepheus then offered Andromeda as a sacrifice, chained to a rock. At the last moment, the hero Perseus, homeward bound with the head of Medusa, rescued Andromeda, his trophy head turning Cetus to stone in the bargain. Other characters in this famous tale, including Pegasus, the hero's mount, appear nearby in the sky.

Antlia

The Air Pump   Size on the sky: 🤏

Northern Hemisphere stargazers can pick out this small constellation in the spring near the southern horizon. It contains two binary stars, one of which can be split with good binoculars.

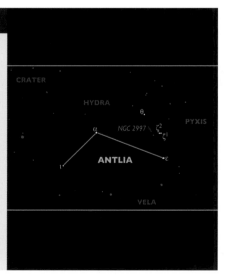

KEY FACTS

The angular southern constellation Antlia contains 3 main stars; the constellation lies about 5 fist-widths south of the bright alpha star Regulus in Leo.

+ months best viewed: March–April

+ seasonal chart location: Spring, southwest quadrant

+ alpha star: Alpha Antlia

As a southern constellation, Antlia's backstory belongs to the realm of science, not mythology. In the 1750s, French astronomer Nicholas-Louis de Lacaille constructed the constellation from his vantage point at the Cape of Good Hope in South Africa. He named it for Irish chemist and physicist Robert Boyle's discovery of the pneumatic pump, originally calling it Antlia Pneumatica. All together, Lacaille charted some 10,000 southern stars. Antlia's alpha star is its brightest one, but it has no proper name. Antlia harbors the galaxy NGC 2997, faintly visible through a small telescope, just inside its corner.

Aquarius

The Water Bearer   *Size on the sky:*

A constellation of the zodiac, Aquarius occupies a lot of celestial real estate along the ecliptic—the apparent path of the sun across the night sky—between Pisces and Capricornus.

KEY FACTS

Its 13 main stars form the spread-out figure of the Water Bearer; within this constellation, the Y-shaped Water Jug asterism is the most prominent feature.

+ months best viewed: September–October

+ seasonal chart location: Autumn, southwest quadrant

+ alpha star: Sadalmelik

One of the fainter constellations of the zodiac, Aquarius is located south of the Great Square of Pegasus, in the "watery" section of the sky with associated constellations such as Cetus the Whale and Pisces. Aquarius has several mythological associations. The ancient Egyptians linked the constellation to the yearly flooding of the Nile River, the most vital event of the agricultural year, and the hieroglyph for water became the zodiacal symbol for Aquarius. To the Greeks, Aquarius was Zeus's cupbearer, Ganymede. In some versions, he pours water or wine from a jug that maintains the celestial river Eridanus.

Aquila

The Eagle   *Size on the sky:*

Named by ancient Mesopotamian stargazers, Aquila the Eagle appears close enough to Earth's Equator to be seen from any terrestrial viewing position.

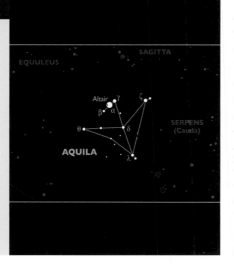

KEY FACTS

This straightforward constellation near the Equator contains 10 stars, with alpha star Altair traditionally representing the eagle's head.

+ months best viewed: August–September

+ seasonal chart location: Summer, southeast quadrant

+ alpha star: Altair

Aquila's alpha star, Altair, is one of the brightest stars in the sky and a good reference point for identifying the constellation. It also forms part of the Summer Triangle asterism along with Vega in constellation Lyra and Deneb in Cygnus. Aquila occurs in the Milky Way in an area of abundant starfields. In Greek myth, the eagle was Zeus's companion and carried the god's thunderbolts for him. He also is thought to have carried the young shepherd Ganymede to the sky to serve as Zeus's celestial cupbearer. Ganymede is immortalized nearby as the constellation Aquarius.

Aries
The Ram   Size on the sky:

Ancient astronomers traditionally placed Aries at the beginning of the celestial zodiac because, millennia ago, the sun was "in" Aries at the time of the vernal equinox in spring.

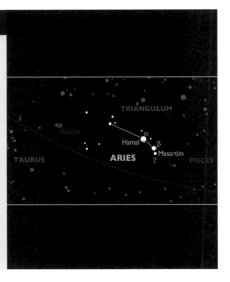

KEY FACTS

The brightest part of constellation Aries contains 4 stars; the "tail" of the ram is Gamma Arietis, or Mesartim, a double star with a wide separation.

+ months best viewed: November–December

+ seasonal chart location: Winter, southwest quadrant

+ alpha star: Hamal

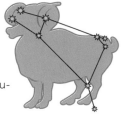

On evenings in late fall and early winter, Aries appears high in the west between the Great Square of Pegasus and the Pleiades asterism in Taurus. The constellation's double star, Mesartim, was one of the first doubles spotted with a telescope—by astronomer Robert Hooke in 1664. In the view of the ancient Greeks, Aries was the source of the Golden Fleece stolen by Jason and the Argonauts. Jason wrested the fleece from the custody of a dragon that had received it from a king. The king had sacrificed the ram in gratitude for saving his two children from an abusive stepmother.

Auriga
The Charioteer   Size on the sky:

The tidy, elegant form of Auriga appears in the heart of the Milky Way and is on Ptolemy's list of the 48 original ancient constellations.

KEY FACTS

The constellation Auriga, found on a line between Orion and Polaris, has 7 main stars. It shares with Taurus a star (lowest on the map) that otherwise would be Auriga's gamma star.

+ months best viewed: December–January

+ seasonal chart location: Winter, center of chart

+ alpha star: Capella

Auriga's shining glory is its alpha star, the brilliant Capella, the sixth brightest star in the sky. Southwest of Capella lies Epsilon Aurigae, an eclipsing binary star veiled every 27 years by a companion star. The constellation lies along the galactic equator, the great circle that passes through the densest part of the Milky Way and contains several interesting star clusters visible through binoculars, notably M36 and M37. The chariot and rider of Auriga may represent Hephaestus of Greek myth, the crippled blacksmith god who built the chariot to accommodate his handicap.

Boötes

The Herdsman   Size on the sky:

The ancient constellation of Boötes is a celestial highlight of early summer. Its alpha star, Arcturus, ranks as fourth brightest star in the sky.

KEY FACTS

The early summer constellation Boötes contains 8 key stars; beta star Nekkar points toward Polaris, the North Star.

+ month best viewed: **June**

+ seasonal chart location: **Summer, center of chart**

+ alpha star: **Arcturus**

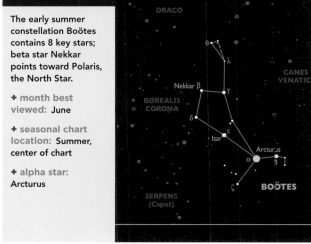

Boötes's alpha star, Arcturus, lies on an arc continuing from the handle of the Big Dipper, giving rise to the mnemonic "arc to Arcturus." In a variation of the myth surrounding the constellation, herdsman Boötes keeps his flock moving about the sky in pursuit of Ursa Major and Ursa Minor. In another, he herds a bear; a loose translation of Arcturus is "bear keeper." In the first week in January, the northern part of Boötes hosts the Quadrantid meteor shower, one of the strongest of the year, producing several dozen meteors an hour during its peak. It occurs at the point where Boötes, Hercules, and Draco meet.

Camelopardalis

The Giraffe   Size on the sky:

Camelopardalis is located in a region of the sky where few stars are visible. No star in this faint constellation has a common name.

KEY FACTS

This constellation contains 5 key stars, including Z Camelopardalis, a cataclysmic variable star that dramatically increases in brightness, then decreases.

+ months best viewed: **December–January**

+ seasonal chart location: **Winter, northeast quadrant**

+ alpha star: **Alpha Camelopardalis**

Camelopardalis is a modern constellation, created in 1613 to fill the celestial gap between the bears, Ursa Major and Ursa Minor, and Perseus. It neighbors Polaris, the North Star, making it a strictly northern constellation, although a faint one. It contains the star cluster NGC 1502, a good target for a telescope, and spiral galaxy NGC 2403, which lies about 12 million light-years from Earth. Although the ancient Greeks called the giraffe a "camel-leopard" (and gave the animal its species name: *camelopardalis*), the constellation possibly was named for the biblical camel that carried Rebecca to her marriage with Isaac.

Cancer
The Crab   Size on the sky:

As a constellation of the zodiac, tiny Cancer the Crab is far less noticeable than most. It has no stars with a magnitude brighter than 4.

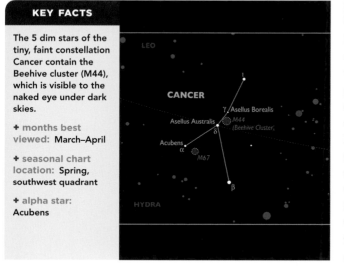

KEY FACTS

The 5 dim stars of the tiny, faint constellation Cancer contain the Beehive cluster (M44), which is visible to the naked eye under dark skies.

+ months best viewed: **March–April**

+ seasonal chart location: **Spring, southwest quadrant**

+ alpha star: **Acubens**

Unassuming Cancer contains some interesting star groups. Viewed with binoculars, the Beehive cluster reveals more than 20 stars, while open cluster M67, with its 500 stars, can be seen through a small telescope. The Tropic of Cancer gets its name from the constellation, although the sun no longer is in it on the summer solstice as it was for early mapmakers. Cancer's story involves Zeus's often jealous wife, Hera. She sent the crab to thwart Hercules, Zeus's son with a mortal, as he battled the multiheaded monster Hydra. Proving no match, the crab perished under the hero's foot, but earned a place in the stars.

Canes Venatici
The Hunting Dogs   Size on the sky:

Location is everything in the night sky: The two stars of Canes Venatici appear just about where two leashed hounds would be expected in relation to Boötes, the Herdsman.

KEY FACTS

A mere 2 stars compose the constellation Canes Venatici, located just below the handle of the Big Dipper. The stars are surrounded by a number of interesting objects.

+ months best viewed: **May–June**

+ seasonal chart location: **Spring, center of chart**

+ alpha star: **Cor Caroli**

An alternative way of viewing Canes Venatici is to look for the two stars running between the legs of Ursa Major, the bear that the dogs are chasing. The small northern constellation was created by Polish astronomer Johannes Hevelius in the 17th century and has been accepted as one of the official 88 constellations. The name of its alpha star, Cor Caroli, stands for the Heart of Charles and was reputedly bestowed on the star by famous English astronomer Edmond Halley in honor of his sponsor, King Charles II. The 500-star globular cluster M3 can be found midway between Cor Caroli and Arcturus in Boötes.

Canis Major

The Larger Dog   Size on the sky:

Canis Major appears just to the southeast of the constellation Orion, not far from the Milky Way. The constellation's alpha star, Sirius, outshines all other stars in the night sky.

KEY FACTS

Canis Major contains 8 key stars, most notably Sirius, brightest in the night sky; smaller stars, star clusters, and nebulae crowd around and within the constellation.

+ months best viewed: January–February

+ seasonal chart location: Winter, southwest quadrant

+ alpha star: Adhara

Sirius, the Dog Star, shows the way to Canis Major, the constellation representing the larger of Orion's hunting dogs. A line traced through Orion's belt and continuing southeast points directly to Sirius within its constellation. The Dog Star figures in the expression "dog days of summer." In late summer, Sirius rises around the same time as the sun, and the two "conspire" to generate extra warmth in the Northern Hemisphere. Several star clusters and nebulae are located in Canis Major. The brightest, open cluster M41, is easily spotted through binoculars and even more impressive through the lens of a telescope.

Canis Minor

The Smaller Dog   Size on the sky:

Orion's smaller hunting dog, Canis Minor, lies directly northeast of its larger companion, Canis Major. The dim constellation lacks any bright deep sky objects.

KEY FACTS

Only 2 stars make up the young dog of Canis Minor, often visualized as a small dog with its head cocked in the direction of northern neighbor Gemini.

+ months best viewed: January–February

+ seasonal chart location: Winter, southeast quadrant

+ alpha star: Procyon

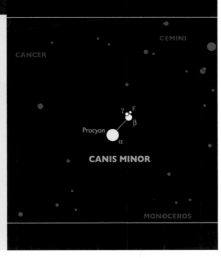

Multiple scenarios explain the dim and subdued Canis Minor. One puts the dog beneath the table of Castor and Pollux, the Gemini twins, waiting for scraps. Another makes him Helen of Troy's favorite pup that allowed her to elope with the Trojan prince Paris. Together with Betelgeuse in Orion and Canis Major's Sirius, Canis Minor's alpha star, Procyon, forms the Winter Triangle, a seasonal asterism that helps orient winter stargazers. Procyon lies only 11.2 light-years from Earth and is the eighth brightest star in the night sky. The constellation also hosts the Canis Minorids, a faint meteor shower of early December.

Capricornus

The Sea Goat   Size on the sky: 🖐

The broad and distinctive triangle of Capricornus stands out in the southern sky, especially in late summer and early fall. Algedi, Arabic for "the goat," is the zodiacal constellation's alpha star.

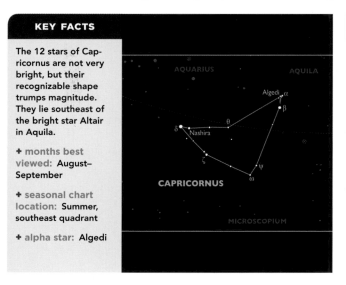

KEY FACTS

The 12 stars of Capricornus are not very bright, but their recognizable shape trumps magnitude. They lie southeast of the bright star Altair in Aquila.

+ months best viewed: **August– September**

+ seasonal chart location: **Summer, southeast quadrant**

+ alpha star: **Algedi**

Long recognized as a goat, Capricornus at some point acquired a fish tail. In one story, the god Pan, a satyr, leaped into the River Nile to escape a monster, and the water transformed him. In an older tale, the animal represents one of Zeus's warriors, discoverer of conch shells with a resounding call that frightened the opposing Titans into retreat. Zeus placed the warrior in the sky with a fish tail and horns to represent his discovery. Alpha star Algedi appears to the naked eye as an optical binary composed of two stars that are closely aligned but separated by more than 500 light-years.

Cassiopeia

The Queen   Size on the sky: 🖐

Visible year-round because of its proximity to the north celestial pole, Cassiopeia's iconic W shape (or M, depending on the season) makes identification easy.

KEY FACTS

The 5 stars of this iconic constellation form the figure of Queen Cassiopeia on her throne within a portion of the Milky Way facing Polaris, the North Star; visible much of year.

+ months best viewed: **October– November**

+ seasonal chart location: **Autumn, northeast quadrant**

+ alpha star: **Shedar**

Chained to her throne, Ethiopian Queen Cassiopeia sits between the constellations of her husband, Cepheus, and her daughter, Andromeda. Boasting of her daughter's good looks, the queen drew the ire of the sea god Poseidon, who felt that his own daughters, the Nereids, had been disrespected. Cassiopeia and Cepheus offered to sacrifice Andromeda to save their kingdom. At the last moment, Perseus saved Andromeda, but Cassiopeia was punished nonetheless, forced to hang upside down half the year. Visible through a telescope, the dense open cluster M52, seen off the leg of the W, contains about 100 stars.

Cepheus

The King   Size on the sky: 🖐

A circumpolar constellation, Cepheus is visible all year in the Northern Hemisphere, but the brightness of its main stars competes with surrounding stars, often making it hard to spot.

KEY FACTS

The body of King Cepheus, facing the open end of Cassiopeia, contains 5 key bright stars. They take the shape of a small house pointing generally toward Polaris, the North Star.

+ **months best viewed:** September– October

+ **seasonal chart location:** Autumn, center of chart

+ **alpha star:** Alderamin

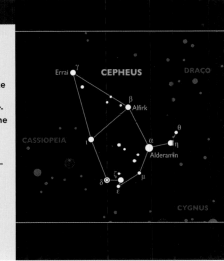

King Cepheus, the hapless consort of Queen Cassiopeia, was forced to put daughter Andromeda at peril because of his wife's unwise boast of their daughter's beauty, which angered the sea god Poseidon. As in his mythological life, the constellation Cepheus plays second fiddle to that of his wife, facing the open end of the W shape of the constellation Cassiopeia. Errai, Cepheus's gamma star, is both a binary star and host to an orbiting planet. Estimated to be almost 1.6 times the size of Jupiter, Errai's planet attests that planets can form in relatively close binary systems, stars that orbit the same center of gravity.

Cetus

The Sea Monster   Size on the sky: 🖐🖐

Very large and faint, the constellation Cetus appears in a sky neighborhood known as the Heavenly Waters in the company of constellations Eridanus and Pisces.

KEY FACTS

The 13 stars of the constellation Cetus are divided between the monster's head and body. The parts are connected by two stars; one of them, Mira, is a variable star.

+ **month best viewed:** November

+ **seasonal chart location:** Autumn, southeast quadrant

+ **alpha star:** Menkar

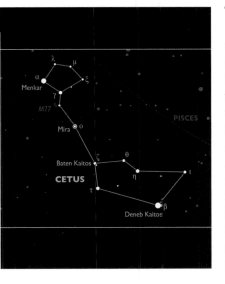

The god Poseidon sent the frightful sea monster Cetus—who gave his name to the mammalian order Cetacea, the whales—to terrorize Ethiopia after Queen Cassiopeia insulted Poseidon's daughters, the Nereids, by boasting that her daughter was more beautiful. Cetus was vanquished by Perseus and turned to stone by the head of Medusa. The constellation is also associated with the whale that swallowed the prophet Jonah in the Old Testament. In autumn, the head of Cetus appears between Taurus and Pisces in the southern sky, with its body bordering Aquarius.

Columba
The Dove   Size on the sky: 🖐

Columba is a smallish constellation, one of the modern ones added in the 16th century to fill out the sky chart. It honors the dove Noah sent out from the ark to search for dry land.

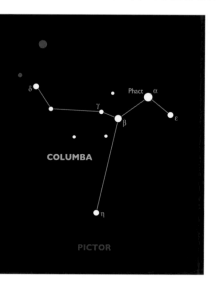

KEY FACTS

The smallish constellation Columba, made up of 8 stars, has the appearance of three legs spiraling from a medium-bright star at the center. Nearby sky objects often outshine it.

+ **months best viewed:** January–February

+ **sky location:** Winter, southeast quadrant

+ **alpha star:** Phact

At the end of the 40 days and nights of epic rain that flooded all the Earth, Noah sent out a dove to see if the waters of the great flood had begun to recede. When the dove returned carrying an olive sprig in its bill, Noah knew immediately that it had. The dove's association with the ark and the flood made it a natural companion of the watery constellations of the Heavenly Waters group. The constellation is located very close to brighter objects in the night sky, making the Dove somewhat difficult to make out, especially for novice sky watchers.

Coma Berenices
Berenice's Hair   Size on the sky: 🖐

Coma Berenices formed part of other constellations—a wisp of Virgo's hair or tuft of Leo's tail—until 17th-century astronomer Tycho Brahe helped designate it a constellation in its own right.

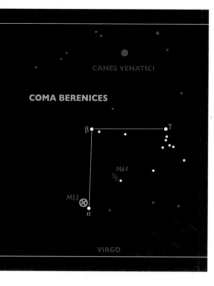

KEY FACTS

Located just north of Virgo and south of Canes Venatici near the Virgo cluster of galaxies, the rough right angle that is Coma Berenices contains 3 stars.

+ **months best viewed:** May–June

+ **seasonal chart location:** Spring, southeast quadrant

+ **alpha star:** Alpha Comae Berenices

A rejuvenated tale of an Egyptian queen provides a namesake and story for Coma Berenices. The consort of Ptolemy III, Berenice sought Aphrodite's aid in the safe return of her husband from war, promising the goddess luxurious hair as an incentive. After the wish was granted, an astronomer of the royal court convinced the ruling couple that a grateful Aphrodite had placed the queen's gift in the stars. The constellation contains the deep sky spiral galaxy M64, known as the Black Eye. It lies between the constellation's two outermost stars, along the line that would represent the base of a triangle.

Corona Australis
The Southern Crown   Size on the sky:

Though an original Ptolemaic constellation, Corona Australis is primarily southern, rising only a few degrees above the horizon during midsummer in the mid-northern latitudes.

KEY FACTS

The crown of Corona Australis, which contains 5 key stars, can be found at the feet of Sagittarius and just west of the bright tail star in Scorpius.

+ months best viewed: July–August

+ seasonal chart location: Summer, southeast quadrant

+ alpha star: Alpha Coronae Australis

For a small, faint constellation, Corona Australis enjoys a wealth of supporting mythology, much in the form of stories referring to a crown of laurel or fig leaves. One version regards it as the crown of Chiron, the Centaur. Another puts Apollo in the picture, having him fashion the crown from the leaves of his love Daphne, who was changed into a laurel tree to escape the insistent Apollo's advances. The constellation hosts an active star-forming region composed of the Coronet cluster and the nearby Corona Australis Nebula. Though visible only through advanced deep sky technology, this region boasts about 30 "newborn" stars.

Corona Borealis
The Northern Crown   Size on the sky:

Despite having one of the more definitive constellation shapes, small and faint Corona Borealis is squeezed in between its bigger, brighter neighbors Boötes and Hercules.

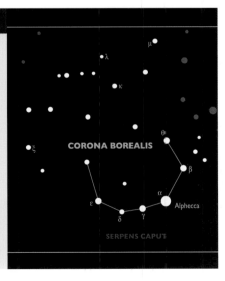

KEY FACTS

An obvious, semicircular crown shape made of 7 stars in a small and faint constellation is located just south of a line traced between the bright stars Vega in Lyra and Arcturus in Boötes.

+ months best viewed: June–July

+ seasonal chart location: Summer, center of chart

+ alpha star: Alphecca

Another Greek crown-related myth provides a plausible tale for the Northern Crown. In it, good-time god Dionysus threw his crown into the sky to impress Ariadne, princess of Crete and his future wife. She had rebuffed him earlier when he courted in the form of a young mortal. The crown toss changed her mind and they married. In American Indian lore, this celestial shape represents a camp circle. Some of the stars in Corona Borealis vary significantly in brightness. One of them is being investigated as possibly being an extrasolar planet; a body larger than Jupiter already has been found.

Corvus

The Crow   Size on the sky:

Corvus, along with Crater, perch in a crook of Hydra, the serpentine megaconstellation. The Crow is located just west of Spica in nearby Virgo.

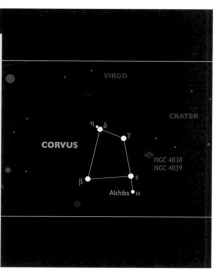

KEY FACTS

The constellation Corvus, located near Hydra's tail, contains 5 stars, 4 of which make up its body, while alpha star Alchiba, just outside the crow's body, represents its downward-facing head.

+ months best viewed: April–May

+ seasonal chart location: Spring, southeast quadrant

+ alpha star: Alchiba

In a myth belying the smarts of members of the Corvid family, a crow is sent for a cup of water by the god Apollo. The bird spied a tempting unripe fig and waited for it to ripen. Anticipating the god's ire, the crow made up a tale about being attacked by a snake, which he carried back along with the cup of water. Apollo saw through the deceit and banished the cup, the crow, and the serpent to the sky. A pair of colliding galaxies known as the Ring-tailed galaxy (NGC 4038 and NGC 4039) can be seen with an 8-inch aperture telescope just outside Corvus, with the "tail" near the border of Corvus and Crater.

Crater

The Cup   Size on the sky:

The small constellation Crater appears near its mythological associates, Hydra and Corvus. To some, Crater first represented a spike on sea monster Hydra's back.

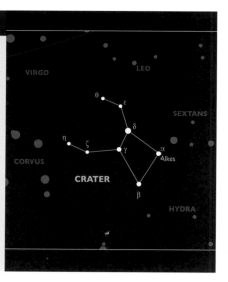

KEY FACTS

The 8 stars of the constellation Crater make a goblet shape: 4 stars form the base of the goblet and 4 more form the cup, which opens toward the constellation Spica.

+ months best viewed: April–May

+ seasonal chart location: Spring, southeast quadrant

+ alpha star: Alkes

The unmistakable goblet shape of Crater can be found in the spring sky right above Hydra and just west of Corvus, an associated constellation. Crater represents the vessel brought to Apollo by Corvus, the crow. Apollo tossed the two into the sky, along with Hydra, when he realized Corvus had lied about the reason he was late in bringing a requested cup of water. Some sky observers also relate the cup to the constellation Aquarius, the Water Bearer, which occupies the same quadrant of the sky in summer. More recent astronomers associate the cup with the Holy Grail or Noah's wine goblet.

Cygnus

The Swan   Size on the sky: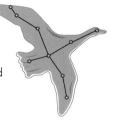

Cygnus appears in a dense and visually rewarding portion of the sky. It is sometimes known as the Northern Cross for its shape and bright stars.

KEY FACTS

The 13 stars of the constellation Cygnus form recognizable wings and a neck and head that in late summer and fall point southward like a migratory bird.

+ months best viewed: August–September

+ seasonal chart location: Summer, northeast quadrant

+ alpha star: Deneb

The ancients associated several bird-centric myths with this prominent constellation. It was seen as Zeus, transforming himself into a swan to seduce Leda, and also as Orpheus, who was murdered and turned into a swan for rejecting a group of maidens. He was placed in the sky next to his beloved lyre, the constellation Lyra. Cygnus also represented one of the Stymphalian birds, the quarry Hercules pursued in one of the 12 labors. Alpha star Deneb, along with Altair and Vega, create the Summer Triangle, and a large diffuse gas emission nebula shaped like and named for North America appears on the Swan's southern border.

Delphinus

The Dolphin   Size on the sky:

The distinct shape of Delphinus, one of the sea creature constellations of the Heavenly Waters sky region, seems to swim toward Pegasus.

KEY FACTS

The constellation Delphinus contains 5 dim stars, 4 of which form the body, which is an asterism known as Job's Coffin; the fifth is the creature's curved tail.

+ months best viewed: August–September

+ seasonal chart location: Summer, northeast quadrant

+ alpha star: Sualocin

The constellation Delphinus holds its place in the sky because he was a favorite of the Greek sea god Poseidon. The little dolphin succeeded in convincing the Nereid Amphitrite, the sea nymph daughter of Nereus, to marry Poseidon after the god himself had attempted and failed to win her attention. Delphinus can be located just west of a straight line traced between Altair in the constellation Aquila and Deneb in the constellation Cygnus. The Dolphin constellation's gamma star is an optical double star, best viewed through a telescope, whose dimmer star has a greenish tinge.

Draco

The Dragon   Size on the sky:

Visible year-round in the Northern Hemisphere, Draco is one of the constellations closest to the north celestial pole. Draco contains the Cat's Eye Nebula (NGC 6543), a dying sun-like star.

KEY FACTS

Draco contains 18 stars; its dragon tail emerges between the bears Ursa Major and Ursa Minor and dips toward sea monster Cepheus, while its head points to Hercules.

+ months best viewed: May–June

+ seasonal chart location: Spring, northeast quadrant

+ alpha star: Thuban

The constellation Draco represented different scaly beasts to different civilizations. To the ancient Greeks, it was the dragon Ladon, who was slain by Hercules. To the Hindus, it was a celestial alligator, and to the Persians, a giant serpent. Draco's alpha star, Thuban, once was the Earth's polestar, before the phenomenon of precession shifted the celestial position of Earth's axis and made Polaris the polestar. The Quadrantid meteor shower erupts from the region where Draco, Boötes, and Hercules meet in the beginning of January. One of the heaviest meteor showers of the year, the Quadrantids last only a few hours.

Equuleus

The Little Horse   Size on the sky:

The second smallest constellation in the area, Equuleus resides in a crowded southern portion of the sky where it is overshadowed by the larger horse constellation, Pegasus.

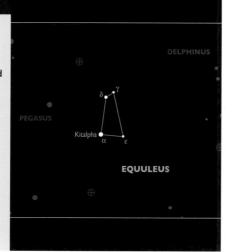

KEY FACTS

The constellation Equuleus has 4 stars, which essentially take the form of the Little Horse's head and neck; the body is left to the imagination. Its alpha star is the only significant feature.

+ months best viewed: July–August

+ seasonal chart location: Summer, southeast quadrant

+ alpha star: Kitalpha

Equuleus is the opening act for Pegasus, rising and setting before its neighbor, a fact that gave it the designation of Equus Prior. The constellation is thought to represent Celeris, the brother-horse of Pegasus. Pegasus is best known as the noble winged steed of Perseus, slayer of Medusa and savior of Andromeda. Celeris belonged to Castor, who along with twin Pollux represents the twins of Gemini. Castor was famed as a skilled equestrian and received the gift of Celeris from the messenger god Hermes. Look for Equuleus between neighbors Pegasus to the northeast and Aquila to the southwest.

Eridanus

The River   Size on the sky:

On a clear night in winter with an open horizon, you have a good chance of seeing Eridanus, although its southern tip disappears below the horizon at about 20° N latitude.

KEY FACTS

Eridanus has 33 stars that make up the long, winding constellation, which starts near beta star Rigel at the hunter Orion's foot and meanders south to the horizon.

+ **months best viewed:** December– January

+ **seasonal chart location:** Winter, southwest quadrant

+ **alpha star:** Achernar

The river of the constellation Eridanus was variously identified as the Euphrates or the Nile. The ancient Greeks saw it as the river into which Phaëthon, son of the sun god Helios, was cast after he failed to control the sun god's chariot and Earth stood in danger of burning up. It is a long river, to be sure, the sixth largest constellation in the sky containing its sixth brightest star, alpha star Achernar. Eridanus bends tightly at its beginning near Rigel and then gradually widens out. It contains a bright star similar to our sun, Epsilon Eridani, which is known to host a confirmed planet.

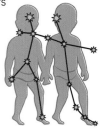

Gemini

The Twins   Size on the sky:

The twins Castor and Pollux stride hand in hand in the southeast winter sky, northeast of the bright star Betelgeuse on constellation Orion's upraised arm.

KEY FACTS

Gemini's 13 key stars are apportioned between the two stick figures of the twins Castor and Pollux, with the constellation's alpha and beta stars forming the heads.

+ **months best viewed:** February– March

+ **seasonal chart location:** Winter, southeast quadrant

+ **alpha star:** Castor

The Gemini twins represent Castor and Pollux, sons of the unions of Leda with a swan-disguised, seductive Zeus and her husband Tyndareus, and brothers of Helen of Troy and Clytemnestra—the exact paternity details remaining murky. Castor and Pollux also served as shipmates with Jason on the *Argo*. The impressive Geminid meteor showers originate from the constellation in the middle of December. Binoculars or a telescope yield views of M35, an open star cluster of hundreds of stars that resides near the three "foot stars" of Castor. The constellation also contains the blue-green Clownface Nebula, visible only through a telescope.

Grus

The Crane   Size on the sky: 🖐

Grus is a breakout constellation, which was separated from the formation Piscis Austrinus, the Southern Fish. It is visible in the Northern Hemisphere only briefly in the fall.

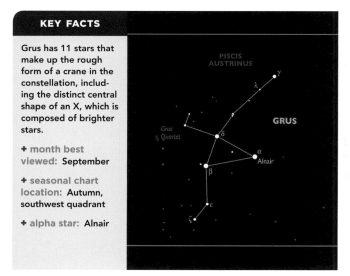

KEY FACTS

Grus has 11 stars that make up the rough form of a crane in the constellation, including the distinct central shape of an X, which is composed of brighter stars.

+ **month best viewed:** September

+ **seasonal chart location:** Autumn, southwest quadrant

+ **alpha star:** Alnair

The largely southern constellation Grus is a modern invention, named in 1603 by German celestial cartographer Johann Bayer in honor of the bird the ancient Egyptians made the symbol of the astronomer. Alternate names for the star group include Anastomas (the Stork), den Reygher (the Heron), and Phoenicopterus (the Flamingo). The constellation harbors the Grus Quartet, a group of four spiral galaxies (NGC 7552, 7582, 7590, and 7599) that are located very close together and exert strong influences on each other. Millions of years from now, the four galaxies will likely unite.

Hercules

The Super Hero   Size on the sky: 🖐

The constellation Hercules contains a he-man's worth of stars, although none of them are exceptionally bright. Look for him overhead in the summer months.

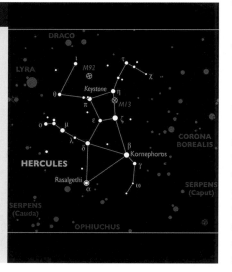

KEY FACTS

It takes 20 key stars to make up the hefty form of Hercules, including the recognizable backwards K at the center; his "torso" contains the 4-star asterism known as the Keystone.

+ **months best viewed:** July–August

+ **seasonal chart location:** Summer, center of chart

+ **alpha star:** Rasalgethi

The half-mortal son of Zeus, Hercules was plagued throughout much of his life by the jealousy of Hera, Zeus's wife and queen of the gods, who resented the tangible reminder of her husband's dalliance. She induced a fit of insanity in Hercules that caused him to murder his wife and children. In grief and repentance, he undertook 12 nearly impossible labors, and at their successful completion was granted immortality. The constellation contains M13, considered the best globular cluster visible in the northern sky. It occurs on one side of the Keystone asterism and can be seen as a blur by the naked eye.

Hydra

The Sea Serpent *Size on the sky:*

Hydra winds across the sky from Cancer to Libra and spans the largest area of any constellation. It used to include today's constellations Sextans, Corvus, and Crater.

KEY FACTS

The constellation Hydra's 17 stars begin with a kite-shaped head, continue past the "heart" at alpha star Alphard, and end with the sea serpent's tail close to the southern horizon.

+ months best viewed: March–April

+ seasonal chart location: Spring, southwest quadrant

+ alpha star: Alphard

In classical mythology, Hydra lost a decisive battle with Hercules, who fought the multiheaded sea serpent as one of his 12 labors. Hydra put up a good fight, regrowing two heads for each one that Hercules lopped off, and it looked pretty hopeless for Hercules until the stumps of each decapitated head were cauterized to stop the regrowth. Constellation Hydra's enormous span belies the relative dimness of its stars; only two of them surpass a magnitude of 3. Alphard, the brightest star in the formation, can be located by looking just east of a line traced from Regulus in the constellation Leo to the star Sirius in Canis Major.

Lacerta

The Lizard *Size on the sky:*

A so-called modern constellation, Lacerta was created by 17th-century Polish astronomer Johannes Hevelius. At first depicted as a small mammal, it later morphed into a lizard.

KEY FACTS

The constellation Lacerta, which zigzags between Cassiopeia, Cygnus, and Andromeda, is composed of 8 stars. It forms a W shape like its brighter, queenly neighbor Cassiopeia.

+ month best viewed: October

+ seasonal chart location: Autumn, center of the chart

+ alpha star: Alpha Lacerte

Lacerta is a constellation without a relevant mythology or backstory. At the time it was created to fill in a gap on the celestial map, competing proposals were offered to name it for Louis XIV or Frederick the Great. Although Lacerta's stars are faint, they remain visible to observers for much of the year and appear almost directly overhead in early fall for those in mid-northern latitudes. Lacerta contains an intriguing deep sky object, BL Lacertae, which is the center of an elliptical galaxy visible only through a substantial telescope. Of great interest, its core may harbor a black hole.

Leo
The Lion   Size on the sky:

One of the most easily visualized constellation shapes, Leo the Lion figured as one of the 13 original zodiacal constellations established by the Babylonians.

KEY FACTS

The kingly constellation Leo has 12 stars, with 8 of them forming the asterism known as the Sickle, which represents the lion's uplifted head.

+ months best viewed: March–April

+ seasonal chart location: Spring, center of chart

+ alpha star: Regulus

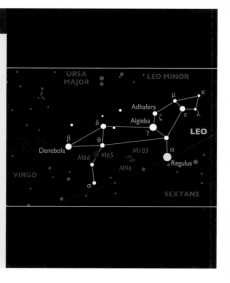

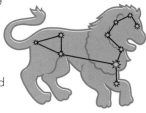

The ancient Egyptians revered the constellation Leo. They associated it with the sun, and may have modeled the Sphinx on its form. Leo also links to Hercules myths as a representation of the Nemean lion that the hero choked to death in the first of his 12 labors. The constellation harbors spiral galaxies M65 and M66, which can be seen with binoculars. Leo conveniently appears at the end of an imaginary line formed from Polaris and through the bowl edge of the Big Dipper. Leo is also the location of the Leonid meteor showers, which are identified with the comet Tempel–Tuttle and peak in mid-November.

Leo Minor
The Small Lion   Size on the sky:

Leo Minor requires a dark night for visibility, but it lies almost directly overhead in the mid-northern latitudes in early spring.

KEY FACTS

Appearing atop Leo, 3 dim key stars make up an indistinct shape of this small constellation; Leo Minor requires a dark night to even be seen.

+ months best viewed: March–April

+ seasonal chart location: Spring, center of chart

+ alpha star: none (beta star: Beta Leonis Minoris)

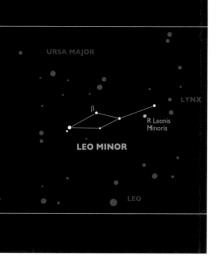

Like other 17th-century additions to the celestial charts, Leo Minor lacks a lot of supporting documentation or mythology. Polish astronomer Johannes Hevelius isolated the formation, which lacks a labeled alpha star, probably as the result of an oversight. Beta star Leonis Minoris is not the constellation's brightest; that distinction belongs to R Leonis Minoris, an oscillating star with more than a five-point range of magnitude during the course of a year. The ancient Egyptians recognized the small Leo Minor group of stars as the hoof prints of a herd of gazelles in flight from neighboring big cat Leo.

Lepus

The Hare   Size on the sky: 🖐

Just as Orion keeps close his celestial hunting dogs, he also has his prey nearby in the form of Lepus, the small constellation representing the Hare.

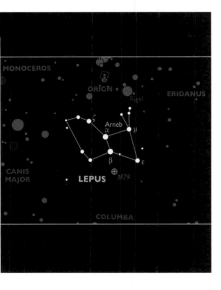

KEY FACTS

The constellation Lepus has 11 key stars that make the small hare shape, including the somewhat distinct ears, which appear just below Rigel in constellation Orion's left foot.

+ **months best viewed:** January–February

+ **seasonal chart location:** Winter, southeast quadrant

+ **alpha star:** Arneb

Lepus the Hare sits crouched and ready to flee just west of Sirius, the bright star in the constellation of a pursuer, Canis Major. Midwinter marks Lepus's highest position in the sky, but it remains close to the horizon for viewers in mid-northern latitudes and dips down below it for part of the year. Lepus contains the globular cluster M79, unusual for the fact that it sits within a southern constellation, while most other globular clusters are found at the center of the Milky Way a significant distance away. It is possible that globular cluster M79 was formed in the neighboring Canis Major dwarf galaxy.

Libra

The Scales   Size on the sky: 🖐

Libra falls within the constellations of the zodiac, the only one named for an inanimate object. It appears between zodiac neighbors Scorpius and Virgo.

KEY FACTS

The constellation Libra has 8 stars that make up a set of scales, the 3 brightest forming a triangle at the top. Binary star Zubenelgenubi can be split into its two stars with binoculars.

+ **months best viewed:** June–July

+ **seasonal chart location:** Summer, southwest quadrant

+ **alpha star:** Zubenelgenubi

A constellation of the northern summer sky, Libra lacks star magnitude when compared to its neighbors. Arab astronomers, and the Greeks before them, considered the alpha and beta stars of Libra to be the claws of the scorpion of neighboring Scorpius. Their names reflect this designation: Zubenelgenubi means "southern claw" and Zubeneschamali means "northern claw." Later, constellation Libra's scales were associated with those held by Astraea, the Greek goddess of justice. Libra's brightest stars can be found along a line traced from alpha star Antares in Scorpius to alpha star Spica in Virgo.

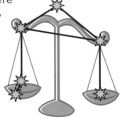

Lynx

The Lynx   Size on the sky: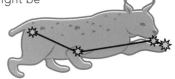

The constellation Lynx came into being in the late 17th century to fill an empty celestial area between Ursa Major and Auriga. Polish astronomer Johannes Hevelius introduced it.

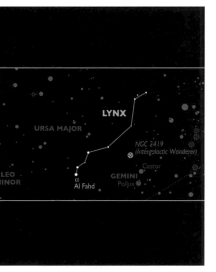

KEY FACTS

The 7 key stars of this inconspicuous constellation form a kinked line somewhat suggestive of a rearing lynx, located to the north of stars Castor and Pollux in Gemini.

+ months best viewed: February– March

+ seasonal chart location: Winter, northeast quadrant

+ alpha star: Al Fahd

Inconspicuous constellation Lynx makes a difficult sighting for casual stargazers. It is said that Lynx earned its name because it requires the observer to possess the sharp eyesight of the northern forest cat to spot it. As a kind of compensation, the constellation contains a number of double stars to reward the observer who uses even a modest telescope. Lynx also harbors a deep sky object in the form of the Intergalactic Wanderer (NGC 2419). The name "Wanderer" refers to the belief that the object is so far from the core of the Milky Way that it might be expected to break away from its current loose orbit around the galaxy.

Lyra

The Lyre   Size on the sky:

Alpha star Vega steals the show in constellation Lyra, appearing directly overhead in the mid-latitude summer sky. It is ranked the fifth brightest star in the sky.

KEY FACTS

The 6 stars of the easily identified constellation Lyra compose the ancient musical instrument; its alpha star Vega ranks among the brightest in the sky.

+ months best viewed: July–August

+ seasonal chart location: Summer, center of chart

+ alpha star: Vega

Lyra, the Lyre, belonged to Orpheus, Apollo's tragic son. He lost his beloved wife, Eurydice, not once but twice, when he blew the chance to retrieve her from Hades. Remaining faithful to her memory, Orpheus was killed by a group of scorned women. An impressed Zeus then sent Orpheus's lyre to the sky. With Deneb in Cygnus and Altair in Aquila, alpha star Vega in Lyra forms an asterism known as the Summer Triangle. In 12,000 or so years, alpha star Vega is slated to become the polestar. The Ring Nebula (M57), located at the edge of Lyra, is formed of gas, and it is visible through a telescope.

Monoceros

The Unicorn   Size on the sky: 🖐

The name "Monoceros" derives from the Greek for "one horned." A constellation created in the 17th century, the Unicorn was first officially recorded by German astronomer Jakob Bartsch.

KEY FACTS

Its 8 stars of fairly low magnitude compose a dim, stylized unicorn with its "head" stars close to Betelgeuse. Brighter neighboring stars and features help locate Monoceros.

+ **months best viewed:** January–February

+ **seasonal chart location:** Winter, southeast quadrant

+ **alpha star:** Ctesias

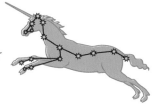

Monoceros appears within the asterism known as the Winter Triangle, which is composed of Betelgeuse in Orion, Procyon in Canis Minor, and Sirius in Canis Major. The constellation is dim in comparison to those very bright stars, but it fills an important gap on the sky chart, and its position on the Milky Way provides many good reference points. One highlight—open star cluster M50—appears between Procyon and Sirius and can be spotted through binoculars. Monoceros may refer to a biblical creature that decided to play in the rain rather than join Noah and the other animals on the ark to escape the great flood.

Ophiuchus

The Serpent Bearer   Size on the sky: 🖐

Southern constellation Ophiuchus interrupts neighboring Serpens, splitting it in two. Entwined mythologies perpetually link them.

KEY FACTS

The constellation Ophiuchus has 14 stars that make up the blocky figure of a man. The Serpent Bearer's alpha star, Rasalhague, represents the head.

+ **months best viewed:** June–July

+ **seasonal chart location:** Summer, center of chart

+ **alpha star:** Rasalhague

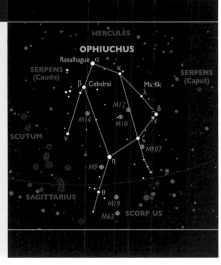

The constellation Ophiuchus likely honors Asclepius, the Greek god of medicine. He appears between the head and tail of Serpens the snake, who taught the god about the healing properties of plants. Asclepius used this knowledge to raise a fatally wounded Orion. A concerned Hades, god of the dead, fearing irrelevance, had Zeus kill Asclepius. Now healer and serpent watch from the sky, easily identified just to the west of the Milky Way. The constellation's location makes it rich in deep sky objects, including a series of globular clusters— M9, M10, M12, M14, M19, M62, and M107—all visible with binoculars.

Orion

The Hunter   Size on the sky: 🖐

Perhaps the most easily recognized constellation in the sky, Orion takes center stage on the celestial equator with more than its fair share of superbright stars.

KEY FACTS

Some 20 stars make up the obvious figure of a hunter. The constellation is punctuated with huge stars of high magnitude such as Betelgeuse, Rigel, and Bellatrix.

+ months best viewed: January–February

+ seasonal chart location: Winter, southeast quadrant

+ alpha star: Betelgeuse

Without a doubt, Orion is a stellar rock star. Among many superlatives, three very bright stars form the hunter's iconic belt; red-tinted, variable Betelgeuse boasts a diameter 300 to 400 times wider than the sun's; and blue-white supergiant Rigel shines 57,000 times brighter than the sun. The great hunter of Greek mythology, Orion appears in the vicinity of his canine companions Canis Major and Minor and other associates, such as Lepus the Hare. Located below Orion's belt, the Great Orion Nebula (M42) is visible as a cloud patch to the naked eye. Its gas cloud churns out new stars at a furious pace.

Pegasus

The Winged Horse   Size on the sky: 🖐🖐

Pegasus shares the sky with other constellations attached to the same mythology: Andromeda, Cassiopeia, Cepheus, Cetus, and the heroic Perseus.

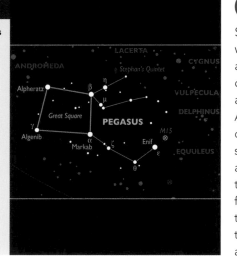

KEY FACTS

The horse figure in this large constellation is made up of 15 stars, with 4 of them forming the Great Square of Pegasus, a prominent asterism.

+ months best viewed: September–October

+ seasonal chart location: Autumn, center of chart

+ alpha star: Markab

Constellation Pegasus is rich in deep sky objects, including the M15 cluster and Stephan's Quintet, a group of five galaxies, four of which are interacting with one another. Pegasus also is a fertile source of exoplanets, planets that circle stars other than the sun. Pegasus shares a star, Alpheratz, and a story with constellation Andromeda (see page 402). Alpheratz forms part of the Great Square asterism with three stars from Pegasus. Pegasus also has associations with the hero Bellerophon, who fought the Chimera with the help of the steed. He had tamed the flying horse with aid from the goddess Athena.

Perseus
The Hero   Size on the sky:

Constellation Perseus, named for a superhero of Greek mythology, has a lot going for it: many bright and easily spotted stars, a superlative meteor shower, and a fascinating double star cluster.

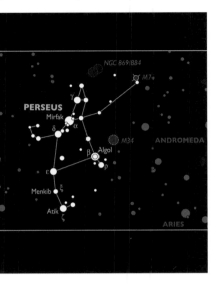

KEY FACTS

Perseus has 14 key stars, with 6 of them at magnitude 3 or lower; the constellation straddles the Milky Way between associated formations Cassiopeia and Andromeda.

+ months best viewed: November–December

+ seasonal chart location: Autumn, northeast quadrant

+ alpha star: Mirfak

Perseus belongs to the same mythological adventure as neighboring constellations Andromeda, Cassiopeia, Cepheus, Cetus, and Pegasus. The hero rescued captive Andromeda, daughter of Cassiopeia and Cepheus, from the predatory sea monster Cetus. The constellation is prominent in the late autumn sky and contains bright and notable signposts. Among them is Algol, the beta star, an eclipsing variable that dims by a full magnitude every three days. Perseus contains double cluster NGC 869 and NGC 884, two separate groups of stars that can be viewed through binoculars. It also is point of origin for the mid-August Perseid meteor showers.

Pisces
The Fish   Size on the sky:

The fish of Pisces are on the small side, but the cord joining them by their tails spans a wide area and crosses the ecliptic, the apparent path taken by the sun across the sky.

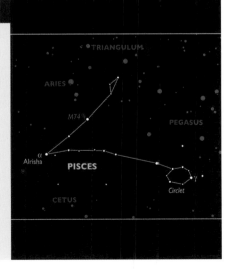

KEY FACTS

The constellation Pisces has 17 key stars; the larger fish in the southern end contains the group of 5 key stars known as the Circle asterism.

+ months best viewed: October–November

+ seasonal chart location: Autumn, southeast quadrant

+ alpha star: Alrisha

Pisces is an ancient constellation of the zodiac that forms a large V in the sky. Its southern asterism, known as the Circle, appears just south of the Great Square of Pegasus. The outside cord attached to the smaller fish is galaxy M74, a faint but elegant spiral galaxy whose arms can be resolved with an 8-inch-aperture telescope. For the ancient Greeks, the two fish of Pisces represented the goddess Aphrodite and her son Eros, who both changed themselves into fish to escape the sea monster Typhon, often equated with Cetus. The cord, tied at alpha star Alrisha, holds mother and son together as they swim.

Piscis Austrinus

The Southern Fish   Size on the sky: 🖐

The key to locating this low southern constellation—low to the horizon even at its highest for northern mid-latitude observers—is its formidable alpha star Fomalhaut.

KEY FACTS

Numerous ancient cultures have visualized these 10 stars as a fish; the name of the alpha star, Fomalhaut, is derived from the Arabic for "mouth of the fish."

+ months best viewed: September–October

+ seasonal chart location: Autumn, southwest quadrant

+ alpha star: Fomalhaut

One of the original 48 constellations Ptolemy identified, Piscis Austrinus is located in the "watery" section of the sky just south of Aquarius and east of Capricornus. In Greek mythology, the Southern Fish was considered to be the sire of the two fish in the constellation Pisces. Alpha star Fomalhaut, considered a young star at between 100 and 300 million years old, ranks as the 18th brightest star to the naked eye and is a mere 25 light-years from Earth. In 2008, the Hubble telescope photographed a planet orbiting Fomalhaut. Named Fomalhaut b, it was the first confirmed extrasolar planet detected via direct imaging.

Puppis

The Stern   Size on the sky: 🖐

Puppis is one of the three breakout constellations created in 1763 from Argo Navis, one of the original constellations on Ptolemy's list of 48.

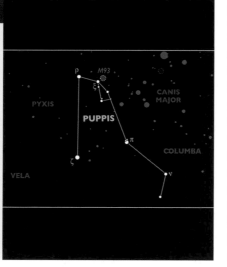

KEY FACTS

The constellation Puppis has 12 stars known as the stern of the largest part of the ship *Argo*; it lies in the rich starfields of the Milky Way.

+ months best viewed: February–March

+ seasonal chart location: Winter, southeast quadrant

+ alpha star: none (brightest: Zeta Puppis)

Puppis, disappointingly to some, doesn't represent a cute little dog, but the stern of the mythological ship *Argo*. The *Argo* carried Jason and the Argonauts on the quest to steal the Golden Fleece, once sported by Aries the Ram, from the clutches of a dragon. When Argo Navis was dismantled during the constellation update of the modern era, its stars were not relabeled within the new constellations and so Puppis was left without a labeled alpha star. Its brightest star is Zeta Puppis, also known as Naos, from the Greek for "ship." A blue supergiant, Zeta Puppis is a spectacular naked-eye star of the second magnitude.

Sagitta
The Arrow   Size on the sky: 👍

Although the shapes associated with many constellations often stretch the imagination, the arrow-like form of the small but distinct Sagitta cleanly hits the mark.

KEY FACTS

The unmistakable arrow shape of the constellation Sagitta contains 5 stars. The arrow's shaft contains M71, a globular cluster visible through binoculars.

+ months best viewed: August–September

+ seasonal chart location: Summer, southeast quadrant

+ alpha star: Sham

The third smallest constellation, Sagitta lies totally within the Summer Triangle asterism formed of Altair in Aquila, Deneb in Cygnus, and Vega in Lyra. Most ancient cultures made associations to the arrow and there are a number of arrow stories to choose from to explain Sagitta's celestial presence. In Greco-Roman myth, it could be the arrow Cupid (Eros) used to spread the love, or the arrow Apollo used to slay the Cyclops, or the one Hercules deployed to dispatch the Stymphalian birds in one of his 12 labors. A relationship to Sagitta, despite the name "arrow," is not widely accepted.

Sagittarius
The Archer   Size on the sky: ✋

Spread across a wide band of the Milky Way, Sagittarius is a treasure trove of celestial superlatives, including two prominent asterisms and eight high-magnitude stars.

KEY FACTS

The complex figure of the constellation Sagittarius contains 22 stars, as well as the clearly defined Teapot asterism and the Milk Dipper asterism.

+ months best viewed: July–August

+ seasonal chart location: Summer, southeast quadrant

+ alpha star: Rukbat

Appearing south of Vega, the Teapot asterism contains the eight central stars of Sagittarius. The Milk Dipper includes the ladle and lambda star handle that form the Teapot's handle. Sagittarius's alpha star is not its brightest; that would be epsilon Kaus Australis. Sigma star Nunki, the second brightest, appeared on the Tablet of the 30 Stars of the ancient Babylonians. The centaur of Sagittarius—half man, half horse—pursues his supposed prey, the neighboring constellation Scorpius, through the sky. Sagittarius is often identified with Chiron of Greek mythology, who paradoxically was a wise and peaceful creature.

Scorpius

The Scorpion Size on the sky:

The narrow outline of Scorpius harbors 2 of the 25 brightest stars as well as outstanding star clusters, including the spectacular Butterfly cluster (M6).

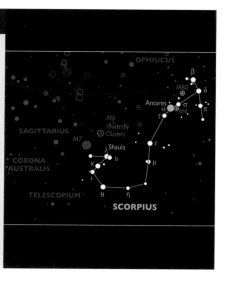

KEY FACTS

A recognizable scorpion shape is made from 17 main stars in the constellation Scorpius, anchored by bright Antares in the core and Shaula at the tail.

+ **months best viewed:** July–August

+ **seasonal chart location:** Summer, southwest quadrant

+ **alpha star:** Antares

A straightforward constellation, Scorpius is located by first-magnitude star Antares, known to the Romans as Cor Scorpionus, or "heart of the scorpion." A red supergiant, about 300 times as large as the sun, Antares's name means "rival of Mars," alluding to its size and color. The Butterfly cluster (M6), one of many open and globular star clusters in the vicinity, reveals its butterfly shape through binoculars. In Greek mythology, Scorpius is tied to the constellation Orion as the animal that inflicted the fatal wound on the hero's leg. The two constellations appear on opposite sides of the sky to keep them separated.

Scutum

The Shield Size on the sky:

This faint modern constellation was created by Polish astronomer Johannes Hevelius in the late 17th century in the feature-rich Milky Way.

KEY FACTS

Only 4 stars compose the diamond-like shield of constellation Scutum that appears between Sagittarius, Aquila, and the lower half of Serpens.

+ **months best viewed:** July–August

+ **seasonal chart location:** Summer, southeast quadrant

+ **alpha star:** Alpha Scuti

Scutum lacks any bright or otherwise notable stars, but its location on the Milky Way puts it in close proximity to several open star clusters, including M11, the so-called Wild Duck cluster. Binoculars alone will reveal this formation that resembles a dense flock of waterfowl, and an 8-inch telescope will show thousands of glittering stars. The Wild Duck appears just southeast of Scutum's northernmost star. The constellation's original name was Scutum Sobiescianum, or "shield of Sobieski," honoring Polish King (and Hevelius's patron) John III Sobieski, who defeated the Ottoman Empire in the 1683 Battle of Vienna.

Serpens

The Serpent   Size on the sky:

The only split constellation in the night sky, Serpens is divided between head and tail, winding over the shoulder of Ophiuchus, the Serpent Bearer, in the middle.

KEY FACTS

Two groups of stars, the triangular head (Caput) and elongated tail (Cauda), make up the two separate parts of the serpent constellation Serpens.

+ months best viewed: June–July

+ seasonal chart location: Summer, southern half of chart

+ alpha star: Unukalhai

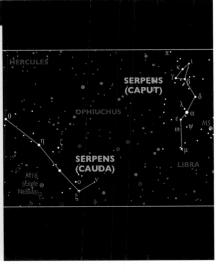

The constellation Serpens formerly encompassed the whole scenario of snake head, tail, and Ophiuchus. They were physically and mythologically united, representing the Greek god of medicine, Asclepius, and the serpent who taught him about the healing power of plants. Serpens Cauda contains the combination nebula and star cluster called the Eagle Nebula (M16), which occurs in an active region of star formation about 7,000 light-years away. Both the nebula and cluster can be spotted with an 8-inch telescope. M16 was famously photographed by the Hubble telescope, which captured a prominent area of star destruction and formation.

Taurus

The Bull   Size on the sky:

Taurus, a zodiacal constellation, is easily located because nearby Orion's belt points directly northwest to Aldebaran, the Bull's alpha star.

KEY FACTS

The renowned constellation Taurus—including the most prominent feature, the horns—contains 13 stars. The southern horn contains Aldebaran, the Bull's "eye."

+ months best viewed: January–February

+ seasonal chart location: Winter, southwest quadrant

+ alpha star: Aldebaran

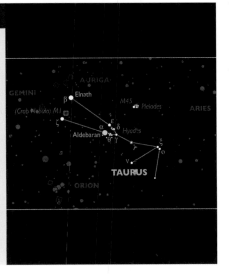

The Egyptians associated Taurus with Osiris, god of life and fertility. Greek god Zeus disguised himself as a bull to capture Europa and carry her to the continent that bears her name. Taurus contains two prominent star clusters, the Pleiades and the Hyades. The Pleiades (M45), an open cluster, is the more famous and displays 6 or 7 of its hundreds of stars to the naked eye. Vying for celebrity is M1, the Crab Nebula. It was the first deep sky object cataloged by Charles Messier, in 1758. Cultures from the American Southwest to China witnessed and recorded the death of the star in 1054 that created the nebula.

Triangulum
The Triangle   Size on the sky: ✋

Despite its nonmythological, utilitarian name—a feature of many of the modern constellations—Triangulum was on Ptolemy's list of the original 48.

KEY FACTS

The 3 stars of this geometric constellation form a distinct triangle located between Andromeda, Perseus, and Aries. All 3 are brighter than magnitude 3.

+ **months best viewed:** November–December

+ **seasonal chart location:** Autumn, southeast quadrant

+ **alpha star:** Mothallah

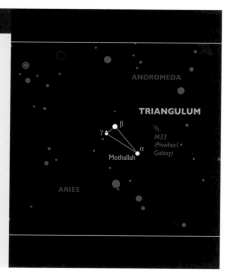

In late autumn, Triangulum is easy to find just east of Andromeda. The Pinwheel galaxy (M33) lies west of the constellation, a spiral galaxy like our own Milky Way. Under good viewing conditions, it appears like a slight glow, although a substantial telescope is needed to view the pinwheel effect. The ancient Hebrews appear to have named this constellation for the small percussion instrument, and the name stuck. The Greeks also took notice that these stars resembled their letter *delta*, and it was known for a time as Deltoton. The theme continues in the name for the alpha star, Mothallah, which means "triangle" in Arabic.

Ursa Major
The Great Bear   Size on the sky: 🖐️🖐️

To various cultures, the formidable assemblage of stars in Ursa Major represented a bear, a chariot, a horse and wagon, a team of oxen, and even a hippopotamus.

KEY FACTS

The complex figure of the constellation Ursa Major contains 20 stars, including the 7-star asterism of the Big Dipper, representing the bear's rear torso and tail.

+ **months best viewed:** March–April

+ **seasonal chart location:** Spring, center of chart

+ **alpha star:** Dubhe

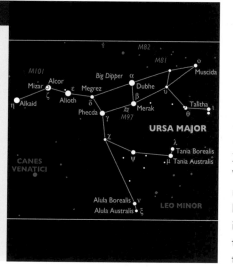

During the year, prominent Ursa Major seems to run in a circle with its back to Polaris, the North Star. The constellation's many named stars include Alcor and Mizar, which appear to be a binary system but are not. Galaxies M81 and M82 lie close together and may be in collision with each other. In Greek mythology, the Great Bear represents Callisto, yet another victim of the goddess Hera's jealousy on finding that her husband, Zeus, had strayed again. When Callisto's son mistakenly tried to kill her while hunting, Zeus placed them both in the sky to protect them.

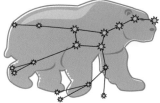

Ursa Minor

The Little Bear   Size on the sky:

The Little Bear plays second fiddle to none, containing the North Star, Polaris, as its alpha star. The constellation and its polestar are visible year-round.

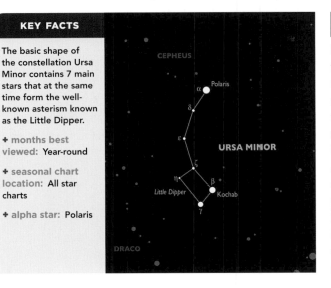

KEY FACTS

The basic shape of the constellation Ursa Minor contains 7 main stars that at the same time form the well-known asterism known as the Little Dipper.

+ **months best viewed:** Year-round

+ **seasonal chart location:** All star charts

+ **alpha star:** Polaris

Navigators, wanderers, and sky gazers throughout history have found their way by the enduring presence of Ursa Minor and Polaris, indicating the celestial pole. This won't always be the case. Over millennia, the celestial pole shifts. It will be closest to Polaris around 2100, and then will start to move away, passing other stars and eventually settling at Vega for a while in about 12,000 years. Ursa Minor appears in the sky as the mythological child of the Great Bear, who represents Zeus's paramour Callisto. Late December brings the Ursid meteor shower, which display about ten meteors an hour at its peak.

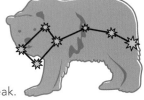

Virgo

The Virgin   Size on the sky:

The only zodiacal constellation representing a woman, Virgo is the second largest constellation and a fertile area for viewing star clusters and galaxies.

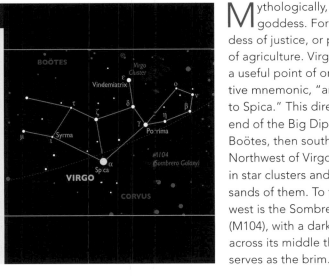

KEY FACTS

The figure of a seated woman in the constellation Virgo contains 13 main stars, with bright alpha star Spica in the vicinity of her left hand.

+ **months best viewed:** May–June

+ **seasonal chart location:** Spring, southeast quadrant

+ **alpha star:** Spica

Mythologically, Virgo appears to represent a goddess. For many she was Dike, the goddess of justice, or possibly Demeter, the goddess of agriculture. Virgo's bright alpha star, Spica, is a useful point of orientation as part of an alliterative mnemonic, "arc to Arcturus, then speed on to Spica." This directs one's gaze south from the end of the Big Dipper's handle to find Arcturus in Boötes, then south again to Spica just below it. Northwest of Virgo is an area dense in star clusters and galaxies—thousands of them. To the southwest is the Sombrero galaxy (M104), with a dark band across its middle that serves as the brim.

Natural Regions of North America

The continental United States and Canada can be divided into nine physiographic regions.

◾ Appalachian Highlands

The oldest mountain chain in North America, the heavily eroded Appalachians extend for about 2,000 miles (3,200 km) from Alabama to Newfoundland, with the highest peak being Mount Mitchell, in North Carolina, at 6,684 feet (2,040 m). This mountain range started forming 480 million years ago, when continental collisions caused volcanic activity and mountain building. Ecologically, this region hosts eastern temperate forests with a great variety of coniferous and deciduous trees and wildflowers, some of them high-elevation specialists.

◾ Coastal Plain

This gradually rising flatland spans some 4,000 miles (6,400 km) in total and covers several distantly related regions along the Gulf of Mexico, the southern Atlantic coast, and the northernmost coasts of the Arctic Ocean. In the past, oceans covered these plains, depositing sediment layers over millions of years, until falling sea levels exposed them. The types of vegetation in these widely separated regions range from subtropical trees and flowers in the southernmost coastal plain to marsh plants along the mid-Atlantic to the largely treeless tundra on the northern coast of Alaska.

◾ Interior Plains

Ranging from the lowlands of the Saint Lawrence River Valley in the east to the mile-high Great Plains in the west, the vast Interior Plains of North America provide fertile soils, especially for productive prairie farms. A shallow sea covered much of this region as recently as 75 million years ago, and sediments from rivers draining the Appalachians and western mountains were deposited in layers throughout the sea. The Great Plains were once covered by vast and diverse expanses of natural grasses, sagebrush, and a varied suite of wildflowers. Much of this ecosystem has vanished, the land brought into use by modern agriculture and extensive grazing.

◾ Interior Highlands

The Ozark Plateau and Ouachita Mountains form the Interior Highlands, which are centered on Arkansas and southern Missouri, with mountains reaching more than 2,600 feet (800 m) high. These ancient eroded highlands were connected to the Appalachians until tectonic activity separated them some 200 million years ago. Ecologically, this relatively small area straddles the southern Interior Plains and the eastern Coastal Plain, with trees and wildflowers representing both regions.

◾ Rocky Mountains

The highest mountain system in North America, the Rockies dominate the landscape for some 3,000 miles (4,800 km), from New Mexico to Alaska, with more than 50 peaks surpassing 14,000 feet (4,300 m). Tectonic activity uplifted the Rockies about 50 to 100 million years ago, making them much younger and less eroded than the Appalachians. The ecological hallmarks of the Rockies are its coniferous forests of pines, firs, and spruces, adapted to high elevations, with wildflower species similarly adapted to elevations and temperatures.

◾ Intermontane Basins and Plateaus

This region is called *intermontane* because it is situated between the Pacific and the Rocky Mountain systems. Pacific mountains block most moisture-bearing clouds coming from the Pacific Ocean, giving desert climates to places like the Colorado Plateau, 5,000 to 7,000 feet (1,500 to 2,100 m) high, and the Great Basin. In Canada and Alaska, the immense Yukon River Valley and the Yukon-Tanana Uplands are part of this region. The deserts feature cacti and a host of other specialist plants of the arid West. Cottonwoods, ashes, and willows line rivers that run intermittently through the dry plains.

◾ Pacific Mountain System

From Alaska to California, mountains and volcanoes tower over the West Coast in an almost unbroken chain. These mountain ranges are geologically young and seismically active, with uplift starting some five mill on years ago. The highest mountain in North America Alaska's Mount McKinley (20,320 feet/6,200 m), is still growing at about a millimeter a year—about the thickness of a fingernail. Distantly separated from the Rockies, this mountain range supports its own distinct varieties of trees and wildflowers adapted to higher elevations.

Landforms are relatively flat, having been eroded and scoured by glaciers over millions of years. The exposed bedrock ranges in age from 570 million to more than 3 billion years old. This is a vast and extensively diverse region of climatic extremes and varied vegetation, from dense boreal forests in the south to frigid tundra in the north, populated by stunted trees, small shrubs, lichens, and ground-clinging herbs.

■ Canadian Shield

The geologic core of North America is the Canadian Shield, which contains the continent's oldest rocks.

■ Arctic Lands

Highlands known as the Innuitian Mountains cover most islands. The icy climate is too harsh for most animals and vegetation, and much of the ground is permanently frozen. Nevertheless, low-growing shrubs, small tundra plants, and lichens manage to survive.

Further Resources

■ Wildflowers

BOOKS

Brandenburg, David M. *National Wildlife Federation Field Guide to Wildflowers of North America.* Sterling, 2010.

National Audubon Society. *Field Guide to North American Wildflowers: Eastern Region.* Alfred A. Knopf, 2001.

National Audubon Society. *Field Guide to North American Wildflowers: Western Region.* Alfred A. Knopf, 1979.

Peterson, Roger Tory, and Margaret McKenny. *A Field Guide to Wildflowers: Northeastern and North-Central North America,* revised ed. Peterson Field Guide Series. Houghton Mifflin Harcourt, 1998.

WEBSITES

Flora of North America
www.efloras.org

Lady Bird Johnson Wildflower Center
www.wildflower.org

U.S. Department of Agriculture, National Resources Conservation Service, Plants Database
www.plants.usda.gov

U.S. Forest Service, Celebrating Wildflowers
www.fs.fed.us/wildflowers

APPS

Audubon Wildflower Identification App
www.audubonguides.com/field-guides/wildflowers-north-america.html

National Science Foundation, Project BudBurst
www.budburst.org/gomobile.php

■ Trees

BOOKS

Dirr, Michael A. *Manual of Woody Landscape Plants: Their Identification, Ornamental Characteristics, Culture, Propagation and Uses,* 4th ed. Stipes Publishing Co., 1990.

Elias, Thomas S. *The Complete Trees of North America: A Field Guide and Natural History.* Gramercy Publishing Co., 1987.

Harris, James G., and Melinda Woolf Harris. *Plant Identification Terminology: An Illustrated Glossary.* Spring Lake Publishing, 2001.

Rushforth, Keith, and Charles Hollis. *National Geographic Society Field Guide to Trees of North America.* National Geographic Society, 2006.

Rutkow, Eric. *American Canopy: Trees, Forests, and the Making of a Nation.* Scribner, 2012.

Sibley, David Allen. *The Sibley Guide to Trees.* Alfred A. Knopf, 2009.

WEBSITES

Flora of North America
www.efloras.org

Lady Bird Johnson Wildflower Center
www.wildflower.org

Missouri Botanical Garden Plant Finder
www.missouribotanicalgarden.org/gardens-gardening/your-garden/plant-finder.aspx

University of Connecticut Plant Database of Trees, Shrubs, and Vines
www.hort.uconn.edu/plants/index.html

USDA Forest Service Fire Effects Information System
www.fs.fed.us/database/feis

Virginia Tech Dendrology Factsheets
www.dendro.cnre.vt.edu/dendrology/factsheets.cfm

■ Rocks & Minerals

BOOKS

Dixon, Dougal, and Raymond Bernor, eds. *The Practical Geologist.* Simon and Schuster, 1992.

Geology Underfoot (series of books). Mountain Press Publishing Company.

Klein, Cornelis, and Anthony Philpotts. *Earth Materials: Introduction to Mineralogy and Petrology,* Cambridge University Press, 2013.

Marshak, Stephen. *Essentials of Geology.* W. W. Norton and Company, 2009.

Price, Monica T. *The Sourcebook of Decorative Stone.* Firefly Books Ltd., 2007.

Roadside Geology (series). Mountain Press.

Wenk, Hans-Rudolf, and Andrei Bulakh. *Minerals: Their Constitution and Origin.* Cambridge University Press, 2004.

WEBSITES

Association of American State Geologists
www.stategeologists.org

Dinosaur Society's Dinosaur News
www.dinosaursociety.com/news/category/
dinosaur-news

Geological Society of America
www.geosociety.org

Mindat.org
www.mindat.org

Paleontology Portal
www.paleoportal.org

U.S. Geological Survey
www.usgs.gov

U.S. Geological Survey Volcano Hazards Program
www.volcanoes.usgs.gov

University of California Museum of Paleontology's
Online Exhibits
www.ucmp.berkeley.edu/exhibits/index.php

Volcano World
http://volcano.oregonstate.edu

■ Weather

BOOKS

Ahrens, C. Donald. *Meteorology Today: An Introduction to Weather, Climate, and the Environment.* West Publishing Company, 2008.

Henson, Robert. *The Rough Guide to Climate Change.* (Rough Guide Reference Series). Rough Guides, 2011.

Williams, Jack. *The AMS Weather Book: The Ultimate Guide to America's Weather.* University of Chicago Press and The American Meteorological Society, 2009.

WEBSITES

American Meteorological Society
www.ametsoc.org/aboutams/index.html

American Meteorological Society's education pages
www.ametsoc.org/amsedu

Atmospheric Optics
www.atoptics.co.uk

Canadian Weatheroffice
www.weatheroffice.gc.ca/canada_e.html

Cloud Appreciation Society
www.cloudappreciationsociety.org

National Weather Association
www.nwas.org/about.php

U.S. National Weather Service
www.weather.gov

U.S. National Weather Service JetStream Online Weather School
www.srh.weather.gov/jetstream

University Cooperation for Atmospheric Research Spark science education
www.spark.ucar.edu

Weatherwise magazine
www.weatherwise.org

■ Night Sky

BOOKS

Aguilar, David A. *13 Planets: The Latest View of the Solar System.* National Geographic Society, 2011.

Daniels, Patricia. *The New Solar System.* National Geographic Society, 2009.

Ridpath, Ian, and Wil Tirion. *Stars and Planets,* 4th ed. Princeton Field Guides. Princeton University Press, 2007.

Trefil, James. *Space Atlas: Mapping the Universe and Beyond.* National Geographic Society, 2012.

WEBSITES

Astronomy magazine
www.astronomy.com

EarthSky
www.earthsky.org

Harvard-Smithsonian Center for Astrophysics
www.cfa.harvard.edu

Hubble Space Telescope
www.hubblesite.org

International Astronomical Union
www.iau.org

National Aeronautics and Space Administration
www.nasa.gov

Sky & Telescope magazine
www.skyandtelescope.com

Space.com
www.space.com/news

About the Contributors

■ About the Authors

Bland Crowder is associate director and editor with the Flora of Virginia Project, whose *Flora of Virginia* was published in December 2012. He is also a freelance writer and editor. He lives in Richmond, Virginia.

Sarah Garlick is a writer and educator specializing in earth and environmental science. She holds degrees in geology from Brown University and the University of Wyoming, and she is the author of the award-winning book *Flakes, Jugs, and Splitters: A Rock Climber's Guide to Geology* (2009). She lives with her family in the mountains of New Hampshire.

Catherine Herbert Howell, a former National Geographic staff member, has written extensively on natural history. She explored the relationships between people and plants in *Flora Mirabilis: How Plants Have Shaped World Knowledge, Health, Wealth, and Beauty* (2009) and covered the importance of birds in culture in the *National Geographic Bird-watcher's Bible* (2012). Howell serves as a master naturalist volunteer in Arlington, Virginia.

Jack Williams was founding editor of the *USA TODAY* weather page in 1982 and the USATODAY.com weather section in 1995. After retiring from USA TODAY in 2005, Williams was director of public outreach for the American Meteorological Society through 2009. Since then, he has been a freelance writer and is the author or coauthor of seven books, including the *National Geographic Field Guide to the Water's Edge* (2012).

■ About the Consultants

George Yatskievych is a curator at the Missouri Botanical Garden. His research involves the wild plants of Missouri and surrounding states. He also has interests in ferns and parasitic flowering plants.

Kay Yatskievych is a retired botanist associated with the Missouri Botanical Garden. She is the author of a field guide to Indiana wildflowers and is working on a catalog of that state's flora.

■ About the Artists

Fernando G. Baptista graduated in Fine Arts from País Vasco in Spain. He worked as a graphic artist for a newspaper in Bilbao, as a freelancer creating reconstructive pieces for museums, and spent six years as a professor of infographics at the University of Navarra. Baptista has won more than 125 international awards. In 2012, he was named one of the five most influential infographic artists of the last 20 years.

Jared Travnicek is a scientific and medical illustrator. He received his M.A. in Biological and Medical Illustration from the Johns Hopkins University School of Medicine in Baltimore, Maryland. Travnicek is a Certified Medical Illustrator and a professional member of the Association of Medical Illustrators.

Jenny Wang was trained in medical and biological illustration at the Johns Hopkins School of Medicine in Baltimore, Maryland. Her areas of focus include scientific illustration, biomedical visualization, and information design. As a nature lover, she is delighted to have contributed to this guide.

Credits for sky and constellation charts and map
Seasonal night sky charts: pp. 394–401, Wil Tirion; Constellation charts and mythological icons: pp. 402–429, Brown Reference Group; Natural Regions of North America: p. 431, NG Maps.

Illustrations Credits

New artwork appearing in this book was created by Fernando G. Baptista, Jared Travnicek, and Jenny Wang.

FRONT COVER: Top Row (Left to Right): Frans Lanting/NG Stock; Renaud Visage/The Image Bank/Getty Images; Yva Momatiuk & John Eastcott/Minden Pictures/NG Stock; Paul Nicklen/NG Stock. Middle Row: Ron Gravelle/NG My Shot; Tim Fitzharris/Minden Pictures/NG Stock; Lincoln Harrison/NG My Shot; Amy White & Al Petteway/NG Stock. Bottom Row: Yva Momatiuk & John Eastcott/Minden Pictures/NG Stock; Paul Nicklen/NG Stock; David Muench/Getty Images; Crisma/Getty Images.

BACK COVER: Top Row: Terry Donnelly; Bret Webster/Science Source; Mike Theiss/NG Stock; Ingo Arndt/Minden Pictures. Middle Row: Frans Lanting/NG Stock; Jack Dykinga; Jim Brandenburg/Minden Pictures; NASA/Science Source. Bottom Row: Carr Clifton/Minden Pictures; Mary Liz Austin; Paolo De Faveri/paolodefaveri.com; Pasieka/Science Source.

INTERIOR: 2-3, Tim Fitzharris/Minden Pictures; 4, Kevin Barry; 6, Carr Clifton; 8, Terry Donnelly; 10, UIG via Getty Images; 14 (UP), Mark Turner/Getty Images; 14 (LO), Judywhite/GardenPhotos.com; 15 (UP), Gerald D. Tang; 15 (LO), Glenis Moore/Science Source; 16 (UP), Gerald D. Tang; 16 (LO), Bob Gibbons/Science Source; 17 (UP), Noveau/Shutterstock; 17 (LO), Visuals Unlimited, Inc./Gerry Bishop/Getty Images; 18 (UP), Arco Images GmbH/Alamy; 18 (LO), Willem Kolvoort/Foto Natura/Getty Images; 19 (UP), Roger Whiteway/iStockphoto; 19 (LO), Robert and Jean Pollock/Science Source; 20 (UP), Stephen P. Parker/Science Source; 20 (LO), Bob Gibbons/Science Source; 21 (UP), Gerald D. Tang; 21 (LO), Michael P. Gadomski/Science Source; 22 (UP), Andy Crawford/Getty Images; 22 (LO), LianeM/Shutterstock; 23 (UP), David Nunuk/Science Source; 23 (LO), karloss/Shutterstock; 24 (UP), Kenneth M. Highfill/Science Source; 24 (LO), Lindasj22/Shutterstock; 25 (UP), Hal Horwitz/Corbis; 25 (LO), Ron Wolf/Tom Stack & Associates; 26 (UP), Sumikophoto/Shutterstock; 26 (LO), Jerry Pavia/Getty Images; 27 (UP), Joshua McCullough, PhytoPhoto/Getty Images; 27 (LO), Clint Farlinger; 28 (UP), Kenneth M. Highfill/Science Source; 28 (LO), Inga Spence/Science Source; 29 (UP), Rod Planck/Science Source; 29 (LO), John Greim; 30 (UP), Geoff Kidd/Science Source; 30 (LO), Gregory K. Scott/Science Source; 31 (UP), Peter Herring; 31 (LO), Maria Mosolova/Getty Images; 32 (UP), Barry Breckling; 32 (LO), Kevin Schafer/Getty Images; 33 (UP), Kenneth M. Highfill/Science Source; 33 (LO), Steve Guttman; 34 (UP), Gerry Bishop/Visuals Unlimited, Inc.; 34 (LO), Michael Wheatley/All Canada Photos/Getty Images; 35 (UP), judywhite/GardenPhotos.com; 35 (LO), William A. Bake/Corbis; 36 (UP), Jeff Lepore/Science Source; 36 (LO), Kenneth M. Highfill/Science Source; 37 (UP), William S. Moye; 37 (LO), Ross Hoddinott/Minden Pictures; 38 (UP), Barry Breckling; 38 (LO), Bill Beatty; 39 (UP), Robert & Jean Pollock/Visuals Unlimited, Inc.; 39 (LO), Laura Berman; 40 (UP), Valerie Giles/Science Source; 40 (LO), Michael P Gadomski/Getty Images; 41 (UP), Roanna Littlefield; 41 (LO), Kenneth M. Highfill/Science Source; 42 (UP), Nature's Images/Science Source; 42 (LO), Dave Welling; 43 (UP), Douglas Craig/iStockphoto; 43 (LO), Artefficient/Shutterstock; 44 (UP), Michael P. Gadomski/Science Source; 44 (LO), Gerald D. Tang; 45 (UP), Michael P. Gadomski/Science Source; 45 (LO), Stephen Dalton/Minden Pictures; 46 (UP), Adrian Bicker/Science Source; 46 (LO), jack thomas/Alamy; 47 (UP), Fotosearch; 47 (LO), David Hall/Alamy; 48 (UP), Mike Theiss/National Geographic Stock; 48 (LO), Mike Comb/Science Source; 49 (UP), David Davis/Science Source; 49 (LO), Will & Deni McIntyre/Science Source; 50 (UP), Luther Linkhart/Visuals Unlimited, Inc.; 50 (LO), Gary Cook/Visuals Unlimited, Inc.; 51 (UP), Brian Barnes/Alamy; 51 (LO), Ron Wolf/Tom Stack & Associates; 52 (UP), Scott Cramer/Getty Images; 52 (LO), Sue Carnahan; 53 (UP), Jerry Pavia/Getty Images; 53 (LO), Roger Hyam/Getty Images; 54 (UP), Visuals Unlimited, Inc./Gerry Bishop/Getty Images; 54 (LO), Scientifica/Visuals Unlimited, Inc.; 55 (UP), John W. Bova/Science Source; 55 (LO), Pi-Lens/Shutterstock; 56 (UP), James Steinberg/Science Source; 56 (LO), 56 (UP) Photo by Jessie Harris; 57 (UP), Douglas Graham/Wild Light Photography, Inc.; 57 (LO), Ventura/Shutterstock; 58 (UP), Nature's Images, Inc./Science Source; 58 (LO), Thomas & Pat Leeson/Science Source; 59 (UP), James Randklev/Getty Images; 59 (LO), Gilbert S. Grant/Science Source; 60 (UP), Scott Camazine/Science Source; 60 (LO), Len Rue Jr./Science Source; 61 (UP), Visuals Unlimited, Inc./John Gerlach/Getty Images; 61 (LO), Richard Bloom/Getty Images; 62 (UP), Gail Jankus/Science Source; 62 (LO), Gerry Bishop/Visuals Unlimited, Inc.; 63 (UP), Brian Gadsby/Science Source; 63 (LO), Martin Ruegner/Getty Images; 64 (UP), Nigel Cattlin/Science Source; 64 (LO), Juan Silva/Getty Images; 65 (UP), Meg Sommers; 65 (LO), Peter Haigh/Alamy; 66 (UP), Gerry Bishop/Visuals Unlimited, Inc.; 66 (LO), David Schwaegler; 67 (UP), Bill Pusztai; 67 (LO), Bob Gibbons/Science Source; 68 (UP), Len Rue Jr./Science Source; 68 (LO), Tim Graham/Getty Images; 69 (Both), Gerald D. Tang; 70 (UP), Lucy Jones/Visuals Unlimited, Inc.; 70 (LO), Gerald D. Tang; 71 (UP), Nature's Images/Science Source; 71 (LO), Visuals Unlimited, Inc./Nigel Cattlin/Getty Images; 72 (UP), Bob Gibbons/Science Source; 72 (LO), Gabriel Bertilson; 73 (UP), Ingrid Russell; 73 (LO), George Grall/National Geographic/Getty Images; 74 (UP), Gerry Bishop/Visuals Unlimited, Inc.; 74 (LO), Gerald D. Tang; 75 (UP), James Steinberg/Science Source; 75 (LO), Gerald D. Tang; 76 (UP), john t. fowler/Alamy; 76 (LO), Gerald D. Tang; 77 (Both), Gerald D. Tang; 78 (UP), George H. H. Huey; 78 (LO), Gerald D. Tang; 79 (UP), All Canada Photos/Alamy; 79 (LO), Ron Wolf/Tom Stack & Associates; 80 (UP), Geoff Kidd/Science Source; 80 (LO), Rich Wagner/WildNaturePhotos; 81 (UP), Gregory K. Scott/Science Source; 81 (LO), Rod Planck/Science Source; 82 (UP), James Steinberg/Science Source; 82 (LO), S.J. Krasemann/Getty Images; 83 (UP), judywhite/GardenPhotos.com; 83 (LO), Steffen Hauser/botanikfoto/Alamy; 84 (UP), Joshua McCullough/Getty Images; 84 (LO), Harley Seaway/Getty Images; 85 (UP), Bill Pusztai; 85 (LC), Nature's Images, Inc./Science Source; 86 (UP), Robert and Jean Pollock/Science Source; 86 (LO), Andreas Riedmiller; 87 (UP), Mark Steinmetz; 87 (LO), Dayton Wild/Visuals Unlimited, Inc.; 88 (UP), Rick & Nora Bowers/BowersPhoto.com; 88 (LO), kpzfoto/Alamy; 89 (UP), Shaughn F. Clements/Alamy; 89 (LO), Martin Shields/Science Source; 90 (UP), Nature's Images, Inc./Science Source; 90 (LO), Scott Camazine/Science Source; 91 (UP), Dr. John D. Cunningham/Visuals Unlimited, Inc.; 91 (LO), Gerald D. Tang; 92 (UP), Howard Rice/Getty Images; 92 (LO), imagebroker/Alamy; 93 (UP), Jeffrey Lepore/Science Source; 93 (LO), Sandra Ivany/Getty Images; 94, Floris van Breugel/NPL/Minden Pictures; 97, Tim Fitzharris/Minden Pictures; 100 (UP), Ron & Diane Salmon/Flying Fish Photography LLC; 100 (LO), Photos Lamontagne/Getty Images; 101 (UP), Barrett & MacKay/All Canada Photos/Getty Images; 101 (LO), Richard Thom/Visuals Unlimited/Getty Images; 102 (UP), Ted Kinsman/Science Source; 102 (LO), Terry Donnelly; 103 (UP), Tim Fitzharris/Minden Pictures; 103 (LO), Tony Wood/Science Source; 104 (UP), Perry Mastrovito/First Light/Corbis; 104 (LO), Ron & Diane Salmon/Flying Fish Photography LLC; 105 (UP), Ron Hutchinson Photography; 105 (LO), David Hosking/Alamy; 106 (UP), Bob Gibbons/Science Source; 106 (LO), David Matherly/Visuals Unlimited/Getty Images; 107 (UP), John Hagstrom; 107 (LO), Stephen J. Krasemann/Science Source; 108 (UP), Philippe Clement/NPL/Minden Pictures; 108 (LO), Michael P. Gadomski/Science Source; 109 (UP), Susan Glascock; 109 (LO), Fred Bruemmer/Getty Images; 110 (UP), Michael P. Gadomski/Science Source; 110 (LO), David Hosking/FLPA/Minden Pictures; 111 (UP), Ted Kinsman/Science Source; 111 (LO), Visuals Unlimited, Inc./Rob Kurtzman/Getty Images; 112 (UP), David Hosking/Minden Pictures; 112 (LO), Bob Gibbons/Science Source; 113 (UP), KENNETH W FINK/Getty Images; 113 (LO), Bob Gibbons/Minden Pictures; 114 (UP), Inga Spence/Science Source; 114 (LO), Thomas & Pat Leeson/Science Source; 115 (Both), Gerald D. Tang; 116 (UP), Susan Glascock; 116 (LO), Dennis Flaherty/Getty Images; 117 (UP), Michael P. Gadomski/Science Source; 117 (LO), Keith Rushforth/Minden Pictures; 118 (UP), David Middleton/NHPA/Photoshot; 118 (LO), David Jensen; 119 (UP), Cora Niele/Getty Images; 119 (LO), Jim Zipp/Science Source; 120 (UP), David Winkelman/David Liebman; 120 (LO), Ron & Diane Salmon/Flying Fish Photography LLC; 121 (UP), Susan Glascock; 121 (LO), David Jensen; 122 (UP), Ethan Welty/Aurora Photos; 122 (LO), Ron & Diane Salmon/Flying Fish Photography LLC; 123 (UP), Tim Fitzharris/Minden Pictures; 123 (LO), Ron & Diane Salmon/Flying Fish Photography LLC; 124 (UP), David Woodfall/Photoshot Holdings Ltd/Alamy; 124 (LO), Michael P. Gadomski/Science Source; 125 (Both), Ron & Diane Salmon/Flying Fish Photography LLC; 126 (UP), Carr Clifton/Minden Pictures; 126 (LO), Colin Marshall/FLPA/Minden Pictures; 127 (UP), Ron & Diane Salmon/Flying Fish Photography LLC; 127 (LO), Jim Brandenburg/Minden Pictures; 128 (UP), Inga Spence/Science Source; 128 (LO), David Liebman; 129 (UP), Kenneth Murray/Science Source; 129 (LO), Adam Jones/Science Source; 130 (UP), James Steakley; 130 (LO), Inga Spence/Science Source; 131 (UP), Geoff Bryant/Science Source; 131 (LO), Frank Zullo/Science Source; 132 (UP), Adam Jones/Science Source; 132 (LO), Ron Boardman/Life Science Image/FLPA/Science Source; 133 (UP), Gerald D. Tang; 133 (LO), Dave Watts/Alamy; 134 (UP), Tim Fitzharris/Minden Pictures/Getty Images; 134 (LO), Geoff Kidd/Science Source; 135 (UP), William Weber/Visuals Unlimited, Inc.; 135 (LO), Joel Sartore/National Geographic/Getty Images; 136 (UP), Ron & Diane Salmon/Flying Fish Photography LLC; 136 (LO), Panoramic Images/Getty Images; 137 (UP), DEA/C.SAPPA/De Agostini/Getty Images; 137 (LO), Dane Johnson/Visuals Unlimited, Inc.; 138 (UP), Michael Orton/Getty Images; 138 (LO), Doug Sokell/Visuals Unlimited, Inc.; 139 (UP), John Shaw/Science Source; 139 (LO), Gerald D. Tang; 140 (UP), Mark Oatney/Getty Images; 140 (LO), Eliot Cohen; 141 (UP), Kent Foster/Science Source; 141 (LO), Phillip Merritt; 142 (UP), Marcos Issa/Argosfoto; 142 (LO), John Glover/Alamy; 143 (UP), De Agostini/S. Montanari/Getty Images; 143 (LO), Peter Chadwick LRPS/Getty Images; 144 (UP), Melinda Fawver/Shutterstock; 144 (LO), E. R. Degginger/Science Source; 145 (UP), Lee F. Snyder/Science Source; 145 (LO), Stuart Wilson/Science Source; 146 (UP), Altrendo Nature/Getty Images; 146 (LO), Ron & Diane Salmon/Flying Fish Photography

(LO), Dr. John D. Cunningham/Visuals Unlimited, Inc.; 239 (UP), John Cancalosi/Okapia/ Science Source; 239 (LOLE), Victor Habbick Visions/Science Source; 239 (LORT), Francois Gohier/Science Source; 239 (LORT), Laurie O'Keefe/Science Source; 240 (Both), Marli Miller/Visuals Unlimited, Inc.; 241 (Both), Marli Miller/Visuals Unlimited, Inc.; 242 (UP), Michael Szoenyi/Science Source; 242 (LO), Marli Miller/Visuals Unlimited, Inc.; 243 (UP), Doug Sokell/Visuals Unlimited, Inc.; 243 (LO), Ted Kinsman/Science Source; 244 (UP), Dennis Flaherty/Science Source; 244 (LO), Bruce M. Herman/Science Source; 245 (UP), Dr. Ken Wagner/Visuals Unlimited, Inc.; 245 (LO), Marli Miller/Visuals Unlimited, Inc.; 246 (UP), mikenorton/Shutterstock; 246 (LO), EastVillage Images/Shutterstock; 247 (UP), Pierre Leclerc/Shutterstock; 247 (LO), Jim Edds/Science Source; 248 (UP), Ken M. Johns/ Science Source; 248 (LO), Marli Miller/Visuals Unlimited, Inc.; 249 (UP), Marli Miller/Visuals Unlimited, Inc.; 249 (LO), Walt Anderson/Visuals Unlimited, Inc.; 250 (UP), Bryan Lowry/SeaPics.com; 250 (LO), Stephen & Donna O'Meara/Science Source; 251 (UP), Explorer/Science Source; 251 (LO), Stephen & Donna O'Meara/Science Source; 252 (Both), Georg Gerster/Science Source; 253 (UP), Francois Gohier/Science Source; 253 (LO), Brenda Tharp/Science Source; 254 (UP), Inga Spence/Visuals Unlimited, Inc.; 254 (LO), Marli Miller/Visuals Unlimited, Inc.; 255 (UP), Robert and Jean Pollock/Science Source; 255 (LO), Marli Miller/Visuals Unlimited, Inc.; 256 (UP), Craig K. Lorenz/Science Source; 256 (LO), William D. Bachman/Science Source; 257 (UP), Douglas Knight/Shutterstock; 257 (LO), Jim W. Grace/Science Source; 258 (UP), ANT Photo Library/Science Source; 258 (LO), Marli Miller/Visuals Unlimited, Inc.; 259 (UP), Marli Miller/Visuals Unlimited, Inc.; 259 (LO), Tim Pleasant/Shutterstock; 260 (UP), Michael Male/Science Source; 260 (LO), Mark Newman/Science Source; 261 (UP), Planet Observer/Science Source; 261 (LO), Michael P. Gadomski/Science Source; 262 (UP), Marli Miller/Visuals Unlimited, Inc.; 262 (LO), William D. Bachman/Science Source; 263 (Both), G. R. 'Dick' Roberts/NSIL/ Visuals Unlimited, Inc.; 264 (UP), Ned Therrien/Visuals Unlimited, Inc.; 264 (LO), Andrew J. Martinez/Science Source; 265 (UP), Thomas & Pat Leeson/Science Source; 265 (LO), Bill Kamin/Visuals Unlimited, Inc.; 266, Mike Grandmaison; 268, Paul Marcellini/PaulMarcellini.com; 269, Mike Grandmaison; 272 (UP), Rene Ramos/Shutterstock; 272 (LO), Michael & Patricia Fogden/Minden Pictures/National Geographic Stock; 273 (UP), Ilya Akinshin/Shutterstock; 273 (LO), Monica Schroeder/Science Source; 274 (UP), Nemeziya/ Shutterstock; 274 (LO), Henry Lansford/Science Source; 275 (UP), Joyce Photographics/ Science Source; 275 (LO), Detlev van Ravensway/Science Source; 276 (UP), Joyce Photographics/Science Source; 276 (LO), Robert and Jean Pollock/Science Source; 277 (UP), Adam Jones/Science Source; 277 (LO), G. R. Roberts/Science Source; 278 (UP), Science Source; 278 (LO), Gregory K. Scott/Science Source; 279 (UP), Mark Schneider/Visuals Unlimited, Inc.; 279 (LO), WimL/Shutterstock; 280 (UP), Brenda Tharp/Science Source; 280 (LO), Gregg Schieve/schievephoto.com; 281 (UP), Ralf Broskvar/Shutterstock; 281 (LO), Mike Hollingshead/Science Source; 282 (UP), Jim Reed/Science Source; 282 (LO), Jim Corwin; 283 (UP), David R. Frazier/Science Source; 283 (LO), Christophe Cadiran/ Science Source; 284 (UP), Jerry Schad/Science Source; 284 (LO), Pekka Parviainen/Science Source; 285 (UP), Jim Reed/Science Source; 285 (LO), Jim W. Grace/Science Source; 286 (UP), Robert & Jean Pollock/Visuals Unlimited, Inc.; 286 (LO), Giselle Goloy; 287 (UP), Mike Hollingshead/Science Source; 287 (LO), Design Pics/Steve Nagy/Getty Images; 288 (UP), Santanor/Shutterstock; 288 (LO), David Hosking/FLPA/Minden Pictures; 289 (UP), Paul Mansfield Photography/Getty Images; 289 (LO), Kamparin/Shutterstock; 290 (UP), Stan Honda/AFP/Getty Images; 290 (LO), Daniel Fredrichs/AFP/Getty Images; 291 (UP), Tim Laman/National Geographic Stock; 291 (LO), Olga Miltsova/Shutterstock; 292 (UP), Kent Wood/Science Source; 292 (LO), Vortex 2/Science Source; 293 (UP), Howard Bluestein/Science Source; 293 (LO), Mike Hollingshead/Science Source; 294 (UP), Jim Reed/Science Source; 294 (LO), Jim Reed; 295 (UP), NOAA; 295 (LO), Anna Omelchenko/Shutterstock; 296 (UP), Chase Studio/Science Source; 296 (LO), Victor Habbick Visions/Science Source; 297 (UP), muratart/Shutterstock; 297 (LO), Lisa Dearing; 298 (UP), Victor Habbick Visions/Science Source; 298 (LO), Gary Meszaros/Science Source; 299 (UP), Science Source; 299 (LO), Mike Hollingshead/Science Source; 300 (UP), Eric Nguyen/Science Source; 300 (LO), Mike Hollingshead/Science Source; 301 (UP), Howard Bluestein/Science Source; 301 (LO), Dr. Bernhard Weßling/Science Source; 302 (UP), St. Meyers/Science Source; 302 (LO), Will & Deni McIntyre/Science Source; 303 (UP), Jim Edds/Science Source; 303 (LO), Planet Observer/Science Source; 304 (UP), Kaj R. Svensson/Science Source; 304 (LO), Syd Greenberg/Science Source; 305 (UP), James Steinberg/Science Source; 305 (LO), AP Photo/Kiichiro Sato; 306 (UP), Kaj R. Svensson/ Science Source; 306 (LO), Jim Reed/Science Source; 307 (UP), Bruce Roberts/Science Source; 307 (LO), Hoa-Qui/Science Source; 308 (LO), Jules Bucher/Science Source; 309 (LO), Jim Edds/Science Source; 320 (UP), David Hosking/FLPA/Minden Pictures; 320 (LO), Dean Krakel II/Science Source; 321 (UP), Alan Copson/Getty Images; 321 (LO), Lowell Georgia/Science Source; 322 (UP), Carr Clifton/Minden Pictures; 322 (LO), Justin Lambert/The Image Bank/Getty Images; 323 (UP), Tony Freeman/Science Source; 323 (LO), Laurent Laveder/Science Source; 324 (UP), Mike Hollingshead/Science Source; 324 (LO), Steve Allen/Science Source; 325 (UP), Mark A. Schneider/Science Source; 325 (LO), Sebastian Saarloos; 326 (UP), Mike Hollingshead/Science Source; 326 (LO), Tim Holt/ Science Source; 327 (UP), Mike Hollingshead/Science Source; 327 (LO), Jamen Percy/ Shutterstock; 328 (UP), Vera Bradshaw/Science Source; 328 (LO), Lizzie Shepherd/Robert Harding World Imagery/Getty Images; 329 (UP), Ron Wolf/Tom Stack & Associates; 329 (LO), Mike Hollingshead/Science Source; 330 (UP), Wesley Bocxe/Science Source; 330 (LO), dotshock/Shutterstock; 331 (UP), Richard W. Brooks/Science Source; 331 (LO), Patrick Endres/Visuals Unlimited/Corbis; 332 (UP), plampy/Shutterstock; 332 (LO), Robert Jakatics/Shutterstock; 333 (UP), David R. Frazier/Science Source; 333 (LO), Jim Reed/ Science Source; 334 (UP), Jim Edds/Science Source; 334 (LO), Sergiy Zavgorodny/Shutterstock; 335 (UP), Bruce M. Herman/Science Source; 335 (LO), Dariush M/Shutterstock; 336 (UP), Science Source; 336 (LO), Mark Horn/Getty Images; 337 (UP), Ted Kinsman/ Science Source; 337 (LO), 4FR/Getty Images; 338 (Both), NOAA; 339 (UP), Mike Berger/ Science Source; 339 (LO), Science Source; 340 (UP), Hank Morgan/Science Source; 340 (LO), Science Source; 341 (UP), NOAA; 341 (LO), Gregory Ochocki/Science Source; 342 (UP), Noel Celis/AFP/Getty Images; 342 (LO), NOAA; 343 (UP), Mike Theiss/National Geographic/Getty Images; 343 (LO), Ken Gillespie/Getty Images; 344, Marc Adamus; 346, Dorling Kindersley/Getty Images; 347, Grant Ordelheide/National Geographic My Shot; 348, Babak Tafreshi/Science Source; 350 (UPRT), Image produced by F. Hasler, M. Jentoft-Nilsen, H. Pierce, K. Palaniappan, and M. Manyin. NASA Goddard Lab for Atmospheres - Data from National Oceanic and Atmospheric Administration (NOAA). NASA/ JPL/GSFC; 350 (LORT), Félix Pharand-Deschênes, Globaïa/Science Source; 350 (UPLE), Davis Meltzer; 350 (LOLE), Paul Leong/Shutterstock; 351 (UP), Pekka Parviainen/Science Source; 351 (LO), Mark Garlick/Science Source; 352 (LORT), Image courtesy NASA; 352 (UP), Detlev van Ravensway/Science Source; 352 (LOLE), Zoltan Kenwell/National Geographic My Shot; 353 (UP), Mark Garlick/Science Source; 353 (LO), NASA, ESA, and the Hubble SM4 ERO Team; 354 (UP), John Chumack/Science Source; 354 (LOLE), Don Dixon; 354 (LORT), Don Dixon/National Geographic Stock; 355 (UPLE), SOHO (ESA & NASA); 355 (UPRT), Kenneth Garrett/National Geographic Stock; 355 (LO), National Geographic Stock; 356 (UP), Science Source; 356 (LOLE), ESA/Science Source; 356 (LORT), NASA/SDO/Steele Hill; 357 (UP), Larry Landolfi; 357 (LOLE), Monica Schroeder/ Science Source; 357 (LORT), Dieter H/Shutterstock; 358 (LORT), Colorization by Eric Cohen/Science Source; 358 (UP), Babak Tafreshi/Science Source; 358 (LOLE), Detlev van Ravensway/Science Source; 359 (UP), Mark Newman/Science Source; 359 (LO), NASA; 360 (UPLE), John Sanford; 360 (UPRT), Detlev van Ravensway/Science Source; 360 (LO), Encyclopaedia Britannica/UIG Via Getty Images; 361 (LORT), Manfred Kage/Science Source; 361 (UP), NASA/JPL/USGS; 361 (LOLE), NASA; 362 (UPRT), Babak Tafreshi/Science Source; 362 (LORT), Philippe Morel/Science Source; 362 (UPLE), Detlev Van Ravensway; 362 (LOLE), Mark Garlick; 363 (UPLE), NASA; 363 (LO), NASA/JPL; 363 (UPRT), NG Maps; 364 (UPRT), NGST; 364 (UPLE), NASA/CXC/SAO (Photo: CXC/M. Weiss); 364 (LO), NASA; 365 (UP), NASA; 365 (LOLE), ESA; 365 (LORT), Victor Habbick Visions/Science Source; 366 (UP), NASA; 366 (LO), Mike Agliolo/Science Source; 367 (UP), Dan Schechter; 367 (LO), James Keenan/National Geographic My Shot; 368 (UP), John Chemack; 368 (LO), Mark Garlick/Science Photo Library/Getty Images; 369 (UPLE), David A. Aguilar; 369 (UPRT), Detlev van Ravensway/Science Source; 370 (UP), Mark Garlick/Science Source; 370 (UP), David A. Aguilar; 370 (LO), Frank Zuillo; 371 (UP), Johns Hopkins Univ. Applied Physics Laboratory/Carnegie Institution of Washington; 371 (LO), NASA; 372 (UPLE), Thomas Tuchan/iStockphoto; 372 (UPRT), Colorization by: Mary Martin/Science Source; 372 (LO), NASA; 373 (UP), NASA/JPL-Caltech/Malin Space Science Systems; 373 (LO), Mark Garlick/Science Source; 374 (UP), NASA; 374 (LO), Atlas Photo Bank/Science Source; 375 (UP), NASA/JPL/Space Science Institute; 375 (LO), JPL; 376 (UP), SPL/Science Source; 376 (LO), NASA, ESA, and M. Showalter (SETI Institute); 377 (UP), NASA/ JPL; 377 (LO), NASA/JPL-Caltech; 378 (UP), NASA; 378 (LO), Chris Butler/Science Source; 379 (UPLE), NASA, ESA, R. Windhorst (Arizona State University) and H. Yan (Spitzer Science Center, Caltech); 379 (UPRT), Colorization by: Jessica Wilson/Science Source; 379 (LO), Mehau Kulyk/Science Source; 380 (UP), NASA, ESA, K. Kuntz (JHU), F. Bresolin (University of Hawaii), J. Trauger (Jet Propulsion Lab), J. Mould (NOAO), Y.-H. Chu (University of Illinois, Urbana), and STScI; 380 (LO), The Milky Way galaxy as conceptualized by Ken Eward and National Geographic Maps; 381 (UP), Adam Evans/sky-candy. ca; 381 (LO), NASA, ESA, and the Hubble Heritage Team (STScI/AURA)-ESA/Hubble Collaboration; 382 (UP), NASA, ESA, and the Hubble Heritage Team (STScI/AURA); 382 (LO), ESA/J. Hester & A. Loll/Arizona State University; 383 (UP), X-ray: NASA/CXC/CfA/R. Kraft et al.; Submillimeter: MPIfR/ESO/APEX/A.Weiss et al.; Optical: ESO/WFI; 383 (LO), ESO/M. Kornmesser; 384 (UP), John Chumack/Science Source; 384 (LO), NASA; 385 (UP), T. A. Rector & B. A. Wolpa, NOAO, AURA; 385 (LO), NASA/JPL-Caltech/R. Hurt (SSC); 386 (UP), NASA Ames Research Center/Kepler Mission; 386 (LO), Jon Lomberg/Science Source; 387 (UP), Royal Astronomical Society; 387 (LO), Celestial Image Co./Science Source; 388 (LORT), José Antonio Peñas/Science Source; 388 (LO), ESA/NASA; 388 (LOLE), Reinhold Wittich/Shutterstock; 389 (UP), Jerry Schad/Science Source; 389 (LO), National Geographic Stock; 390 (UP), Daniel McVey/National Geographic My Shot; 390 (LO), Babak Tafreshi/Science Source; 391 (UP), Peter Lloyd/National Geographic Stock; 391 (LO), The Art Archive/British Library/Via Art Resource, NY.

Index

Boldface indicates illustrations.

NATIONAL GEOGRAPHIC

ILLUSTRATED GUIDE
TO
Nature

CELEBRATING
◄**125**►
YEARS

The National Geographic Society is one of the world's largest nonprofit scientific and educational organizations. Founded in 1888 to "increase and diffuse geographic knowledge," the Society's mission is to inspire people to care about the planet. It reaches more than 400 million people worldwide each month through its official journal, *National Geographic*, and other magazines; National Geographic Channel; television documentaries; music; radio; films; books; DVDs; maps; exhibitions; live events; school publishing programs; interactive media; and merchandise. National Geographic has funded more than 10,000 scientific research, conservation and exploration projects and supports an education program promoting geographic literacy. For more information, visit www.nationalgeographic.com.

For more information, please call 1-800-NGS LINE (647-5463) or write to the following address:
National Geographic Society
1145 17th Street N.W.
Washington, D.C. 20036-4688 U.S.A.

For information about special discounts for bulk purchases, please contact National Geographic Books Special Sales: ngspecsales@ngs.org

For rights or permissions inquiries, please contact National Geographic Books Subsidiary Rights: ngbookrights@ngs.org

ISBN: 978-1-4262-1174-4 (trade)
ISBN: 978-1-4262-1309-0 (regular)
ISBN: 978-1-4262-1255-0 (deluxe)

Printed in China

13/RRDS/1

Published by the National Geographic Society

John M. Fahey, *Chairman of the Board and Chief Executive Officer*
Declan Moore, *Executive Vice President; President, Publishing and Travel*
Melina Gerosa Bellows, *Executive Vice President; Chief Creative Officer, Books, Kids, and Family*

Prepared by the Book Division

Hector Sierra, *Senior Vice President and General Manager*
Janet Goldstein, *Senior Vice President and Editorial Director*
Jonathan Halling, *Design Director, Books and Children's Publishing*
Marianne Koszorus, *Design Director, Books*
Susan Tyler Hitchcock, *Senior Editor*
R. Gary Colbert, *Production Director*
Jennifer A. Thornton, *Director of Managing Editorial*
Susan S. Blair, *Director of Photography*
Meredith C. Wilcox, *Director, Administration and Rights Clearance*

Staff for This Book

Barbara Payne, *Editor*
Paul Hess, *Text Editor*
Sanaa Akkach, *Art Director*
Catherine Herbert Howell, *Developmental Editor*
Miriam Stein, *Illustrations Editor*
Linda Makarov, *Designer*
Uliana Bazar, *Art Researcher*
Carl Mehler, *Director of Maps*
Marshall Kiker, *Associate Managing Editor*
Judith Klein, *Production Editor*
Galen Young, *Illustrations Specialist*
Katie Olsen, *Production Design Assistant*

Production Services

Phillip L. Schlosser, *Senior Vice President*
Chris Brown, *Vice President, NG Book Manufacturing*
George Bounelis, *Vice President, Production Services*
Nicole Elliott, *Manager*
Rachel Faulise, *Manager*
Robert L. Barr, *Manager*

Dig Into More Nature-Related Books From National Geographic!